Acclaim for John Huntington's
Show Networks and Control Systems
(Formerly *Control Systems for Live Entertainment*)

"We recommend this book to engineers and students who ask us, 'How do I learn how to do what you do?' We also tell our newly hired engineers that they should be conversant with the entire contents of this book because it covers the core knowledge commonly needed in the design of show control systems."
— **Glenn Birket**, P.E.; President, Birket Engineering, Inc., designers of show and ride control systems since 1984.

"Today's entertainment technology students training to work on tomorrow's cutting edge are just as likely to need to know IP addresses and data protocols as knots and drawing conventions. John Huntington's book provides an invaluable introduction to networking for entertainment systems, helping students build a foundation and vocabulary for systems integration across lighting, sound, media, scenery, and more."
—**David Boevers**, Associate Professor, Carnegie Mellon University School of Drama.

"Given the increasing complexity of entertainment systems, today's tech needs access to an abundance of fresh information, and this book provides it in a well-organized, easy-to-digest manner. This is the best book you will find on the subject of entertainment control systems and show networks because Huntington not only teaches it but also uses it extensively in the real world. If you're serious about your profession in the entertainment production industry, you'll read this book now."
—**Richard Cadena**, author of *Electricity for the Entertainment Electrician & Technician* (Focal Press 2009) and *Automated Lighting: The Art and Science of Moving Light*, 2nd Edition (Focal Press 2010); founder of Academy of Production Technology; PLASA technical editor; and freelance lighting designer.

"John's book, *Show Networks and Control Systems*, is a treasure chest of information that leads us to a creative world of possibilities, with many great avenues of escape when we find ourselves stuck in a cul-de-sac of ideas."
—**Jonathan Deans**, Sound Designer for Cirque du Soleil; Broadway; and events, shows, and installations worldwide.

Show Networks and Control Systems describes the technology on the leading edge of the entertainment industry and looks out into the future of entertainment networks. All serious entertainment technologists will study this book and everyone will learn something from it."
—**Gary Fails**, President, City Theatrical

"The new generation of entertainment technology is almost wholly focused on making individual systems communicate and cooperate with each other so that the show operates like a single coordinated machine. This book is on the cutting edge of that new wave, and it presents these indispensable techniques and concepts in clear, concise terms and examples. It's easily the best resource for anyone needing a better understanding of where entertainment control technology is today and where it's headed in the future."
—**Scott Fisher**, President, Fisher Technical Services.

Show Networks and Control Systems
Formerly *Control Systems for Live Entertainment*

John Huntington

Zircon Designs Press

Copyright © 2012 by John Huntington.

Zircon Designs Press
Brooklyn, NY, USA.
www.zircondesigns.com

All rights reserved. Except as permitted under the U.S. Copyright Act of 1976, no part of this publication may be reproduced, distributed, or transmitted in any form or by any means, or stored in a database or retrieval system, without the prior written permission of the publisher.

Written and laid out by the author in Adobe® Framemaker® 10.

LCCN: 2012943024

ISBN: 978-0615655901 (Zircon Designs Press)

Version 1.0

Foreword

Having read the previous editions—or, should I say, used the previous editions as my guide, reference, idea shaker, and knowledge source—it makes perfect sense for a newer version to be released, as technology changes very quickly and I need to stay on top of the innovation curve.

Networking, entertainment control, and show control influence art for live and real-time performances. In fact, these technologies are arts unto themselves, especially when we make comparisons to some of the other disciplines within our entertainment world that attempt to create natural moments, scenes, and displays through their given technology. All that being said, it is not what you know but how you use the knowledge that matters. Like all creative thoughts and practices, you need to learn the fundamental skills in order to have a final result.

John Huntington's book offers all the information you should know, but also—and more to the point—all the information that you didn't know, which is where we all find the dead end, or more likely the deathtrap, in our otherwise glorious journey. Whether it's in a suitcase, backpack, briefcase, workbox, or an Ipad, I highly suggest you have a copy nearby when you are at work, in addition to the copy you have at home. Just knowing you have the answer somewhere within reach sometimes is enough to help you survive. For me, it opens up my mind to ideas—and perhaps inventions—that I would have missed if I were always in the dark.

—**Jonathan Deans**, Sound Designer for Cirque du Soleil; Broadway; and events, shows, and installations worldwide, 2012.

Table of Contents

Foreword v

Table of Contents vii

Preface xv
    I Wonder xv
    Why Does This Book Exist? xv
    What's New in This Edition? xvi
    For Whom Is This Book Written? xvii
    How Should This Book Be Used? xvii
    What Is Included? xviii
    Conventions xviii
    Disclaimers xix
    Website xix
    Special Thanks to My Production Team xix

Chapter 1: Introduction 1
    What Is Entertainment Control? 1
    What Is Show Control? 2
    What Is a System? 2
    What Is a Network? 3
    What Is a Standard? 4
    Moving On 9

Part 1: Entertainment Discipline Overview 11

Chapter 2: Lighting 13
    Lighting Control Equipment 13
    Lighting Control Approaches 20

Chapter 3: Lasers 23
    Laser System Control Equipment 23
    Laser Control Approaches 25

Chapter 4: Audio 27
    Audio Control Equipment 27

Audio Control Approaches 31

Chapter 5: Image Presentation **35**
Image Presentation Control Equipment 35
Image Presentation Control Approaches 37

Chapter 6: Stage Machinery **41**
Machinery Control Systems 41
Sensing/Feedback 42
Control 44
Drive Devices 46
Emergency Stop 47
Commercial Entertainment Machinery Control Systems 48
Machinery Control Approaches 48

Chapter 7: Animatronics **51**
Animatronic Effects 51
Animatronic Control Systems 52
Animatronic Control Approaches 54

Chapter 8: Fog, Smoke, Fire, and Water **57**
Fog and Smoke Equipment 57
Fire and Water Control Systems 58
Fog Smoke Fire and Water Control Approaches 58

Chapter 9: Pyrotechnics **61**
Pyrotechnic Control Systems 61
Pyrotechnic Control Approaches 63

Part 2: Entertainment Control. **65**

Chapter 10: Entertainment Control Basics **67**
Show Types 67
Cues 68
Cueing Methods 69
Operational Modes 72
Commands/Data 73
Data Relationships 74
Feedback 78
Control Structures 80
System Control Architectures 81
Physical Topologies 83
Variables 85

    Logical Operators 86

Chapter 11: Electrical Control System Basics **89**
    Sensors and Switches 89
    Contact Closures 93
    Contact Nomenclature 94
    Isolation 97

Chapter 12: Numbering Systems **99**
    Base 10 (Decimal) Notation 99
    Base 2 (Binary) Notation 100
    Base 16 (Hexadecimal) Notation 101
    Binary-Coded Decimal Notation 103
    Math with Binary and Hex 103
    Bitwise Operation 104
    Converting Number Bases 105
    Sample Numbers in Different Formats 108
    Storing Numbers in Computers 108

Chapter 13: System Design Principles **111**
    Principle 1: Ensure Safety 111
    Principle 2: The Show Must Go On 115
    Principle 3: Simpler Is Always Better 117
    Principle 4: Strive for Elegance 118
    Principle 5: Complexity Is Inevitable, Convolution Is Not 118
    Principle 6: Make It Scalable, and Leave Room for Unanticipated Changes 118
    Principle 7: Ensure Security 118
    System Troubleshooting 119

Part 3: Data Communication and Networking **123**

Chapter 14: Data Communication **125**
    An Introduction to Communications Layering 125
    Character Encoding 126
    Data Rate 127
    Bandwidth 127
    Multiplexing 127
    Communications Mode 129
    Error Detection 129
    Flow Control 132
    Electricity for Data Transmission 133
    Transmission/Modulation Methods 140
    Light for Data Transmission 141

Radio for Data Transmission 143

Chapter 15: Point-to-Point Interfaces 145
Parallel Interfaces 145
Serial Interfaces 146
TIA/EIA Serial Standards 147
Practical Serial Connections 151
Serial Connection Example 152
High-Speed Serial Point-to-Point Interconnects 154
Moving On 156

Chapter 16: Networking Basics 157
Open Systems Interconnect (OSI) Layering Scheme 157
Packet Switching 159
Encapsulation 159
Packet Forwarding Schemes 160
Network Types 161
Ethernet 162
Ethernet Implementations 166
Ethernet Hardware 169
IEEE 802.11 "Wi-Fi" 175
Ethernet in Our Industry 179

Chapter 17: Show Networks 181
TCP, UDP, and IP 181
Transmission Control Protocol (TCP) 182
User Datagram Protocol (UDP) 183
Internet Protocol (IP) 183
Dynamic Host Configuration Protocol (DHCP) 185
ipconfig/ifconfig Command 186
Example Network Using DHCP and ipconfig 187
Link-Local Addresses 189
Static/Fixed IP Addresses 189
ping Command 190
Example Network Using Fixed IP Addresses and ping 191
Subnets and Network Masks 194
Example Network with Two Subnets 197
Address Resolution Protocol (ARP) 200
Ports and Sockets 203
Testing Networks 204
Moving On 209

Chapter 18: Advanced Show Network Topics ... 211
    Broadcast Domain ... 213
    Network Topology and Broadcast Storms ... 216
    Default Gateway ... 219
    Layer 3 Routing ... 220
    Virtual LANs (VLAN) ... 221
    Example Network with VLANs and Managed Switches ... 226
    Other Network System Protocols ... 232
    IPv6 ... 233

Part 4: Standards and Protocols Used in Entertainment ... 239

Chapter 19: Digital MultipleX (DMX512-A) ... 241
    DMX's Repetitive Data Approach ... 241
    Addressing ... 242
    Universes ... 243
    Controlling Equipment Other Than Dimmers ... 243
    Physical Connections ... 245
    Data Transmission ... 247
    DMX Distribution and Termination ... 249
    DMX Patching/Merging/Processing/Test Equipment ... 250
    Alternate Start Codes ... 251
    Enhanced Function ... 251
    DMX Over a Network ... 252
    Network Over DMX? ... 257
    DMX in the Entertainment Control Market ... 257

Chapter 20: Remote Device Management (RDM) ... 259
    Basic Structure ... 259
    RDM Message Structure ... 260
    The Discovery Process ... 263
    RDM Messages ... 265
    RDM and Networks ... 268
    RDM in the Entertainment Control Market ... 268

Chapter 21: Architecture for Control Networks (ACN) ... 269
    A Bit of Blue Sky Thinking ... 269
    Background and Mandate ... 270
    Overview ... 270
    ACN's Acronym Soup ... 271
    Protocol Structure ... 274
    Identifiers and Addresses ... 274
    Discovery ... 276

  Control of Devices 278
  ACN Implementations 279
  ANSI E1.31: Streaming ACN (sACN) 280
  ACN in the Entertainment Control Market 280

Chapter 22: Musical Instrument Digital Interface (MIDI) 281
  Basic Structure 281
  Physical Connections 282
  MIDI Messages 283
  Channel Messages 283
  System Messages 285
  Active Sensing 287
  MIDI Sync 287
  System-Exclusive Messages 287
  MIDI Running Status 291
  General MIDI 292
  MIDI Processors/Routers/Interfaces 292
  Recommended MIDI Topologies 294
  Common MIDI Problems 294
  MIDI over Networks 296
  MIDI in the Entertainment Control Market 299

Chapter 23: MIDI Show Control (MSC) 301
  MSC Command Structure 301
  Recommended Minimum Sets 306
  MSC Commands 306
  Limitations of MIDI Show Control 312
  MSC in the Entertainment Control Market 314

Chapter 24: MIDI Machine Control (MMC) 315
  MMC Systems 315
  Command/Response Structure of MMC 315
  MMC Motion Control 316
  MMC Message Structure 316
  Common MMC Commands 318
  MMC in the Entertainment Control Market 319

Chapter 25: SMPTE and MIDI Time Code (MTC) 321
  Background 321
  Time Code Addresses 321
  Traditional Audio/Visual Synchronization 321
  Live Entertainment Time Code Applications 323
  Time-Code Types 324

Practical Time Code For Live Shows 325
SMPTE Time-Code 327
SMPTE Time Code Hardware 329
SMPTE Linear Time Code in the Entertainment Control Market 330
MIDI Time Code 330
MIDI Time Code in the Entertainment Control Market 332
Other Time Codes 333

Chapter 26: Open Sound Control (OSC) 335
OSC Overview 335
OSC in the Entertainment Control Market 338

Chapter 27: Other Control Protocols 341
Open Control Alliance (OCA) 341
MIDI Visual Control (MVC) 343
Network Time Protocol (NTP) 343
Simple Network Management Protocol (SNMP) 344
Virtual Network CoNtrol (VNC) 344
Industrial I/O Systems 345
Legacy Video Connection Standards 347

Part 5: Show Control 355

Chapter 28: Show Control 357
Evolution of Show Control 357
Why Show Control? 360
What Is a Show Controller? 361
My Show Control Design Process 362
Question 1: What are the safety considerations? 362
Question 2: What type of show is it? 363
Question 3: Is the show event-based, time-based,
or a hybrid? 363
Question 4: What is the control information source? 363
Question 5: What is the type of user interface required? 363
Question 6: What devices must be controlled/connected? 365
Other Concerns 366
Finally 367
Moving On to Some Examples 367

Chapter 29: A Theatrical Thunderstorm 369
The Mission 369
Design Considerations 369
The Systems 370

Show Control Script 372
Approach 1 373
Approach 2 374
Approach 3 376

Chapter 30: Put on a Happy Face 379
The Mission 379
Design Considerations 379
The Systems 380
Approach 1 382
Approach 2 384

Chapter 31: Ten-Pin Alley 387
The Mission 387
Design Considerations 388
The Systems 389
The Approach 393

Chapter 32: Comfortably Rich 397
The Mission 397
Design Considerations 397
The Systems 398
Show Control Script 400
Approach 1 401
Approach 2 403

Chapter 33: It's an Itchy World after All 407
The Mission 407
Design Considerations 408
The Systems 408
The Approach 412

Conclusion 421
Contact Info and Blog 422

Acknowledgments 423

Appendix: Decimal/Hex/Binary/ASCII Table 427

Glossary 435

Index 453

Preface

I WONDER . . .

"I wonder . . ." is the beginning of a question that has served me well throughout my life. I seem to be asking "I wonder why . . . ?" or "I wonder how . . . ?" about 1,000 times a day, because I'm driven by an interest in what goes on "behind the scenes" in nearly every area I encounter: how electricity is generated, how corn chips are manufactured, or how magic tricks are done. This book is a result of that continual process of inquiry, focused on the area that has always fascinated me the most: show business. More specifically, this book is about entertainment control systems and networking, a subset of the fascinating intersection of art and science we call entertainment technology.

WHY DOES THIS BOOK EXIST?

In 1990, I was a technical editor at *Theatre Crafts* and *Lighting Dimensions* magazines (now *Live Design*), having just graduated from the Yale School of Drama, where I wrote my thesis on entertainment control systems. It was clear to me at that time how these entertainment control technologies underlie nearly everything we do on shows, and one day, researching an article for *Theatre Crafts*, I went to Times Square to the Drama Book Shop (the Amazon.com of the day for books on entertainment) to see what information was available on the topic. On the shelves, I found a wide variety of theatre technology titles, that explained scenic construction, lighting design principles, paint techniques, and even sound in those early days of digital audio. But I wasn't able to find a single book that explained the then widely-used DMX512 digital lighting control standard, and I was even more shocked to find that the majority of the lighting technology books didn't even *mention* it. I also couldn't find anything about the details of show control technologies like SMPTE Time Code or MIDI, or anything at all on networking, but that was less surprising, since those technologies were developed in other industries and adapted for our field. DMX, on the other hand, was developed by and for *our* industry, and was as critical to the success of shows then as it is now. After that day, I started a process that led to the first edition of this book in 1994.

As of this writing, we seem to be in a similar situation regarding networks and their ever-increasing use on shows. Searching Amazon and Google Books for terms like "networks" and "live entertainment," I found mostly references back to my own book, along with a couple others that don't really give a complete picture of the ways networks are used on shows. I find this a bit of surprising deja-vu, since network technologies increasingly lie at the core of entertainment control systems, which are the technological backbone of every modern show. While not every entertainment technician needs to be a networking expert, I do believe that everyone working in entertainment technology has to have at least a cursory understanding of these critical technologies, and that situation led to the new edition of this book.

WHAT'S NEW IN THIS EDITION?

Focal Press published the first three editions of this book, which together sold more than 7,000 copies. However, at the end of 2011, they decided that they were no longer interested in future editions. With the increasing impact and ubiquity of networking in our market, I felt a new edition with an expanded networking focus was important, so I got a sabbatical from my position as a Professor at New York City College of Technology (City Tech), and figured out how to self-publish using Amazon's Createspace™. Since I did all the production and layout work myself for the previous edition in Adobe® Framemaker®, it was a relatively straightforward (although enormous!) task to undertake the process I had gone through on each previous edition: I completely reorganized the content; checked, updated, and expanded the information; and added a lot more stuff (see below). And, to reflect the changes in the market—and the self-publishing status—I also changed the title.

New Title

The title of previous editions of this book was *Control Systems for Live Entertainment*, and the book was always intended to cover the basics of all the control systems used on shows. However, many in the market thought of this as only the "show control book". While I certainly do cover the advanced topics of show control in depth (this is still one of the only books to do so), there's always been a lot of other information in here—especially about networks—that is increasingly important to everyone working in live entertainment technology. So, I wanted to add "network" to the title in some way. In addition, many of the technologies and techniques covered here are used in shows and installations like museums, themed-retail attractions, and theme parks. I would call all of those forms of entertainment "shows", but they aren't necessarily "live". And so, the new title: *Show Networks and Control Systems*.

Glossary

Due to popular demand, there is now a Glossary, starting on page 435.

More Graphical Guidance

I have added several graphical aids to help you navigate through the information; see "Conventions," on page xviii.

New Organization

Due to the impact of networks on our market, I've completely reorganized this edition. I moved the discipline overviews to the introduction, and expanded the datacom and networking sections significantly.

FOR WHOM IS THIS BOOK WRITTEN?

While writing this book, I have kept in mind two groups of readers. First, I've targeted those working in or studying the entertainment technology field: technicians, designers, technical managers, and so on, with a special focus on entry-level readers. (Note that I left product designers and engineers off this list—they should be reading the same technical source materials I read.) Second, I've targeted technically literate folks outside the entertainment industry who want to learn more about entertainment control. This group may include an electrical engineer who wants to get some background on and context for the live entertainment standards or protocols that he has been reading about, or a computer artist interested in interfacing her software with show production elements. As in previous editions, I focus here not on gear (which constantly becomes obsolete), but on open techniques and standards, because I think it's critical that people know not only the "what" but the "why" of entertainment technologies.

Though I've tried not to be extremely technical, this is a field of ever-increasing technological complexity. In order to make the content of the book manageable, I assume the reader knows the general difference between analog and digital, what a Volt is, the basics of computing, and so on. However, you do *not* need to be an expert in all these areas, since, in entertainment technology, we deal mostly with higher-level systems integration, and we do so very practically. It's great for an entertainment technician to have the skill to do component-level electronic circuit-board repair, or packet-level analysis of network traffic. However, it's even more important for the average technician to learn how to put together reliable systems and make sure the show actually happens—experts can always be found when necessary. For those who want to pursue the core technologies further in depth (something I strongly encourage), I have suggestions listed in the book resource sections on my website (`http://www.control-geek.net`).

HOW SHOULD THIS BOOK BE USED?

I have attempted here to present the information in a form readable straight through by motivated, independent readers, while also making the structure modular enough to be useful for working professionals and educators (I teach classes based on the book myself). For those of you working through this information on your own, please, if you get either bored or bogged down in technical details, feel free to skip around the book; I've included a lot of cross references to help you with that.

For those using this book in a class, I think this subject, while fundamental, is best suited for upperclassmen. In addition, in my more than a decade of teaching, I've found two ways to deal with information-intensive topics like this. The first is that you can present everything up front, methodically working through the information and building in complexity. The problem with this approach is that some people get bored and drift off, or get completely lost. The other approach is to throw people into a situation, and have them formulate questions which we can then answer. This approach is frustrating to some people.

So, in my teaching and in this book, I've tried to do a little of both. If you're bored, please skip ahead; if you're frustrated, there's the glossary starting on page 435, an extensive index starting on page 453, and the Table of Contents, on page vii.

WHAT IS INCLUDED?

As I mentioned earlier, this book's intended audience is end users, not product developers. For this reason, I include enough detail of standards and protocols to enable users to understand these technologies and their implications for system design. But I have intentionally left out a lot of the nitty gritty technical details, because—fortunately for us—as protocols have gotten more complex, this complexity has been increasingly buffered from the user by product developers. In short, I've waded through those technical details so I hope you won't have to!

CONVENTIONS

While the book uses fairly standard heading levels and so on, there are a few conventions:

- If a term is **bolded**, then it is defined in the glossary which starts on page 435. I marked the first major usage of the term in the book.
- If you see something like $F2_{16}$, or 1010010_2, the subscripted number indicates the base of the number which precedes it (see "Numbering Systems," on page 99).
- Key terms are called out in the margin with a symbol like this:

- Dangerous stuff is indicated with this symbol in the margin:

- If you see a small "TL" next to or inside of a graphic, that indicates that it was drawn by Tom Lenz, who handled graphics for the second edition. Aaron Bollinger created all the other excellent graphics throughout the book.

> Text set apart in a grey box is something interesting but not directly related to the immediate topic, or a historical or personal anecdote.

- Some network icons are from Cisco's free to use symbol collection.[1]
- Uncredited photos are by the author.

---

1. http://www.cisco.com/web/about/ac50/ac47/2.html

DISCLAIMERS

And now for the *"It's not my fault!"* disclaimers: While I've made every effort to ensure that the information in this book is accurate, *DO NOT* implement anything in any product or system based only on the information in this book. The goal here is understanding; if you want to go to the next level—*implementation*—you need to obtain information from the appropriate standards or other organizations (see my web page at `http://www.controlgeek.net` for contact information). They have the newest, most up-to-date information and are worthy of your support.

Additionally, safety is the responsibility of system designers and operators. I include some general safety principles in this book; these are based on the way I do things and not necessarily in compliance with any industry standard. It is *your* responsibility to ensure safety in any system with which you deal!

To counteract obsolescence as much as possible, I have not dwelled on specific equipment or systems—they age very quickly. This book focuses on the underlying concepts of entertainment control systems, which will survive a lot longer than the newest, greatest piece of gear. Pieces of gear depicted in pictures or mentioned in the book are simply representative products which appear to me to be in widespread use. No endorsement of any sort is intended.

WEBSITE

Standards body and manufacturer contact information, errata on this edition, my blog, and much more is available at my website: `http://www.controlgeek.net`

SPECIAL THANKS TO MY PRODUCTION TEAM

Literally hundreds of people helped me with this book—see "Acknowledgments," on page 423. But I want to extend a special thanks to my self-publishing production team: Netta Rabin[2] for help with the book layout and for designing the cover; Aaron Bollinger[3] for creating all the excellent graphics and putting up with all the changes; and Cindy Nowicki[4] for copy editing the manuscript, and finding mistakes that three other copy editors missed.

---

2. `nettarabin.com`

3. `Piratechnical.com`

4. `cindynowicki.wordpress.com`

Chapter 1: **Introduction**

WHAT IS ENTERTAINMENT CONTROL?

Humans have been wired by evolution with a strong desire to hear and tell stories, and in show business, storytelling has been our job for thousands of years. Entertainment technology provides a powerful set of storytelling tools that have extended beyond the traditional performing arts like theatre and dance into a wide array of venues and types of shows, including concerts, the circus, theme parks, corporate meetings, special events, cruise ship shows, themed retail, location-based entertainment, museum exhibits, mega churches, and so on.

"Entertainment control systems," or just **entertainment control**, means the control system used for any element in the show environment; for example, the control signals used between a lighting-control console and a dimmer rack; the network used to synchronize and control a number of video servers; the data sent between a pyro controller and a flash pot firing system; or anything else related to the field.

Evolution of Entertainment Control

In the 1980s when I started my career, most lighting control systems were relatively straightforward and easy to understand for anyone with a basic electronic background. They used one wire per dimmer, each wire sending a simple analog voltage to represent the level to which that dimmer should be set. Sound systems were analog and there was little automation. Film was cumbersome to use on live shows, and video was prohibitively expensive in general; display devices were always too dim and never had enough resolution. Machinery systems were only used in the largest venues, with relatively straightforward control systems.

In modern show systems, however, there is a wide array of highly sophisticated, computerized controllers invisibly (or not) buried inside the equipment found all around a performance venue. These devices all need to be controlled and, increasingly, that control data is carried over common computer-industry networks. But with networked lighting control systems, there is no simple circuit to trace down with a voltmeter in order to find out why a dimmer is not working—without the right knowledge and test equipment, the core of the control system is effectively invisible. Modern sound systems can now have complex computer control and sample synchronization issues few would have even imagined in the 1980s. While you're unlikely to see film used on a show any more, sophisticated, networked digital video is everywhere and cost effective. Powerful machinery systems can now be found even on small shows. As a result, networking and control technologies have become ubiquitous and critically important to the success of shows, and even an entry-level technician or an undergraduate entertainment technology student needs at least a basic understanding of these subjects.

WHAT IS SHOW CONTROL?

Show control, a much maligned and misused phrase, simply means connecting two or more separate entertainment control systems together. A computer controlling fog machines that regulate the amount of fog in a harbor scene doesn't amount to show control; a system that *links* the control of the fog machine with an audio playback system generating maritime sound effects does. The signals sent between a lighting console and a group of dimmers is an entertainment control system, but it is not show control. If the lighting cues were triggered from a system that also took positional data from the moving scenery that was being lit, this would be show control. The key is that you don't have a *show control system* unless control for *more than one* production element is linked *together*. So show control is part of entertainment control, but not all entertainment control is show control (a lot more on this is in Part 5, "Show Control," on page 355).

WHAT IS A SYSTEM?

Almost everything we do in entertainment control deals with systems of components, devices, or even systems made up of other systems. But just what is a system, anyway? According to the American Heritage Dictionary, a **system** is, "a group of interacting, interrelated, or interdependent elements forming a complex whole."[1] This is a workable definition for our market, since any control system has to interact, it is by definition interrelated, and it's interdependent in that the whole is greater than the component parts.

Outputs, Connection Methods, Inputs, and Processors

When breaking a system down to its simplest level, we see outputs, inputs, signal connection methods, and (optionally) processors. From these simple building blocks, very sophisticated systems can be built.

---

1. "system." *The American Heritage® Dictionary of the English Language*, Fourth Edition. Houghton Mifflin Company, 2004. 14 Mar. 2007. <Dictionary.com http://dictionary.reference.com/browse/system>.

2 • *Chapter 1: Introduction*

An output sends some sort of signal, state, information, or data to an input, using some sort of connection method:

$$\text{From Output} \xrightarrow{\text{Connection Method}} \text{To Input}$$

Outputs always connect to inputs: A lighting console control output would connect to a dimmer control input, a machinery computer output would connect to a motor drive input, and so on. The device that takes the input might then "process" the information in some way or even pass it on to another output. Processors could of course be **analog**, but with the ever-declining cost of computer hardware and increase in digital power, the vast majority of processing devices are now **digital** and contain some sort of intelligence.

WHAT IS A NETWORK?

Networks form the backbone of many entertainment control systems, so we need to cover them in depth. But what is a network anyway? Point-to-point interfaces (discussed starting on page 145) are designed to connect two—or maybe a few—devices together. Point-to-point connections require direct connections from each device to every other device, so if you needed to connect a large number of devices, you would end up with an unruly mess, such as the one pictured at right. While this structure allows any device to communicate with any other, it creates a cabling and interfacing management nightmare.

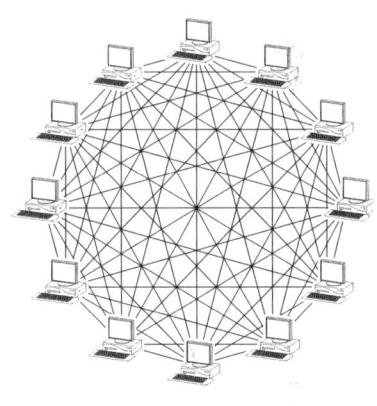

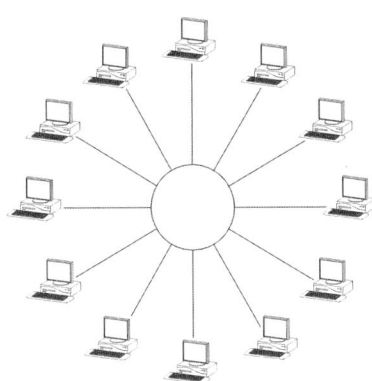

The solution for efficiently connecting large numbers of devices is to use a **network**, where we use a common physical infrastructure that still allows each device to have access to all the others (see figure at left). Connections between devices then simply become virtual pathways, and the resulting system can be dramatically cheaper, more powerful, simpler overall, and far easier to manage. Any device connected by the network is called a **node**; if the node is a computer that actually has data to communicate, it may be referred to as a **host**.

Evolution of Entertainment Networking

Our audiences demand performance and reliability from our show systems that far exceeds their own experience with similar technologies (i.e., cell phones), so our "show must go on" industry needs robust, secure networks. At the same time, the physical environment backstage is often not kind: show equipment gets abuse that can exceed what might be found in many factory or military environments (think about dragging networking cabling through the mud after a circus). In addition, we need networks that are **deterministic** (able to deliver data in a predictable amount of time), because if that light cue goes in 100 milliseconds today and 2 seconds tomorrow, a show could be ruined.

For all these reasons, in the early 1990s, when networks began to emerge in our industry, we pushed for industrial-grade, high performance (i.e. expensive) systems, and tried to steer away from office-grade networks, which were designed with (low) cost and connectivity as primary concerns, and reliability and performance somewhere further down the criteria list. In the end, we were not able to compete with the value offered by the economies of cost and scale found in the massive business networking equipment market, but—fortunately for us in entertainment—the larger information technology (IT) industry found an increasing number of "mission critical" applications similar to our own. They solved many of the problems that made these office networking technologies unsuitable for our use, and early in the 21st century, one clear winner emerged as the victor in the battle for network domination: Ethernet. (covered extensively in Chapter 16, on page 157). Ethernet has become so integral to entertainment control that it's likely to be the only network most entertainment technicians encounter in their career.

WHAT IS A STANDARD?

Standards form the backbone of our industry, but just what is a standard? Dictionary.com has 27 definitions[2] for **standard**, but probably the most pertinent one for us is the first: "something considered by an authority or by general consent as a basis of comparison; an approved model." In control systems, this generally means that a group of some sort gets together and agrees upon a useful working practice, protocol, measurement, or connection method, and we then call the resulting agreement a standard. Standards can be official or unofficial, private or public, strict or loose—although, generally, the stricter the standard, the more consistent (standardized) the outcome. Standards have existed in various forms throughout history, but have generally been tied closely to the promotion of com-

---

2. "standard." Dictionary.com Unabridged. Random House, Inc. 04 Mar. 2012. <Dictionary.com http://dictionary.reference.com/browse/standard>.

merce in one form or another. There are three general types of standards: proprietary, de facto, and open.

Proprietary Standards

As a manufacturer or designer, you could design a control approach yourself: pick the connector, transmission media, and signaling levels and then create the necessary control protocols, messages, data formats, and so forth. With this approach, you have complete control over the way a system works and you can optimize the connection method specifically for your applications and for your product's idiosyncrasies. However, you also have complete *responsibility* for getting everything working. You have to design every part of the system, debug it, develop test equipment and procedures, and then support your system after the product is released and used in ways you never imagined. Your customers are locked into a complete solution from you or a competing system from your competitors; they can't take your best products and mix and match them with the best devices from another manufacturer. Such a control method is a **proprietary standard**—it is typically owned by one organization, and can only be used with the permission (implied or formal) of that organization. Some companies see this situation as a competitive advantage (I disagree).

De Facto Standards

A practice or technique may be used by so many in the marketplace that it becomes a **de facto standard**—one that is widely used, but not (necessarily) officially standardized. De facto standards could be based on one manufacturer's way of doing things or just accepted practice. For example, "stage left" and "stage right" aren't formally defined by a standards-making body, but instead are based on widely-used practice. These standards are informal and, therefore, not generally officially maintained or administrated by anyone.

Open Standards

Open standards are typically developed by volunteers under the auspices of a not-for-profit organization, and are open for any and all to use without license or restriction. Often, the need for an open standard becomes clear in the market after a number of proprietary or de facto standards duke it out, and users (who pay the bills) start screaming for interoperability. With a successful open standard, a customer, designer, or systems integrator can choose the best equipment for their application from any compliant manufacturer and still be more or less assured that all the equipment will work together. Components, systems, cables, and connectors can be standardized, sourced from multiple manufacturers, and made in bulk, reducing cost. Safety across an industry can be improved. New equipment, such as processors and test equipment, can become economically viable, since the combined market size created by the standard gives even small companies a big enough market to survive. All of this generally reduces the cost of doing business, something in which everyone is interested. Additionally, it's my opinion that successful open standards cause industry growth and expanded markets and the development of new equipment never imagined by the standard's creators. There are many examples of successful open standards that have expanded markets in our business: DMX512, MIDI, SMPTE, TCP/IP, and Ethernet are just a few. For all these reasons, open standards form the core of this book.

Common Attributes of Successful Standards
I've found that successful standards have a few things in common:

1. Successful standards are *pulled* into existence by the market; they are not pushed. They fill a clear commercial demand in the market, especially one driven by users. Often, this means that multiple, noninteroperable systems already exist in a market segment and users are screaming for interoperability.
2. They are limited in scope and ambition and optimized for some task. Consensus standards-making processes ensure this, since only a minimal level of functionality will be agreed upon by all the parties involved.
3. They are open for all to use. All of the standards (of which I'm aware) that have caused our market to grow have been open.
4. They leave clear room for expansion and allow shortcomings to be corrected.

I would argue that DMX, MIDI, and (full-duplex, switched) Ethernet are successful standards in our market that meet all of these criteria. All three were pulled (not pushed) into existence at the demand of users. They all provide functionality optimized for a particular task; they are all open for anyone to use; and they all left room for expansion. Ethernet is probably the standard that has had the most impact on our market, and it is fast becoming our universal control data *transport* standard because it does one and only one thing extremely well: transport bits. The bits could be lighting data, streaming audio, video files, or porn—Ethernet doesn't care.

Standards Development
Open standards are typically developed by a group of manufacturers, consultants, designers, and/or concerned users. This process, which often takes a tremendous amount of time, argument, and compromise, may result in either an open standard or a "recommended practice."

Open standards-making groups generally work under the auspices of a not-for-profit engineering society (e.g., the Audio Engineering Society [AES] or the Soci-

The Limitations of Standards

I have long been a proponent of open standards. However, I have learned over the years that there are limits to what should be standardized and how much a standard should attempt to do. In the early 1990s, I was working as a systems engineer for the (now long lost and legendary) lighting company, Production Arts Lighting (PAL); we were providing a lighting control system for a cruise ship show room. Looking at the drawings, it became apparent that on the backstage wall stage left was to be a mess of different electrical boxes: one sound box full of XLRs and some volume controls; one rigging control box for the connection of a rigging system remote; and a box of ours for house light control, remote focus unit connection, and so on. At that young age, this seemed like a stupid thing to me: Why, when the cruise ship line was spending so much, could they not have a standardized, nicely integrated, and well thought out panel—especially one right there on the stage where so many would see it?

Over the years, I've come to realize that not only were a group of individually constructed and coordinated panels cheaper, but, in fact, *better*. At PAL, we knew lighting systems, and we wired things a certain way, used certain brands and types of connectors, labelled things in a specific way that made the most sense for the lighting application, and used a specific kind of panel that was made efficiently and in a way with which we were comfortable working. Each company on the job had developed their own methods, and had arrived at a very effective way of working, optimized for their process and needs. To force all those vendors to work together on an integrated panel would have cost the client an enormous amount of money in coordination time, and, in the end, would have forced each vendor to compromise in some way, leading to a solution that was mediocre for all and optimized for none.

On the other hand, it wasn't a total free-for-all; some coordination (standardization) was done, where it would benefit the client. For example, to give a consistent look and fit everything into the small control room, the client specified the brand of 19" rack they wanted us to use. This was beneficial to the client, while not being a particular burden to anyone, since all the racks followed the very effective 19" rack standard. (The 19" standard has been very successful because it *only* defines how panels will mechanically fit, allowing anyone's panel to screw onto anyone's rack. It does not attempt to dictate on what side the power button should be—or if there should be one, what specific indicator lights mean, or how large or small a unit should be.)

Drawing the line between what can be effectively standardized and what should be customized is something I've been fascinated with for many years. For example, I was originally a strong advocate for universal, far-reaching control protocol standards with wide application. With such protocols, a designer would only have to learn and implement one approach and they would then be able to communicate with anything. What I didn't realize is that, for a variety of reasons, "universal" far-reaching standards efforts rarely achieve the designers' ambition. Instead, they typically end up either dying off, getting watered down (which may actually make for an improvement), or providing limited functionality to a small subset of the originally intended market. In those situations, it probably would have been more effective to simply address a market subset from the beginning. For example, MIDI Show Control (Chapter 23, on page 301) was designed with the ability to control nearly anything in a show. MSC did fill (and continues to fill) some specific cue-based control market niches very well, such as triggering "Go" commands on lighting consoles, or triggering cue-based sound FX playback systems. But MSC never did find universal, all-for-everything application, for both political and technical reasons. Why did this fine effort not reach its full potential, and why did others, like Medialink, a commercial universal control network, (covered in the first edition of this book) or the Audio Engineering Society (AES) AES-24 standard, which attempted to provide universal control for any kind of audio equipment (covered in the second edition of this book) fail? It's certainly not for technical reasons, since each was developed by smart engineers who were capable of solving all the technical challenges. I would argue instead that they failed for political and commercial reasons, and, most of all, because they tried to offer too much to too many.

ety of Motion Picture and Television Engineers [SMPTE]), which may exist solely for the purpose of developing and promoting standards. Alternatively, a standard may be developed within a trade association (e.g., the **Professional Lighting and Sound Association** [PLASA]), which exists to promote the common interests of an industry (a good standard can spur growth for the whole market). These engineering societies or trade associations typically work under the rules of national or international umbrella organizations, such as the American National Standards Institute (ANSI) or the International Electro-technical Committee (IEC). Open standards are generally free to use, but the standards documents themselves might be sold to help defray some of the administrative development costs.

Many of the most widely used standards on the Internet actually do not follow this structure at all; they were instead developed under the auspices of the free-wheeling Internet Engineering Task Force (IETF). The IETF does not belong to ANSI but it is overseen by the Internet SOCiety (ISOC), a not-for-profit membership organization whose role is to promote the growth of the Internet. IETF standards start as "Internet drafts," and then, after a comments and revisions process (typically conducted over an e-mail list), the document is submitted to the Requests For Comments (RFC) editor, who publishes the official RFC and assigns it a number. RFCs are never changed—they simply become obsolete by the issuance of subsequent RFCs. RFCs can also eventually be formalized into "standards track" RFCs by the IETF; those RFCs can eventually be turned into Internet Standards (STD). This process, though seemingly chaotic, has produced many of the technologies found at the core of the Internet.

Standards-Making Process
The standards process is typically started with the formation of a "working group" or a "task group" made up of concerned participants whose work is open to the public for review or collaboration. Generally, a subset of the group develops a draft of a proposed standard using as precise language as possible, and then releases the draft to the public for a period of review and comment. Under procedures established by the sanctioning body, the group resolves the comments, objections, or suggestions and, if all goes well after several rounds of this process, the group issues the final standard. The group then takes on the responsibility for maintenance, support, and future upgrades of the standard.

Due to the compromise inherent in the collaborative creation process, successful open standards generally end up defining "lowest common denominator" functionality. Cutting to the basic, minimal core of what the market really needs at that moment is the fastest route to a successful standard. Even so, the standards process often takes many months or even years, with members essentially (or actually) volunteering their time for the good of the industry.

MOVING ON

Now that we've covered some basics, let's move on to the rest of the subject, which I've broken down into five parts. Part 1, "Entertainment Discipline Overview," on page 11, gives a brief overview of each of the technical disciplines you would typically find on a show. Part 2, "Entertainment Control," on page 65, covers techniques and technology used across all the disciplines. Part 3, "Data Communication and Networking," on page 123, goes into depth into the backbone data transport techniques used in just about any show system, while Part 4, "Standards and Protocols Used in Entertainment," on page 239, goes into details of protocols used on shows. Part 5, "Show Control," on page 355 ties everything together discussing how to connect different kinds of systems together, with lots of examples.

Part 1: **Entertainment Discipline Overview**

To give some background on the field, we will go through a brief introduction to each entertainment discipline and list the most widely used control standards and protocols used in that discipline. Because many of the protocols and standards are used by more than one department, specific details can be found in Part 4, "Standards and Protocols Used in Entertainment," on page 239.

If you know all this stuff already, please feel free to skip ahead to Part 2, "Entertainment Control," on page 65.

Chapter 2: **Lighting**

The basic function of lighting is illumination, of course, but it can also provide stunning effects, help to set a scene, or focus and direct the audience's attention. Light without control isn't much use for most shows, and because of this, control has long been a critical part of lighting design and system functionality. Gaslights offered the first centrally controllable luminaires; they were dimmed using mechanical valves—less gas produced less light. "Conventional" lighting fixtures are, of course, now powered by electricity and have their intensities controlled by electronic, solid-state valves known as **dimmers**. And, of course, intensity is now only one part of the lighting control equation; color, angle, focus, video image, and many other parameters now are all part of the designer's palette.

The lighting market was basically created by the availability of open control standards. Few can now imagine a world where it's not easy, inexpensive, and convenient to connect virtually any control console to any manufacturer's dimmer, color scroller, video server, or a wide array of other devices.

LIGHTING CONTROL EQUIPMENT

There are generally three parts to a lighting control system: a control **console**, the control data distribution system, and the controlled devices, such as dimmers, color scrollers, and moving lights.

Lighting Consoles

Early controllers mimicked the positions of the operating handles on manual dimmers: a row of sliders would be set as a "scene" or "preset," and the operator would manually **cross-fade** between the presets. In this preset operation, every lighting level of each scene would have to be meticulously entered into the system and every parameter would be recorded into each cue. In other consoles, only the changes to each scene need be entered; these consoles are known as "tracking" consoles. Consoles typically have touchscreens, encoder wheels, and recordable presets that speed up the programming time for moving lights, LEDs, media servers, and similar devices that need sophisticated control.

Lighting control consoles fall into four basic categories (of course, like anything else, the lines between the categories are often blurry): multi-scene preset, submaster-based rock-and-roll/club consoles, fully computerized systems, and moving-light controllers.

Courtesy Strand Lighting

In simple terms, multi-scene preset consoles (like the one shown in the photo) allow one scene to be preset while another is being used to control the levels of the lights onstage. At the cue "go" point, the live scene is cross-faded to the preset scene; the cross-fade can be executed either by the on-board computer or manually.

Like a multi-scene console, a rock-and-roll/club console still allows control of each individual channel, but instead of (or in addition to) using preset cues, the show can also be run from "submasters," into which individual looks can be recorded or patched. This approach allows the designer quick, spontaneous access to different looks, which can be selected randomly. More recent higher-end rock-and-roll consoles are able to operate in both cue-based and manual modes.

Fully computerized "conventional" consoles allow "go-button" operation for well-defined, sequential shows, such as theatrical productions or corporate events. Each preset look is recorded along with the transition time to the next look, as a cue. In this way, the show lighting can be performed exactly the same way each night, provided the operator and

Courtesy ETC®

stage manager execute the cues at the appropriate points. Such consoles generally still retain some manual fading functions, so that talented operators can match lighting fades to onstage action; submasters may also be provided for a variety of uses.

Moving lights have many more control parameters than conventional luminaires; controlled parameters on a moving light may include intensity, color, multiple axis positions, gobo,[1] shutter, iris, focus, or even framing shutters. While some traditional consoles offer moving light functionality for a limited number of units, consoles designed specifically for the application are typically used for bigger shows.

For permanent installations, the cost and space requirements of a large console can be prohibitive, and having so many physical controls available to an untrained user may cause confusion and unneeded service calls. To solve this problem, many manufacturers have developed rack-mount versions of their systems with limited user controls; some have even gone so far as to put all the intelligence of a typical lighting console into the software of a standard personal computer. With these standard computer-based systems, the show can generally be run from either the computers directly or show information can be downloaded into a rack mount unit for playback.

---

1. A gobo, or "template," is a stencil-like metal piece used to create an image for projection by a luminaire.

Precedence

There are times where a controlled lighting device such as a dimmer or moving light might need to take control signals from two different sources. For example, the houselights in the venue might need to be operated from a permanent, architectural control system during daytime operations of a venue, but then be controlled by the show lighting system during the show. Alternatively, a console could have the red side lights programmed into two different looks; in either case, the system has to decide what to do. There are generally two approaches to resolve this issue: **highest takes precedence** (HTP), or **latest takes precedence** (LTP). In an HTP system, the highest control value for a parameter wins. So, if the red back lights are at 50% in fader 7, and 75% in fader 34, then they will be set to 75%, because that is the highest value. But if we were running in an LTP scheme, and I raised fader 7 to 100%, and then raised 34 to 50%, then the latest action—fader 34—would take precedence and the light would change from 100% down to 50%.

Dimmers and Patching

In the early days of electronic dimmer control, it was desirable to have the largest dimmer possible. Load circuits from the individual luminaires were "hard patched" into the dimmer's output through telephone operator–style patch cords hanging from a panel, a series of sliders over a matrix of copper bus bars, or through a number of other methods. As dimmer prices began dropping in the late 1970s, and computer control became widespread, it became cost-effective to have many small dimmers—one dimmer for each light (or maybe for two lights)—rather than a few large dimmers with many lights patched in.

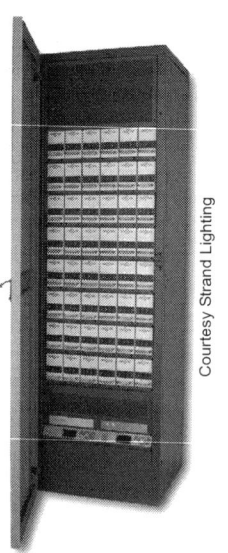

In "**dimmer-per-circuit**" systems, patching is done on the control side (rather than the load side) of the dimmer. In computerized consoles, this "soft-patching" is accomplished entirely within the realm of the controller. Designers and operators do not directly control the dimmers in these systems; they deal with virtual "channels," to which any number of dimmers can be electronically patched. Portable systems often still include some hard patching capability to provide maximum flexibility.

Early dimmers were operated manually: a handle on the dimmer was directly coupled to the wiper on a resistance plate or autotransformer. The advent of electronically-controlled dimmers allowed remote, centralized control to become a reality. In an electronic dimmer, the alternating current (AC) waveform is "chopped" to vary the overall average voltage output. There are a number of dimming techniques: forward phase control, reverse phase control, and "sine wave."

Forward Phase Control

Most dimmers use a pair of silicon controlled rectifiers (SCRs) to switch the power line—one SCR for each half of the AC waveform. The SCR is turned on at some point in the half cycle via a control signal on its gate; the SCR, by its nature, turns itself off at the AC waveform's zero-crossing. By varying when the SCR is turned on, a controller can alter the light's output level. This control method is known as forward phase control (FPC). Turning on the SCR in the middle of the AC half cycle generates a big spike, which makes noise, both mechanical (filament "sing") and electrical (RFI/EMI). To counteract such noise, a toroidal inductor, or "choke," is wired in series with the output of the SCR to soften the dimmer's turn-on transition. The time it takes for the current to start flowing, known as the dimmer's "rise time," is proportional to the size of the choke. Forward phase control is inexpensive to implement and has proven very reliable, but it has a few disadvantages: the choke adds weight and size to the dimmer module, and even good dimmers are never silent. In addition, the rise time is load dependent, and regulation of the dimmer's output can be no better than one half cycle delayed—once the SCR is turned on, only the absence of flowing current (an AC zero-crossing) can turn it off.

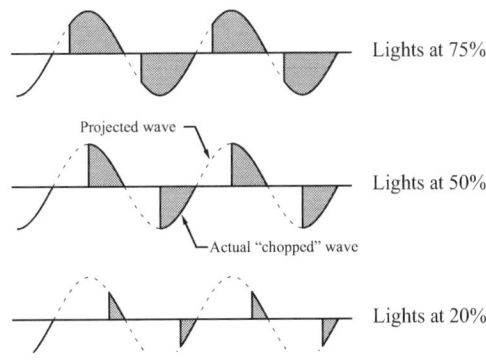

Reverse Phase Control

In the early 1990s, manufacturers started implementing reverse phase control (RPC) into their dimmers. Through the use of power devices such as insulated-gate, bipolar transistors (IGBTs) or metal-oxide silicon field-effect transistors (MOSFETs), the waveform can actually be turned off in a controlled fashion at any point *during* the half cycle. RPC lets the wave-

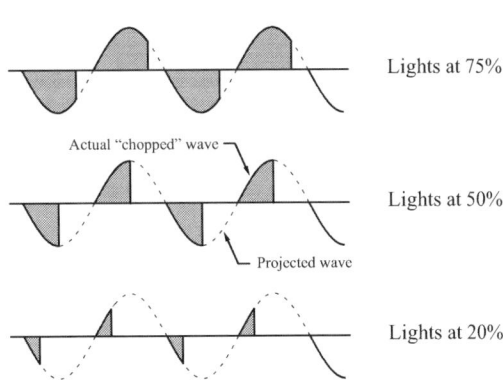

form ramp up in its normal sinusoidal fashion, starting at the zero-crossing; the power device is then turned off at the appropriate point in the half cycle. This approach has a number of advantages: Since there is no sharp turn-on spike, RPC reduces both electrical and physical noise, as well as the need for the bulky, heavy choke used to smooth out the waveform. RPC dimmers can actively sample the dimmer's output and shut off at any point; this allows dimmers to be designed that are virtually immune to short circuits and overloads, since regulation occurs within every half cycle.

The advent of chokeless dimmers enabled manufacturers to offer viable "distributed" dimming systems. Conventional dimmers are generally centrally located in racks, with large power feeds and individual load cables running to each light in the system. In distributed systems, bars of small numbers of dimmers can be distributed throughout the lighting rig, each with a small power and control feed. Since the dimmers themselves generate so little noise, they can be placed in audience or stage areas.

Sine Wave Dimming

With the availability of sophisticated, fast power switching devices, so-called "sine wave" dimming takes the "chopping" approach of FPC or RPC one step further by using high speed pulse-width modulation (PWM). With this approach, an electronic switching device is turned on and off at an extremely high rate (typically 40–50 kHz), creating a variable pulse-width output. The "duty cycle" of the pulses are varied to generate the desired output. If the dimmer is at its maximum, the pulses would be turned on continuously, creating a 100% output. If the dimmer is set to a lower level, the power line is switched to the "on" position in proportion to the level of the control input, and this generates the desired output. This chopped up output is then filtered to create a very smooth, variable amplitude sine wave output; manufacturers claim less than 1% distortion in the output.

Input wave

Pulse-width modulation

Output wave

Sine wave dimmers are more complex, and need more processing power than FPC or RPC dimmers. Correspondingly, they are significantly more expensive than conventional dimmers, but have significant advantages, especially for noise-critical applications. Since they create a more or less true sine wave output, they generate no more filament sing than if the lamp were connected to a constant source of power such as a wall outlet, even when the lamp is dimmed. Well-built sine wave dimmers also generate much less radio-frequency (RFI) and electro-magnetic interference (EMI); the weight and cost of the choke is eliminated, and they generate much less of a harmonic load onto the power line (a big issue in large systems). Additionally, with a sine wave output, these dimmers can be used to power virtually any type of device.

Moving Lights

The moving light has had a huge impact on the lighting industry. With these automated luminaires, parameters such as pan, tilt, focus, gobo, color, and so on are all remote-controllable. Control approaches for early moving lights were often proprietary, but, as often happens, open standard control protocols now dominate the market.

LED Fixtures

In the early 2000s, LED fixtures gained widespread use in the entertainment lighting market. LEDs have existed commercially since the 1950s, but are now inexpensive, bright, and adaptable enough to be viable for entertainment lighting purposes. In fact, the entertainment market is a significant niche for the larger LED lighting market, since we are often looking for very strong colors (something most people don't want in their living room).

Video Servers

Video has historically been a separate department, but with media "convergence" starting in the early 2000s, many lighting systems incorporate and control sophisticated video presentation systems. Systems which play back prerecorded media are typically referred to as "media servers," and all this technology is covered in Chapter 5.

Other Lighting Equipment

The development of open control standards in the lighting industry led to a huge explosion of new types of lighting devices, all of which need control. Gobo rotators, which take a projected cut-out image and rotate it; color scrollers (see photo), which move a scroll of colored gel in front of a light; and dozens of other types of devices all have found a place in lighting control systems.

LIGHTING CONTROL APPROACHES

Now that we've covered various types of lighting devices that might be found in a lighting control system, we can move on to the heart of the control system itself.

Analog (0–10V) Control

Along with the advent of easily remote-controllable dimmers came the first de facto dimmer control "standard": analog **direct current** (DC) voltage control. The concept is quite simple: One wire is run from the console to each controlled dimmer and a DC voltage corresponding to the level of the dimmer is sent down the wire. 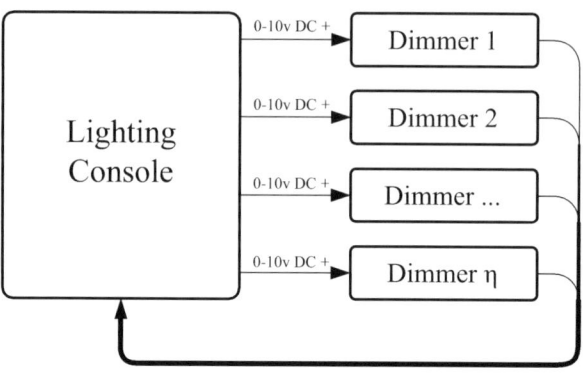 Zero volts represents an off condition; 10 volts (generally) represents a dimmer output of 100% or "full on." In analog-controlled dimmers, this control voltage is received by trigger circuitry, which generates SCR firing pulses at the proper times to achieve the desired light output. While digital control using DMX (discussed below) is used in the vast majority of installations, analog control can still be found in small systems in some low-budget venues, such as clubs, discos, and small theatres. If you're interested in using this low-tech, simple system, please consult the PLASA Standard ANSI E1.3 – 2001 (R2011), 0 to 10V Analog Control Specification.

DMX512-A

The "ANSI E1.11-2008 Entertainment Technology USITT DMX512-A, Asynchronous Serial Digital Data Transmission Standard for Controlling Lighting Equipment and Accessories" started humbly, but is now the most successful and widely-used entertainment lighting control standard in the world. It has become so successful, in fact, that it is now used outside the lighting world as well—it is covered in depth in Chapter 19.

RDM

ANSI E1.20 - 2010, Remote Device Management over USITT DMX512 Networks implements remote control functionality over existing DMX infrastructure. It's covered in Chapter 20, "Remote Device Management (RDM)," on page 259

Proprietary Ethernet and IP Control Solutions

A typical production "tech table" may have several console monitors for the lighting designer, and a remote console data entry unit. Backstage, we may want a few DMX outputs on one side of the stage for some dimmer racks, a few more DMX outputs over the stage for some moving lights, and a remote monitor for the stage manager. In the catwalks, we may want a few DMX outputs and a connection for a remote focus unit. Each of these devices would typically need a different type of cable run separately to or from the main lighting controller and some sort of patching and distribution in the control room. With an Ethernet-based system, a single Ethernet network can be run throughout the performance facility, and the appropriate lighting "node" is simply connected wherever you need control, a monitor, or a remote entry or focus station. Many lighting manufacturers make proprietary systems for exactly these purposes, but most of these implementations are not compatible with any other. Additionally, many consoles these days have a standard Internet Protocol address (see Chapter 17 for more information), and can now be controlled over standard network connections using strings of characters, just as if the remote system were typing commands into the system directly.

Art-Net

One manufacturer, Artistic Licence, made its Ethernet implementation free for all to use, and their approach is discussed starting on page 252.

ACN and sACN

In late 2006, after many years of work, PLASA completed a sophisticated, open, interoperable, network-based lighting control protocol, known now as E1.17 – 2009, Entertainment Technology — Architecture for Control Networks (ACN). I hope that ACN (and its subset, Streaming ACN) will have a major impact on the lighting control industry, and perhaps beyond, because ACN was designed for control of any type of equipment. sACN is covered starting on page 255, and ACN is covered in Chapter 21.

MIDI and MIDI Show Control

MIDI Show Control (see Chapter 23) is a widely used way for interconnecting lighting consoles so that multiple consoles can trigger from a single go button. Additionally, standard MIDI messages (see Chapter 22, on page 281) can be used to trigger submasters and other similar console features.

Chapter 3: **Lasers**

Laser systems provide spectacular effects for a wide variety of productions, from city-scale light shows to concerts and corporate events. The basic concept behind a laser light show is very simple: Using one or more colored, very narrow beams of light (provided by one or more lasers), a pattern or graphical picture is drawn very rapidly—so fast that we are no longer able to perceive a moving laser dot, but instead perceive a continuous line or image. These effects are generally provided as a service by laser light show companies, and, traditionally, these companies custom-build or integrate their own proprietary systems.

The lasers used in laser light shows are powerful enough to be dangerous, and can cause skin damage, blind spots, or even permanent blindness, and damage equipment such as cameras. Always consult an expert before working with any laser system, and be sure they consult with the appropriate authorities—laser use for shows is tightly regulated, especially in the US.

LASER SYSTEM CONTROL EQUIPMENT

Laser systems generally break down into series of standard component types. First, of course, there are the laser light sources, which produce powerful, collimated beams. Lasers for light shows were traditionally gas lasers producing individual colors, but in the early 2000s many light shows moved from these large, fragile, water-cooled monsters, to compact, durable, Diode-Pumped Solid-State (DPSS) "white" lasers.

Next in the optical chain is traditionally a color-splitting and blanking device, which acts as a sort of remote-controlled electric prism. This device is called a polychromatic acousto-optic modulator (PCAOM), and allows only a certain color to pass at any given instant, with all the remaining light sent out as a "waste" beam. These devices work at blindingly fast speeds, allowing the subsequent components of the laser system to deal only with a particular color at any given fraction of a second. Solid state diode lasers and DPSS lasers offer the advantage of being "internally modulated," meaning that color selection and blanking can be achieved internally by turning the laser beam itself on and off quickly and the PCAOM is not needed.

The final stage of a typical laser projector is the "image engine," which drives the beam(s) into some desired graphical form. For "vector" (line-based) graphical images, typically two fast-moving, servo positioners (often called "galvos", see photo) with mirrors attached are used to "scan" the beam (one for the X axis and one for the Y). For raster (video) imagery, RGB solid-state lasers can be used with various technologies; these systems can produce a projected video image, and some have a very large depth of field, allowing the image to be in focus over uneven projection surfaces.

Courtesy Pangolin Laser Systems

So far, we've only talked about hardware, but a crucial part of a laser system is, of course, the software, which takes a graphic image from the mind of the designer and turns it into raw signals that control the projector components.

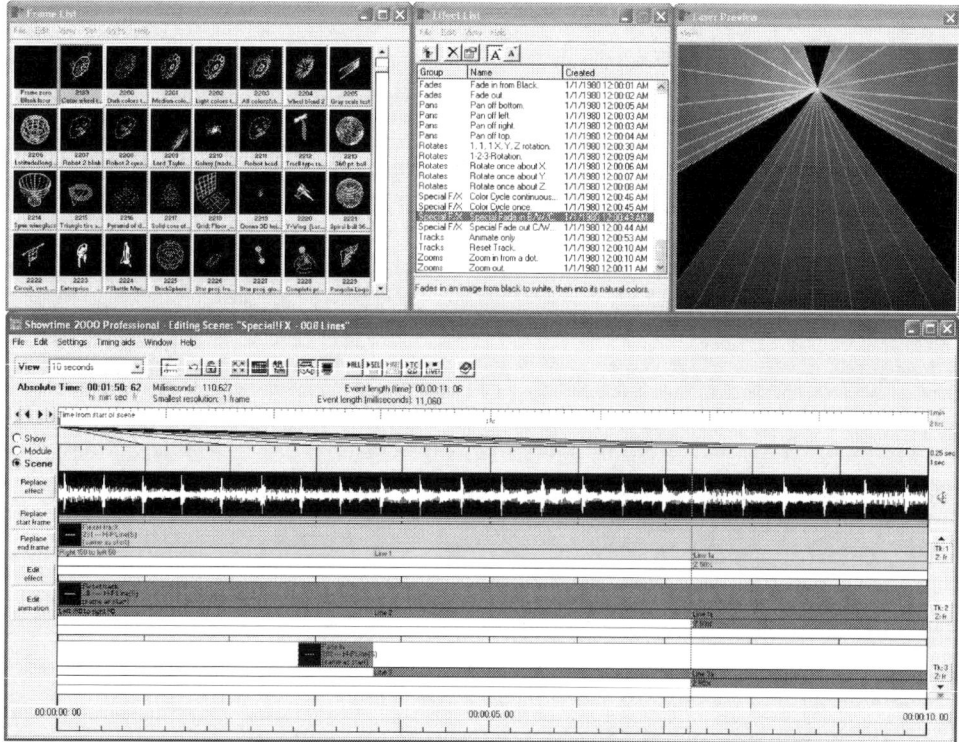

Courtesy Pangolin Laser Systems

This kind of control software can control a laser projector directly, or signals from the software can be digitally recorded and played back.

LASER CONTROL APPROACHES

Many laser light show systems are custom built and use proprietary control protocols. However, there are a number of standard methods used in a variety of systems, and we will cover a few of them here.

ILDA Standards

The International Laser Display Association (ILDA) has standardized various aspects of laser systems and they have developed "The ILDA Standard Projector," which standardizes subjects such as scanner tuning, connector and pin out specifications, and recorded track assignments. See `http://www.laserist.org/` for more information.

DMX512-A

The continuous nature of DMX (Chapter 19) makes it well suited to some aspects of laser control, but it is generally neither precise nor fast enough to control devices, such as scan heads, directly. However, many laser systems generate DMX signals to control conventional lighting equipment, relays for general purpose usage, or other similar devices. In addition, some laser control systems can accept DMX input for live control over aspects of the laser system, or for the selection of pre-programmed graphics from a standard lighting console.

SMPTE Time Code, MIDI

Many laser systems can accept or generate various show control signals, such as MIDI (Chapter 22), MIDI Time Code, or SMPTE (Chapter 25).

Chapter 4: **Audio**

Audio plays a critical role in many productions: It can be used to amplify a performer's voice so that he or she can be heard in venues larger than would otherwise be possible; it can connect the audience emotionally to musical performances; or it can set an aural scene through the use of sound effects.

Performance audio systems traditionally break down into two basic types: reinforcement and playback. Reinforcement systems use microphones, mixers, amplifiers, and speakers to bring live music, singing, speech, or other sounds to a larger audience than would otherwise be possible acoustically. These systems are used in musical or other live productions—everything from rock and roll to Broadway theatre. Playback systems present prerecorded music, dialogue, or effects to an audience, such as those found in legitimate and corporate productions, special events, theme parks, product launches, and even rock-and-roll concerts (although the audience may not know it). With the explosion of digital sound technology, more and more productions are using hybrid reinforcement/playback systems—the (human) reinforcement mixer for a Broadway show is able to press a single button and initiate complete sound-effects sequences. Conversely, a playback operator for a corporate presentation could also control a few channels of microphones. However, because sound and audio is so complex and is significantly affected by environmental conditions, performer variability, and other factors, many audio systems, while digital, are entirely manually operated—although the (human) mixers may use considerable computer horsepower to make their mixing jobs easier.

AUDIO CONTROL EQUIPMENT

In the world of lighting, there is an increasing number of intelligent devices distributed throughout a performance facility, each of which needs control communication. In the audio world, however, the opposite is taking place: With the advent of inexpensive yet high-power **Digital Signal Processing** (DSP) technologies, self-powered loudspeakers, and other innovations, more functionality is being put into less physical units, and it is now possible to have an extremely sophisticated performance audio system made up of very few components. However, a wide variety of audio devices still need control, and a few of those types are covered here.

Audio Playback Devices

The digital explosion has affected audio playback perhaps more than any other area in audio. Analog tape is long gone for virtually all show applications. CD players, and other similar digital devices, are generally now used only for testing, or very simple playback on shows. The vast majority of show audio is now being played back from some sort of hard disk-based playback system; one form of this is a linear, multitrack recording and playback deck.

Courtesy JoeCo Ltd

Another form of hard disk playback is software running on a standard PC, or even a fully computerized audio "server."

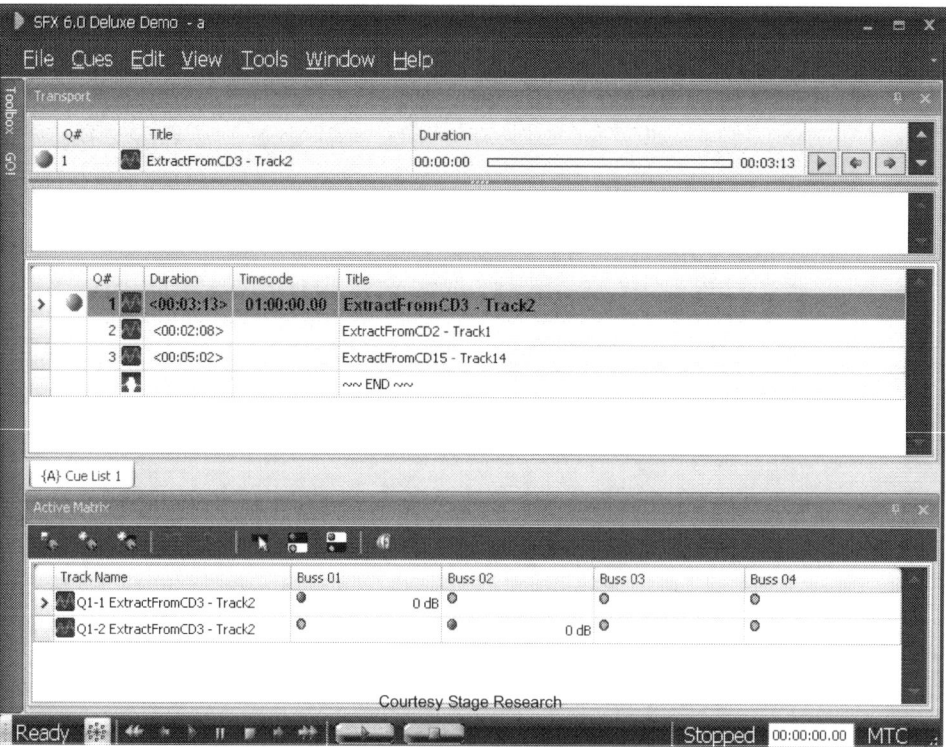

All offer near-instantaneous, random-access playback capabilities, with cue durations limited only by the amount of available disk (and/or memory) space.

Mixers and Matrices

The core of any traditional audio system is the "mixer," often called the "console" or "desk," which routes, distributes, processes, and manipulates audio. Traditional performance audio mixers come in two basic flavors: front of house (FOH) or monitor. Performance-oriented FOH consoles typically combine a large number of inputs into a smaller number of outputs; monitor consoles typically route a smaller number of inputs to a large number of outputs. Of course, like everything else these days, the line between these two types of mixers is getting blurry; many mixers are now available that do both tasks and are reconfigurable with the push of a button.

There are an ever-increasing number of digital systems that offer sophisticated automation and recall capabilities; some even allowed mixing DSP functions to be moved backstage, near the microphones and speakers, while the human operator mixes on a remote, all-digital "control surface."

Other digital mixing systems do away with the control surface altogether, and use software "front ends." These systems are generally not designed for seat-of-the

pants live operation (unless used with complimentary control surfaces), but often work in a "live assist" capacity on complex live shows, or to completely automate "canned" shows. These systems are less like a traditional mixer and more like a digital "matrix" (named for the schematic appearance of the unit's signal flow), which allows any or many inputs to be routed (sometimes proportionately) to a number of outputs. Some of these units are designed specifically for show applications, are cue-based, capable of sophisticated timed and triggered transitions between programmed presets, and can be controlled in real time.

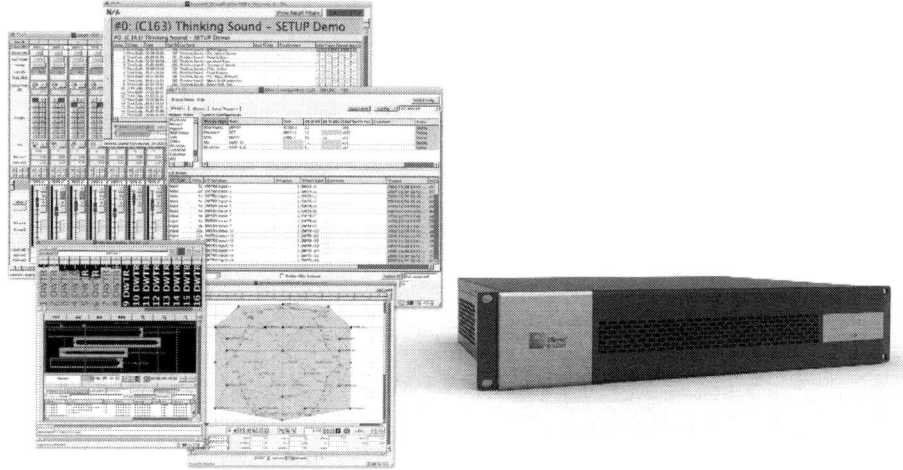

Courtesy Meyer Sound Laboratories

Other matrix systems are designed primarily for the permanent installation market and are more flexible in their configuration, but less flexible in their transitional and cueing ability, but are therefore generally cheaper.

Audio Processors

Using audio "processors" such as equalizers, signal delays, crossovers, reverberation, and effect units, designers can control many aspects of audio signals. Many of these processors are capable of interfacing with control systems in a variety of ways, or even come with sophisticated software that runs on an external PC for control or configuration.

Amplifiers

Amplifiers are the workhorses of audio (equivalent to lighting dimmers in many ways), taking the small signals from mixers or processors and increasing the magnitude of those signals to drive loudspeakers. In the past, amps were usually set to one level using local manual controls; gain manipulations were done by varying the level of the input signal. Now amplifiers are increasingly capable of remote-controlled operation, and many manufacturers have proprietary (and non-interoperable) systems, which allow centralized control and monitoring of nearly any amplifier parameter, no matter where in a performance facility amps are located. None of these systems integrate with one another (these companies apparently view this as a competitive advantage—I disagree).

Self-Powered Speakers

Self-powered speakers combine equalization, crossover, protection, and amplification functions in the speaker cabinet, minimizing complex signal connections and adjustments and length-limiting high-power speaker-cable runs. With so much intelligence in every speaker and speakers distributed all over a performance facility, remote control and monitoring become even more important.

AUDIO CONTROL APPROACHES

The wide variety of control equipment available in the audio field often requires a customized approach for applications. However, there are a number of standard interconnection methods that we can introduce here, and cover more completely in later sections.

Contact Closures

Playback and other kinds of audio devices often offer simple GPI (General Purpose Interface) or contact closure interfaces (see "Contact Closures," on page 93, for more information). Contact closures offer the designer a crude way to interface with the device, and many times a crude interface is better than none at all (or is all that is needed). These interfaces rarely can do more than simple "play" and "stop" types of commands; trying to tell a unit to cue up to a certain sound or "back up three cues" using contact closures may prove difficult. Serial interfaces offer much more control.

Serial Standards

Many devices in audio, particularly those that have roots in the broadcast market, can be controlled using simple ASCII commands (page 126) sent over an RS-232 or 422 serial port interface (Chapter 15). Each manufacturer makes up their own command set; for example, a playback device may be designed to start playing when it receives the ASCII character "P." While this means that each device may have a different command set, the fact that the manufacturer put a serial port on the device at least gives a system designer a fairly straightforward and powerful way to interface with it, and control can be quite sophisticated with a well-designed unit. Since every manufacturer designs their own control language, few standards can be covered here, but some of these units follow de facto standards.

IP and Ethernet

Many audio playback and processing devices contain an Ethernet port (Chapter 16) over which ASCII commands can be sent, typically using TCP/IP (Chapter 17). There are no widely accepted standards for this type of control, but many manufacturers provide a proprietary command set which can be easily implemented. Some manufacturers implement their proprietary control approaches using "Open Sound Control" (Chapter 26). As of this writing, work is also underway on the Open Control Alliance (page 341), which is intended to provide an open command set and control architecture.

MIDI

MIDI has had a major impact on the music and sound industry and is used widely for sound control. See Chapter 22, on page 281, for more information.

A Quick Note on Word Clock

In many applications, audio samples from all the equipment in the system must lock together exactly in time; in other words, all the samples must be exactly in phase. For this reason, a separate "word clock" signal is often distributed around part of an audio system to "sample lock" all the audio gear. Further discussion of this topic is outside the scope of this book.

Audio Transmission Networks

There are a number approaches for transmission of high-quality digital audio throughout a facility or venue. Because these approaches are designed for transmission of audio rather than control, they are generally outside the scope of this book. However, many of the systems feature some control capability, so we will briefly mention them here.

Audio-Video Bridging (AVB)

As of this writing, an exciting new, open audio and video distribution system is just becoming available: **Audio Video Bridging** (AVB). AVB is being developed under the auspices of the developers of Ethernet: IEEE. The system is based on standard Ethernet, but uses special Ethernet switches which can actually measure the propagation time of frames of data through its system, and pass that information along to the receiver for later reassembly. The AVnu Alliance is a trade association promoting the use of AVB and certifying equipment, and is a great resource for more information: `http://www.avnu.org/`.

Cobranet

Cobranet, a widely used, proprietary technology from Cirrus Logic, is one of the more widely used Ethernet-based audio transmission systems. In addition to transmitting a huge number of audio channels, it can also transmit some control information, typically the SNMP Protocol—see "Simple Network Management Protocol (SNMP)," on page 344. For more information on Cobranet, see `http://www.cobranet.info/`.

Ethersound

Ethersound, a proprietary system developed by Digigram, is a low-latency, Ethernet-based audio transmission system. Ethersound also offers some control capabilities. For more information, see `http://www.ethersound.com/home/index.php`.

Audinate Dante

Audinate is a company that makes Dante, a widely used proprietary network scheme that runs on standard Ethernet using UDP/IP and either unicast or multicast broadcasting. For more information visit `http://www.audinate.com`.

Audio Engineering Society AES-50

AES-50 and Sony's related HyperMac protocol, are high-performance, high-quality transmission systems, designed to deliver audio in a point-to-point manner. The systems can also carry Ethernet frames as part of their data streams. For more information, see `http://www.aes.org/`.

MADI

MADI is an audio interfacing system used by the company RME, and is increasingly being used in the entertainment market. The system offers 64 channels of 24 bit audio at sample rates of up to 48 kHz, or 32 channels at 96 kHz. Data can be transmitted over distances more than 100 meters, over either coaxial cable and BNC connectors or fiber. For more information, see `http://www.rme-audio.com/`.

Chapter 5: Image Presentation

The quality and sophistication of large scale, high-resolution imagery is constantly increasing, while the cost is simultaneously decreasing, and spectacular imagery is now being fully integrated into all sorts of live productions, from tiny shows to mega-spectaculars. In the past, large-scale images of any kind would likely have been created and presented using film, but video has now displaced film for just about everything in a live show except legacy applications.

IMAGE PRESENTATION CONTROL EQUIPMENT

Much of the equipment used in a typical imagery presentation system is there to create, recall, display, or process the imagery. However, we are concerned here with *controlling* all that gear, so let's start by covering the types of image presentation equipment that can be controlled.

Video Servers

The primary source for video playback is the **video server**, which typically plays back a digital video file from a hard disk. These systems can either be software running on standard PCs, hard

Courtesy doremi labs, Inc.

disks, and video cards, or can be purpose-built systems (see photo). Many servers are capable of a wide variety of real-time control over parameters, such as geometry and playback speed, and work with and manipulate in real-time multiple layers. External control is an important feature of video servers, since they are often integrated with other performance elements.

Computer Presentation Control Equipment

With the digital video explosion, the line distinguishing video playback from computer graphics can become quite fuzzy. In a corporate event, it's quite possible that a video server will be used to play back prerecorded media, and a CEO or other presenter may want to use a computer presentation program such as PowerPoint® or Keynote for his or her presentation.

However, other professional solutions (pictured) are available that provide imagery edge-blended across multiple screens in a more professional way, and allow the content to be "authored" in the venue and updated in real time with no rendering time. These professional systems typically are built on standard PC's connected using a network.

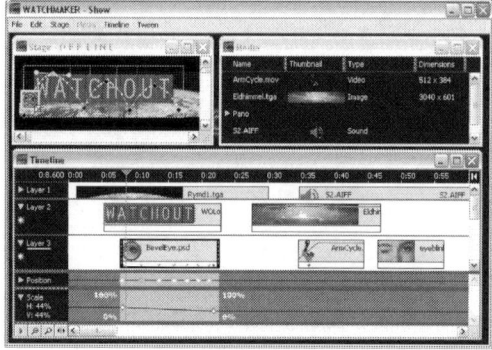
Courtesy Dataton AB

Digital Disk Players

Digital video disk or digital versatile disk (DVD) were—and Blu-Ray players are—a staple of the consumer playback market, and while you might find one occasionally on a show, they have been mostly replaced by video servers.

Video Routers/Switches

Video switchers, like audio mixers, are typically used to mix multiple video sources—cameras, video servers, effects generators—into fewer program outputs. Highly sophisticated switchers are available, and external control (typically serial or network-based) is usually available on professional models. There are two general types of switchers, those for live production applications (such as a live broadcast) and those for installation purposes, where the level of required control doesn't need to be as sophisticated or flexible.

Photo by Dorian Meid

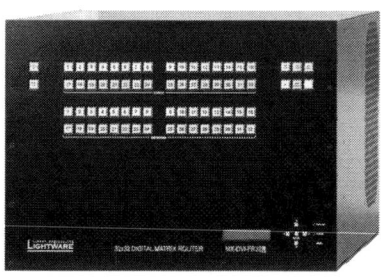
Courtesy Lightware USA

A **routing matrix** or "crosspoint matrix" is a similar type of device, but, unlike a switcher, which blends or switches many inputs into a few outputs, a routing matrix can patch many inputs to many outputs simultaneously and independently. A number of audio/video routing matrices—capable of switching both audio and video—are remote-controllable, and have many applications in the show environment.

Display Devices

Professional video monitors are generally capable of external control, if only for simple things such as power on/off and input selection. Video projectors and other large-format display devices, such as LED displays, can accept a wide variety of input sources, and many have network connections for control of items such as power and lamp on/off, input selection.

Film Presentation Equipment

You're unlikely to see film on a live show anymore, but you might encounter it in legacy applications, so here's a brief introduction. Still "slides" were a staple of corporate presentations for many years, and large-format slides were often used to create high-resolution images for large-scale events, either outdoors (i.e., projecting on the side of a building) or indoors (i.e., projecting huge and highly detailed background images). Kodak made the de facto standard 35 mm slide projectors, which used a circular "carousel" to sequence a series of slides, typically 80 or more. Large format projectors each had their own media standards, and the media could be static or changed in a random-access way, or even "scrolled."

Motion-picture projectors range from 8 mm home projectors, to the commercial 35 mm projectors found in movie theaters, to 70 mm custom systems such as IMAX® or Showscan.® Most motion-picture projectors, however, have not been designed for external control, since in traditional film applications, the media contains all the information needed for the presentation (image and sound). In a traditional movie theater application, the projector is simply started and rolled until the show is over; houselights can be dimmed automatically based on sensors that read foil-tape tabs on the film or by other automatic systems. Over the years, however, some film projectors have been built that can generate or chase external control signals, such as time code.[1]

IMAGE PRESENTATION CONTROL APPROACHES

Image presentation systems can generally accept one or more of a variety of open control protocols, with the most common and powerful available sent over a network or serial link.

IP and Ethernet

TCP/IP (Chapter 17, on page 181) is a protocol stack used at the core of many networks, and in our market, it is often used in conjunction with Ethernet (Chapter 16, on page 157) as the backbone of many presentation systems. In addition, many

---

1. When I worked for Associates and Ferren in the 1980s, we made time-code controllable 35mm film projectors for bands such as Pink Floyd and Roger Waters.

of these systems can accept commands over the network.

Serial Standards
Many types of image presentation equipment can be controlled using simple ASCII commands—see "Character Encoding," on page 126—or other control data sent over an EIA serial connection, such as 232 or 422—see "TIA/EIA Serial Standards," on page 147. Proprietary implementations from a few manufacturers became de facto standards and are good examples of typical control implementations. Two of these are covered later, in the "Pioneer LDP/DVD Control Protocol," on page 347 and the "Sony 9-Pin Protocol" section, on page 350.

Time Code
SMPTE Time Code's roots are in video editing, but it has gone well beyond its original applications and become ubiquitous for a variety of time-based entertainment applications—everything from synchronizing audio decks to controlling huge, time-based live shows. Many kinds of image presentation equipment can either read or generate SMPTE or MIDI Time Code; both are covered in Chapter 25, on page 321.

DMX512-A
DMX was designed to control lighting equipment, but its success as an open lighting control protocol, along with the increasing integration of video and lighting systems, has led to the somewhat peculiar situation where DMX is now widely used as a control protocol for highly sophisticated video servers. DMX is covered in detail in Chapter 19, on page 241.

MIDI
MIDI is sometimes used to control VJ (video jockey) systems. See Chapter 22, on page 281, for more information.

Contact Closures/General Purpose Interfaces
Some playback and other kinds of video and film presentation devices offer general purpose interface (GPI) or "contact closure" interfaces. These give the system designer a crude way to interface with the device, and many times a crude interface is better than none at all, or is all that is needed. These interfaces can rarely do more than simple "play" and "stop" types of commands. Contact closures are covered in detail in the "Contact Closures" section, on page 93.

A Quick Note on Video Sync
In video studios, video frames from all equipment must lock together exactly in time; in other words, all the frames must be exactly in phase. For this reason, a

separate video "sync" signal (outside the scope of this book) is often distributed around a studio or installation to synchronize all the video gear. This signal is often called the "house" or master sync, and equipment synched in this way is often referred to as being "gen-locked."

Chapter 6: Stage Machinery

"Stage Machinery" covers a broad range of automation control systems that are used in scenic-related applications on shows, including stage elevators, scenic automation systems,[1] automated rigging systems, mechanized props, or theme park "show-action equipment." Since such systems offer control over motion, they are often called **motion-control** systems, a term borrowed from the industrial automation industry.

Stage machinery systems generally fall into two categories, though the line between the two can, of course, be blurry. Systems in the first category, often called "stage equipment," are usually permanently built into a performance space, and include machines such as orchestra lifts, automated rigging systems, permanent turntables, and elevators. The second category is typically installed on a per-show basis, to move scenic pieces across a show deck on a track, "fly" them above the stage, or move them up and down across the stage itself. Each moving piece in a system is typically known as an "axis"[2] or "effect."

Stage machinery, of course, is inherently dangerous. Always consult an expert before attempting to do anything in this field, and be sure to check any applicable regulations.

MACHINERY CONTROL SYSTEMS

For our purposes of study, we can break down machinery control systems into four basic parts: sensing/feedback, control, drive, and emergency stop (see block diagram below). "Sensing/feedback" components provide information to the control system, such as the position or current velocity of a scenic platform. The "control" part of a system usually takes inputs from encoders and instructions from operators and sends control outputs to the "drive" part of the system, which is what actually regulates and/or delivers the power to the devices that provide the

---

1. Many scenic automation operators call their systems "show control" systems, but this is incorrect unless they connect more than one show discipline together (see "What Is Show Control?," on page 2, for a definition).
2. The "axis" term comes from the industrial motion control market, where they are often controlling systems with X and Y or X/Y/Z axes.

motion. Emergency stop systems, of course, stop machinery in case of an emergency (such as a performer in the wrong place at the wrong time).

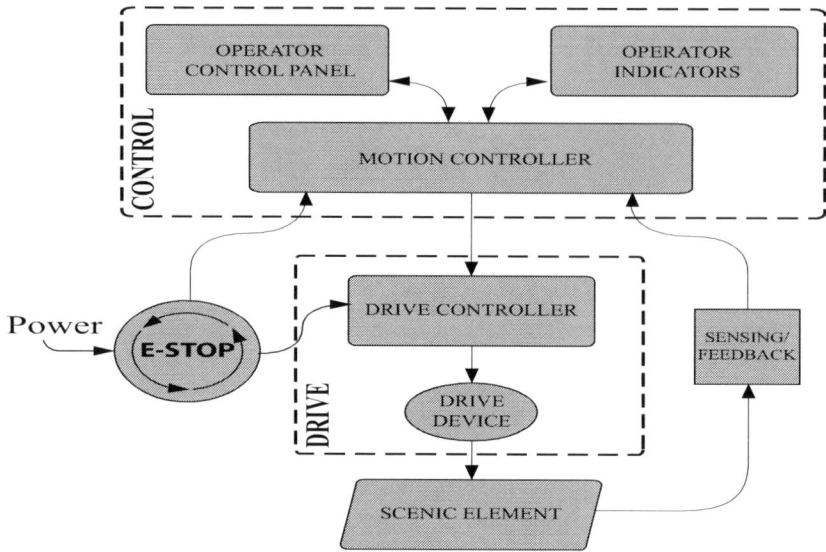

Let's look at each part in a bit more depth.

SENSING/FEEDBACK

Most motion-control systems use a closed-loop structure (page 79), and we generally close the motion loop with positional and velocity feedback.

Positional Feedback

There are three basic kinds of positional feedback: limit switches,[3] analog, and digital. Each approach indicates to the control system where a machine is within its range of travel; the system can then compare the actual position with a "target" position indicated by the operator, and determine how and in which direction to move the machine.

Limit switches are simply mechanically actuated switches (more in "Limit Switches," on page 90), and are the crudest and most limited form of positional feedback, but they are simple to implement and extremely reliable. With limit switches, the only way to change a desired target position is to physically move the switch (or the cam or other device actuating the switch). For a simple system,

---

3. Actually a form of digital (on/off) feedback, but we will treat this separately here.

limit switches are adequate; in larger, more sophisticated systems, limit switches are generally used only for "end-of-travel" (EOT) applications.[4]

With analog positional feedback, a proportional analog signal is used to encode the position of the controlled device—a particular signal indicates a specific position. The two primary analog feedback devices are potentiometers and resolvers. Potentiometers ("pots") are simple resistive devices that output an absolute DC voltage proportional to shaft position: five volts could mean 15 feet across a stage, 5.9 volts could indicate 18 feet. A resolver is a sort of rotary transformer, with one winding on its shaft and others in the resolver housing. At a certain angular position, a particular AC voltage is induced to the resolver's outputs, and this signal can be decoded into positional information. Both pots and resolvers are absolute devices: a certain position of the device's shaft always equals a certain output (even when they have just powered up), and this is a beneficial aspect for critical systems. This also means that the entire motion of a machinery system must be geared down to one (or a few) revolutions of the resolver or pot. For this reason, and the fact that pots and resolvers are more difficult to accurately interface with digital systems at high resolution, they are rarely used any more.

The most common devices for digital positional feedback are rotary shaft encoders or linear encoders (more in the "Encoders" section, on page 90). Since much of scenic motion starts with a rotary drive mechanism, such as a winch drum, and the fact that commonly available linear encoders are generally far too short for large-scale scenic encoding applications, rotary shaft encoders are found more frequently in entertainment applications than are linear encoders.

Velocity Feedback

All the methods we've discussed so far indicate to the control system the *position* of the scenic element; for true motion control, we need to know not only where an object is, but also how fast it is going; we need velocity feedback. As you might expect, there are both analog and digital methods for generating velocity data.

To generate an analog velocity signal, a generator, or "tachometer," is used. At a particular rotary velocity, a specific voltage is output; the control system can continuously compare this voltage to the target velocity and adjust the system accordingly. Older systems often used digital feedback for positional information and (analog) tachometers to close the velocity loop. Analog tachometers are rarely used anymore, because digital systems can actually derive the object's velocity from the change in its position over a given small increment of time.

---

4. Better to hit a limit switch and stop the system than hit the physical "end of travel"!

CONTROL

The feedback technologies discussed in the previous section indicate the position, velocity, or other information about the devices that need to be automated. Now, let's look at the actual parts of the system that control the motion of the machinery.

User Interface

We must, of course, tell a machinery system to do something before it can do it, so controls and operator indicators are typically used (more in Chapter 11, on page 89).

Motion Profiles

One of the key jobs of a motion control system is to create and maintain a **motion profile**, which determines how a motion-control axis will accelerate, what top speed it will reach, and how it will decelerate. This acceleration (change in velocity) is typically either programmed in as a "slope" (slope of the velocity curve if graphed out) or as an acceleration time. "Trapezoidal" profiles have a constant acceleration and deceleration slope; "S-curve" profiles use more gently changing slopes.

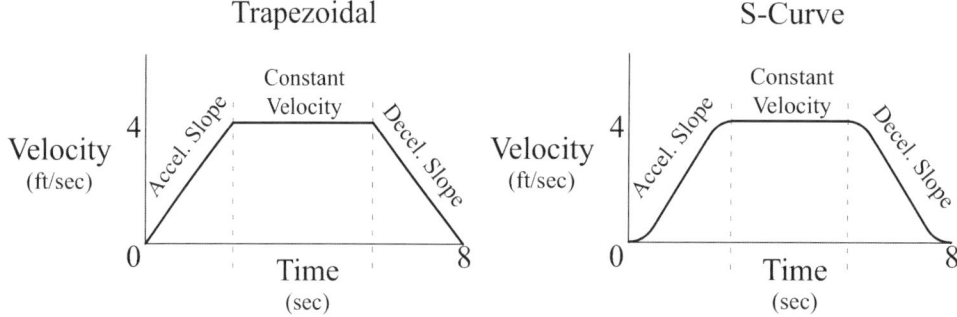

Machinery Controllers

The operator input and positional and velocity information must all be acted on by some controller, which tells the drive side of the system what to do. Since the factory automation and process control markets are so much bigger than the entertainment market, little equipment has been designed with the needs of our industry in mind. As a result, most off-the-shelf systems are not able to operate in cue-based modes (critical for our industry, of course). So, off-the-shelf units are frequently used at the core of entertainment motion-control systems, with an additional computer used as an operator "front-end" to handle operator interface, cue editing, data storage, and execution.

Programmable Logic Controllers (PLC)

Originally designed as a substitute for relay-logic control systems (such as those found in controls for elevators and other large machinery), programmable logic controllers (PLCs) have become remarkably sophisticated—math, HMI, and sophisticated motion-control capabilities are all built-in.

Photo: Andrew Gitchel

PLCs: are low-cost, heavy-duty industrial computers with pre-wired, optically isolated inputs and relay, transistor, solid-state-relay, or even analog outputs. In a typical application, a PLC's inputs are connected to control or limit switches, encoders, and other input devices. The outputs of the PLC are connected to motors, motor drives, indicators, solenoids or solenoid valves, contactors (which can drive heavy loads), or any other device needing control.

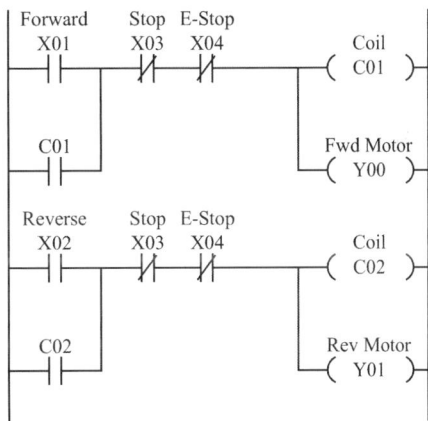

The internal program that runs the PLC is written for each application by the user or system integrator, generally in a "ladder logic" language, which is a sort of software-executed relay logic schematic (graphically, the program looks something like a ladder, hence the name—see the figure). The PLC is programmed using various types of "elements," which are analogous to relay contacts and coils (from its roots in relay logic). After the PLC reads the input conditions, it evaluates the status of the inputs, processes that data based on the user program, and then generates the outputs.

Dedicated Motion Controllers

While PLCs are general-purpose industrial controllers, dedicated motion controllers are also available that are intended specifically for motion control applications. They connect directly to operator interfaces, encoders, and drive systems, and can work based on an internal program written by a user or respond to external control commands. The programs used in these controllers are optimized for this task; and can include direct commands for things like acceleration or target position, rather than coding all that into a complex ladder logic program.

DRIVE DEVICES

"Drive" here is intentionally vague, as many machinery and motion controllers can control a variety of drive technologies: electric devices, electric motors, and electrically-actuated hydraulic or pneumatic valves.

Relays/Contactors

Relays and contactors, discussed in the "Relays/Contactors" section, on page 95, use a small control current to energize an electromagnet, which then pulls a set of contacts together. These contacts can be designed to handle very large amounts of current and, therefore, we can think of them as "drive" devices, since large, high-power devices such as motors or other heavy loads can be "driven" from small control signals.

Solenoids

A very simple type of electro-mechanical drive device is the solenoid, which is an electromagnet, which, when energized, moves a piece of metal. These devices are often used in stage effects applications needing simple, short actuation, and since they are electrical, they are easy to interface with electrical controllers.

Motor Drives

Although entire books exist on the subject of electric motors and drives, the entertainment control engineer should understand some of the basics of drives used in entertainment machinery. There are two basic types of motion-control drives: open-loop and closed-loop "servo" systems.

A DC servo "amplifier" is its own closed-loop system. A tachometer coupled to the motor provides velocity feedback to the drive. This information is then used by the drive itself to close the velocity loop—the drive is simply instructed by the controller how fast to go and in which direction. In a DC motor, voltage corresponds to speed, and current determines torque. By varying these two parameters, a DC servo drive can have quite smooth and tight control over a motor's velocity. AC servo drives are also available (as are nonservo DC drives).

A "stepper" motor works much differently: the motor has several windings at different angular orientations, and applying current to these windings in proper sequence causes the motor to spin. Since sequencing is a task well suited to computers, stepper-motor drives are very easy to interface with digital control systems; a certain number of sequential pulses issued to a motor will equal a certain position. If all is functioning correctly, firing a certain number of sequential pulses should cause a stepper to turn to a certain position; for this reason, stepper motors are generally run as open loop. Stepper motors are generally used in small, low-power designs, and are useful for entertainment applications, such as props or internal effects in moving lights.

AC motors are generally much cheaper than DC motors, but AC drives are more complex. In an AC motor, frequency determines speed, and current still corresponds to torque. Three-phase drives are mostly run open-loop. Because of volume manufacturing, high-power three-phase drives have become available at very low cost and are now the dominant form of motor control for most electrical high-power entertainment machinery applications.

Fluid Power

Almost all the concepts we've just discussed have hydraulic (oil) or pneumatic (air) equivalents; these systems can be used where tremendous power is necessary but not much space is available (since fluid-power motors can be significantly smaller than their electrical equivalents) or where linear motion is desired. Hydraulic and pneumatic systems are, of course, often computer-controlled, and use many of the same principles detailed above

for electrical systems. Valves, which can be controlled electrically or mechanically, are used instead of electrical drives, and come in both solenoid (on-off only) and proportional (continuously variable) versions (see photo). In a solenoid valve, the solenoid is mechanically coupled to the valve actuator and, when energized, the valve either allows or stops fluid flow. Proportional valves are used when variable fluid flow (which usually corresponds to variable speed) is desired. These valves, like electric motor drives, often accept a variable analog or digital control signal.

EMERGENCY STOP

Emergency stop systems are a critical safety component of all stage machinery systems. See "Emergency Stop Systems," on page 112, for more.

Chapter 6: Stage Machinery • 47

COMMERCIAL ENTERTAINMENT MACHINERY CONTROL SYSTEMS

Since few of the components or devices mentioned above are "plug and play," most entertainment machinery systems are designed and installed by a company, which packages standard and custom components into a system suitable for the application.

Controllers for Installed Applications

A number of rigging companies offer standardized controllers for automated control over permanent stage equipment, such as turntables, orchestra lifts, and fly systems. They often have PLCs or dedicated motion controllers at their core and a custom designed operator interface. Most have some sort of user interface and the requisite emergency stop button.

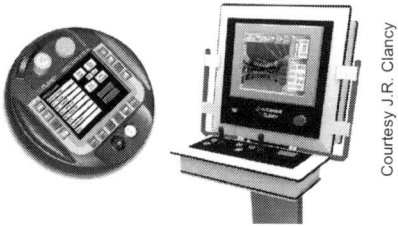

Controllers for Production Applications

Many companies now offer machinery control systems custom-designed for the demands of live production and performance, and these systems are used on Broadway shows, concerts, in theme parks, and other similar venues. Many commercial systems use standard desktop PCs as front ends, with dedicated motion controllers or PLCs located with the drive systems in remote "drive racks" (see photo). With any machinery control system, safety is of course of utmost concern; well-designed systems always include extensive emergency stop ("E-stop") and other safety features.

Courtesy Hudson Scenic Studio

MACHINERY CONTROL APPROACHES

The process control and factory automation industries dwarf the entertainment technology industry, and so, as in many other areas, we hitch a ride on their technologies. Factories and refineries need distributed control connection methods for a huge number of sensors, switches, and actuators, and these devices may be distributed throughout a large facility. Many manufacturers have developed proprietary networking solutions for these applications, traditionally called "field busses," or I/O (input/output) systems. Ethernet is increasingly used in automation applications because it offers a near universal connectivity standard; however, no one approach tends to dominate entertainment machinery applications. So, all that can be provided here is a brief introduction to several commonly used systems; more details are offered in specific sections later in the book.

Drive System Interconnection

Some drive devices still use a simple, but very effective, scheme: ±10V DC, where +10V indicates maximum forward velocity and –10V indicates maximum reverse. No voltage, of course, means no motion. ±10V control is very common for entertainment drive systems, although there is an increasing number of both open and proprietary digital control schemes.

IP/Ethernet

Ethernet (Chapter 16) is being widely used in many different disciplines throughout the entertainment industry. A huge number of remote input/output or "I/O" systems are now available that use Ethernet as a backbone; most use proprietary control data approaches.

Courtesy Opto22

Field Busses

There are a number of industrial "field bus" technologies used in stage machinery systems; because these can also be used as general purpose I/O, they are listed in the "Industrial I/O Systems" section, on page 345. Commonly used systems include: AS-Interface (AS-i) (page 345) .DeviceNet (page 345), EtherCAT (page 346); Modbus and Modbus TCP (page 346), and Profibus (page 347).

Interfacing with Machinery Control Systems

While the majority of machinery systems are triggered completely manually for safety reasons, some machinery control systems can communicate with other devices over a show network using proprietary data sent out via Ethernet. In addition, any of these systems could also be built to accept other signals, such as MIDI Show Control (Chapter 23) or SMPTE or MIDI time code (Chapter 25).

These external systems should never compromise the safety of the system, and there are a number of approaches to keep humans in the control loop while adding the sophistication of control offered by the computer (see "Humans in the Loop," on page 113).

Chapter 7: Animatronics

In **animatronics,** or "character animation," machines designed to look like living characters perform to prerecorded sound tracks.[1] Animatronics incorporates aspects of stage machinery, audio, and computer control systems. Animatronic control systems can generally be broken down into the same component categories as stage machinery (Chapter 6): sensing, control, drive, and E-stop. However, in animatronics, these technologies are used on a smaller scale than in stage machinery.

ANIMATRONIC EFFECTS

In an animatronic system, there are two basic types of effects: on/off and proportional. On/off effects tend to be more common, due to their lower cost and ease of use.

On/Off

On/Off effects[2] are usually controlled using relatively inexpensive solenoid valves, pneumatic cylinders, or motors that create movement when turned on or off for some period of time. Typical on/off functions might be "open/close eye-

---

1. The term "animatronics" or "audio animatronics" was coined by Disney early in the development of its theme parks, but is now in widespread usage.
2. Sometimes referred to as "digital."

lids" or "move arm right/left." This approach is cost effective, but has limitations—on/off effects usually have only fixed speed control (through needle valves in pneumatic systems), and precisely repeatable positioning in the middle of an axis movement is difficult to achieve.

Proportional

Proportional[3] animatronic effects use some sort of variable control. Instead of having an on/off command, such as "move arm right," an analog control maps the range of the arm's motion to a continuous scale (i.e., 0–10V, or 0-255). In this way, speed, acceleration, and deceleration can be controlled, and the arm can be programmed to start moving slowly, then move quickly, and then move slowly into a precise position.

Open/Closed Loop

Most animatronic effects are controlled in an open loop system, to minimize cost. Higher-end characters (like those you might see in major theme parks) can use closed-loop systems with positional, velocity, or advanced feedback systems; these kinds of feedback systems would typically be used with proportional effects.

ANIMATRONIC CONTROL SYSTEMS

Animatronic control systems are generally operated in two phases: programming and playback. During the programming period, the character's movements are precisely matched to the audio track; during the playback phase, the movements are played back in sync with the sound track. Good programming is key to achieving sophisticated and believable motion in animatronic systems, and some systems use a programming panel, consisting of button panels to turn on and off effects, and knobs, sliders, or joysticks to run the proportional effects. High-end (typically custom) systems may even record this performance data from humans, using sensors attached to their bodies, or allow interactivity with the audience. In any case, as the character's moves are "performed" by a human programmer, the control systems digitally sample and record the performance data in sync with the audio soundtrack.

Generally, animatronic control systems allow each effect (arm, mouth, eyes, and so on) to be recorded separately, thereby allowing the programmer to record each "track," just a musician may perform each part of a song in a recording studio. An animatronic programmer might start by programming the character's mouth, then move on to eyes, head, and so forth. If a segment of a track's performance (a few

---

3. "Proportional" is sometimes, and incorrectly, referred to as "analog" for historical reasons.

eye blinks) needs to be "punched in" and rerecorded, it can be done later, and the performance can be built up by programming each set of movements individually.

The performance can also be tweaked and polished "off line," and programming systems often feature a number of post-production editing features (see screen capture below).

A/V Stumpfl Wings Platinum example program created by Michael Baier and James Hergen

For instance, it may be difficult for a human performer to exactly anticipate a musical soundtrack, or an animatronic system may have some latency, so someone may record a good performance that is slightly late. In many manufacturer's systems, the programmer can simply grab the recorded data tracks and drag them back slightly in time. For actions such as mouth movement, it is also possible to have the programming system create a movement based on the sound track; for example, whenever the volume on the voice track exceeds a certain threshold, the mouth moves. This track can then be tweaked and cleaned up after processing by the programmer.

Animatronic control systems either rely entirely on a computer with interfacing hardware, which is used both for programming and playback, or use a computer to create a program, which is then downloaded into playback hardware (see photo). With the latter structure, the (relatively) expensive programming interface is only needed for the programming period.

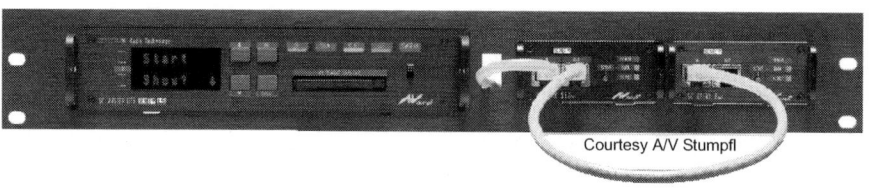
Courtesy A/V Stumpfl

ANIMATRONIC CONTROL APPROACHES

Depending on the effect being controlled, there are a number of control methods used in animatronics for either internal control or interfacing with the outside world.

Contact Closures

Contact closures (page 93) are used extensively in the world of animatronics, both for internal control and external interfacing.

0–10V DC

For proportional effects, analog control is often used, with the effect's range of motion mapped to a 0–10VDC scale (see the "0–10V DC" section, on page 54). For instance, 0 volts may represent a leg fully retracted, and 10 volts may tell the leg to fully extend. Any voltage in between would represent some position between those two extremes.

DMX512-A

DMX (Chapter 19) is well suited to some forms of animatronic control, because it is a simple standard which sends continuous information at a fairly high data rate. It also allows an animatronic controller to control lighting equipment in simple applications.

MIDI and MIDI Show Control

Some systems use MIDI (Chapter 22) as a core control signal, using MIDI notes to either trigger digital on/off or proportional effects (with velocity representing the proportional value). This approach is very cost-effective, and the MIDI messages can easily be recorded into standard musical MIDI sequencers and other devices. In addition, any system that outputs MIDI should also be able to output MIDI Show Control (Chapter 23) commands to control other devices in the show environment.

SMPTE and MIDI Time Code

Many animatronic control systems can "chase" SMPTE or MIDI Time Code. Both are described in detail in Chapter 25.

Chapter 8: Fog, Smoke, Fire, and Water

This category contains a variety of production elements, some usually found only in theme parks: fog, flammable gas, water fountains, and such. Most of these systems (other than fog) are generally purpose-built, with a custom interface to other show systems. This category is sometimes also called "process control," but this term has other specific meanings in industrial control.[1]

FOG AND SMOKE EQUIPMENT

The Professional Lighting and Sound Association's (PLASA) *Introduction to Modern Atmospheric Effects*[2] defines fog as "a mixture of liquid droplets in air that reduces visibility and reflects light," and "smoke" is "small, solid particles produced by burning and dispersed in the air. In the context of atmospheric effects, 'smoke' is used to refer to any aerosol made of solid particles rather than liquid droplets." While smoke effects are occasionally used in performance, fog effects are typically more easily controlled, and will be the focus here.

Courtesy Look Solutions USA, Ltd.

A large variety of off-the-shelf fog machines are available that use glycol-based solutions, frozen liquid or gaseous carbon dioxide, liquid nitrogen, or other materials. In theatrical-style live shows, fog generation usually falls under the jurisdiction of the electrics department, and in these situations, fog machines are often interfaced to lighting control systems.

---

1. In industry, "process control" means the control of processes in chemical, wastewater, refinery, and other facilities.
2. *Introduction to Modern Atmospheric Effects*, 4th edition. See http://tsp.plasa.org/tsp/documents/published_docs.php

FIRE AND WATER CONTROL SYSTEMS

Most fire and water systems are custom-designed and built from the ground up to create spectacular effects for theme park or similar attractions. Water systems are generally custom and use standard process, industrial, and machinery control components, such as PLCs (page 45). Animatronic control techniques are sometimes used for fountain control as well.

Because of the obvious danger, safety concerns are of utmost importance with any flame effect, and safety standards, such as the National Fire Protection Association's "NFPA 160 Standard for Flame Effects Before a Proximate Audience" should always be followed. Flame systems are typically custom-built from a variety of industrial control system components, and use, at their core, industrial burner-management systems, originally designed for use on large boilers. The primary purpose of flame control systems is to assure that a source of ignition (such as a pilot flame or spark) exists prior to the release of gas, since one of the biggest hazards is not flame itself, but the release of unignited gas that finds an unintended source of ignition (e.g., a smoker in the audience). Flame effect control systems can also monitor "gas sniffers," which detect gas leaks before they reach the "lower explosive limit" for combustion, intrusion alarm systems, and building systems (such as the fire alarm system), so that, in the event of a fire elsewhere in the facility, or carbon monoxide or high temperature above a defined threshold, the entire fire effect system will be immediately brought to its safest state.

Emergency stop systems (page 112) for flame effects also have to be thought through very carefully: The "safe" state of every control output must be known and understood, and the system must be designed so that every possible failure within the system leads to a safe state. In addition, humans should always be in the control loop of any potentially hazardous system; see the "Humans in the Loop" section, on page 113, for more information. Obviously, flame control is an area for expert consultation and even experienced designers typically send their work out for peer review.

FOG SMOKE FIRE AND WATER CONTROL APPROACHES

There are no entertainment control standards specifically designed for control of these elements, but several methods are used frequently.

Contact Closures
Many systems make extensive use of contact closures (whether electro-mechanical or solid state) for a variety of purposes. See "Contact Closures," on page 93, for more information.

DMX512-A
DMX512-A (Chapter 19) was not designed to control dangerous elements, but is often used to control fog machines and similar devices.

0-10V
Some systems can use 0-10V proportional control, like an old lighting system; see the "Analog (0–10V) Control" section, on page 20

Industrial I/O Systems
Industrial I/O systems are used in many custom fire, water, or other systems; see the "Machinery Control Approaches" section, on page 48.

Chapter 9: **Pyrotechnics**

This chapter is about chemical pyrotechnics: flash pots, sparkle devices, concussion mortars—things that go boom (pyro substitutes such as air cannons would go into other categories, depending on their construction).

Pyrotechnics are defined by the National Fire Protection Association as "controlled exothermic chemical reactions that are timed to create the effects of heat, gas, sound, dispersion of aerosols, emissions of visible electromagnetic radiation . . . ."[1] Generally, "pyro" refers to smaller, indoor displays for production applications, and "fireworks" are the devices used in large outdoor displays.[2] Safety is of utmost importance in any pyro or fireworks application, and standards such as NFPA's "1126 Standard for the Use of Pyrotechnics before a Proximate Audience" and "1123 Code for Fireworks Display" should always be followed whenever pyro is used.

In addition, humans should always be in the control loop of any potentially hazardous system, such as pyro; see the "Humans in the Loop" section, on page 113, for more information. Remember, the job of the pyro operator is to *not* fire an effect if a dangerous situation occurs, such as an audience member or performer who is in the wrong place, or when flammable materials are placed near the effect.

PYROTECHNIC CONTROL SYSTEMS

The fundamental principle of electrical pyrotechnic control is very simple: Controlled electrical current is used to heat a small wire, which triggers an explosive reaction of chemicals positioned near the wire. Varying levels of sophistication are used to implement this simple principle, from switch closures and "button boards" to computer-based, time-code controlled triggering systems.

---

1. "NFPA 1126 Standard for the Use of Pyrotechnics before a Proximate Audience," 2011 Edition, p. 1126-6.; you can read it online for free here: `http://www.nfpa.org/about-thecodes/AboutTheCodes.asp?DocNum=1126&cookie_test=1`
2. Fireworks are defined in NFPA 1123, Code for Fireworks Display as "Any composition or device for the purpose of producing a visible or an audible effect for entertainment purposes by combustion, deflagration, or detonation."

A typical pyro control system consists of a control system and a number of effects devices, such as "mortars," or "flashpots." The controller has a series of switches, each with "arm" and "test" positions, for the various effect circuits. When the switch on the controller is moved to the "test" position, a very small current (thousandths of an ampere at a low voltage) is applied to the electric match. If the circuit is good, continuity will exist on the firing circuit; if the firing circuit has a problem, there will not be continuity, and in either situation, the controller can determine whether or not the circuit is ready to be fired. Once the operator has determined that the circuit is good and it is time to place the pyro effect into standby, the switch is moved to the arm position. For manual controllers, the fire command, for safety reasons, is typically issued using a key switch; when this switch is turned, voltage is applied to the electric match, causing the effect to fire.

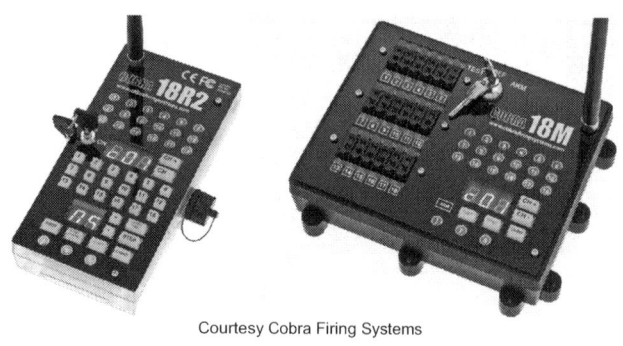

Courtesy Cobra Firing Systems

Computerized pyro and fireworks controllers generally feature either stand-alone or PC-based controllers, and communicate bidirectionally (and sometimes wirelessly) through a proprietary digital protocol to "firing modules." A typical firing module is addressable, and handles firing and continuity testing for a number of pyro effects. Continuity checks are done locally on request from the master controller, based on a comparison with the firing instructions; the results are communicated digitally back to the controller.

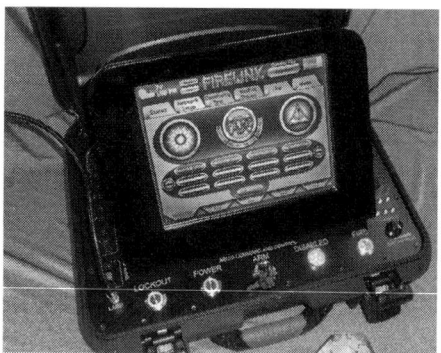

Courtesy Birket Engineering, Inc.

The computerized approach has a number of advantages for large, complex systems. With a digital system, the operator can be a safe distance from the action, with only small control cables (or a radio link) leading to the firing site. In addition, since the system is computerized, it is easy to add automated cue control and precise timing and synchronization. The computerized system can even contain pre-programmed information on shell caliber, event trigger time, and pre-fire time.[3]

Safety is, of course, of utmost concern in pyrotechnics, and computerized pyro controllers uses extensive error detection (page 129), and the system generally defaults to nonoperation in case of a failure (fail-safe, see "Fail-Safe Design," on page 112).

PYROTECHNIC CONTROL APPROACHES

Many pyro systems use a proprietary control structure, but many also use open standards.

Contact Closures

Of course, at its lowest level, pyro control involves contact closures, which are described in detail in the "Contact Closures" section, on page 93. They can also be used for external interfacing.

SMPTE and MIDI Time Code

Most computerized pyrotechnic systems can either chase their own internal time code scheme or external SMPTE or MIDI Time Code (Chapter 25).

MIDI Show Control

MIDI Show Control (Chapter 23) commands are supported by some systems for event-based control. External control allows pyrotechnics to become an integral part of any production, and humans can always be incorporated in the loop, downstream of the controller, to ensure that the effects are ready to fire safely.

---

3. Time offset to compensate for the time it takes the shell to travel before exploding.

Part 2: **Entertainment Control**

This section offers an overview of control concepts that are common to all of entertainment control. If you are experienced in our industry, the terms and concepts discussed here may seem obvious. However, I encourage you to at least skim through the section, since some concepts may be new, or new to you.

In addition, you may wonder why I discuss here, in formal and abstract ways, concepts such as "cues," which we take for granted in our industry. Much of this book is about implementing systems in computers, and these systems, of course, can't take anything for granted. So we need to make these concepts explicit in order to understand and implement them.

Chapter 10: Entertainment Control Basics

When we design a control system we have to make some decisions about how the system should work based on what exactly we are controlling, what we want the system to do, and some basic characteristics of the system's application. The entertainment control basics covered in this chapter form the foundation of show systems, and are what make our systems so unique in the larger world of control technology.

Before moving forward, a couple notes. I present many concepts here in an either/or, black/white way. In reality, the distinctions may not be so clear cut. However, I've found that making sharp distinctions is a helpful way to explain these abstract concepts; I leave it to you to figure out the shades of grey relevant to your situation. Finally, another thing to consider is that control systems can be made up of combinations of any or all of the basics we discuss here. While it's possible, for example, for a linear show to use time-based cueing methods, shows are often done using event-based cueing. Open loop systems might use either relative or absolute systems. Distributed systems might use either star topologies or bus, and so on. Each situation can be different.

SHOW TYPES

In order to plan our control systems, we have to think a bit about the type of show. For our purposes, I'm going to categorize all shows into two basic types: linear, and nonlinear; each has specific impacts on the way the control systems are structured.

Linear Shows
A **linear show** is one in which there is a single, fixed storyline, that is normally performed in the same sequence in each show. This may take the form of a live theatrical-style show, where the performers act out the same script in each show.

Nonlinear Shows
In a **nonlinear show**, multiple, separate components of a show (or even multiple shows) can all run independently, in a dynamic order, or even simultaneously. For our purposes here, a group of linear segments (such as a concert where the run-

ning order of songs changes every night) would be considered nonlinear, because the dynamic running order demands nonlinear flexibility from the control system.

One example of a nonlinear show would be an interactive game-like attraction where there are multiple possible branches to the story being performed. Another example would be a haunted house attraction, where the audience members themselves trigger various aspects of the show as they walk through. Unless audience throughput is very tightly controlled and regulated (which would detract from the experience), there is no way to know when audience members will trigger the cues in any particular area, and the areas may trigger in any order.

CUES

The **cue** is the basic building block of show production, and it's a concept we deal with so intuitively that we may not have ever really thought about a definition. Computers and networks, however, like things to be defined precisely, and some readers may not be familiar with our field, so we will begin with a general introduction.

In theatrical-style shows, a human "stage manager" is usually in charge of running the show, and uses a headset system to communicate with individual system operators in each "department" (lighting, sound, machinery, video, and so on). Warnings are given for each cue; for example, "Sound cue 13 stand by." If ready, the sound operator then replies, "Sound standing by." At the appropriate point in the production, the stage manager initiates the cue by saying, "Sound cue 13 go." The sound operator "takes" the cue, generally by pressing a button on a computer controller, or executing some other action. When the cue is completed, the sound operator replies, "Sound complete" (of course, with experienced crews, the "standing by" and "complete" steps are often omitted).

Alternatively, the stage manager can control "cue lights" to signal performer entrances or to communicate with run crew members who cannot easily or safely use a headset (for example, a person flying scenery over the stage would probably not wear a headset because it could get tangled in the rigging). The stage manager turns on a cue light to indicate that an operator should stand by for the next cue. When the stage manager turns the cue light *off*, the operator executes the cue. This model of control is often applied in nontheatrical shows as well, or operators may take many of their own cues and take cues from the stage manager only when larger, interdepartmental coordination is necessary.

While we deal intuitively with the concept of cues, they can be fairly complex or confusing for a computer. A cue can be a number of different things, depending on the production element to which one is referring and its current state and context. For example, a cue can be self-contained: once started it runs until completion (i.e., a pyro explosion). A cue can also modify something that is continuing: a lighting cue is generally a transition from one light "look" to another, but "light cue 18" can refer to both the *transition* from light cue 17 *and* the resulting new look 18, which will continue onstage after execution of the cue transition.

This concept of cues, of course, can also be used internally in entertainment control systems, since those of us backstage know that technical show elements are rarely performed or controlled completely "live," but instead are often a series of predefined states or transitions executed using computerized control consoles at specific points (cues) throughout the show. In addition, things such as lighting consoles, sound controllers, machinery systems, and so forth, can often be set up for external triggering by a control system.

CUEING METHODS

Cues or control states in entertainment control systems must somehow be triggered, or executed. Generally, triggering approaches fall into three categories: event based, time based, or hybrids of the two.

Event Based

A live performance is generally event based. A stage manager may call cues based on an event's occurrence; for example, the speaking or singing of a certain line or a performer's movement. For instance, when a performer screams, "Run for your life!," the stage manager gives a "go" command over the headset to lighting and sound operators. The lighting operator, who has been standing by for this cue, presses the "next cue go" button on her control console, triggering the explosion lighting effects, and the sound operator triggers the sound effects in a similar fashion.

A tremendous amount of intelligence exists in this human-controlled, event-based system. The explosion effect is triggered on the event—the actor's line—regardless of when in time he says it, where he is standing, what day of the week it is, or the phase of the moon. Event-based systems allow performers to vary their timing, improvise, and make mistakes. In our example, if the performer is having a bad night and simply runs screaming off the stage without saying his line, the stage manager can improvise and still trigger the explosion effects by giving the go commands. This type of intelligence is not easily built into automated systems. If a voice-recognition system was designed to trigger the effects off the actor's "Run for your life!" line and the performer forgot to say it, the system could get hopelessly out of whack.

Here are some examples of events:

- A performer sits on an onstage sofa.
- The sun appears over the horizon.
- A specific measure and beat of music is performed.
- A performer speaks a line of text.
- An operator presses a button on a control console.
- An audience member passes a sensor.

Cue Lists

Event-based systems are most typically represented in computer systems by "Cue Lists", with the cues listed out vertically. Here, the "standing by" cue is indicated—this indicates which cue will be executed when the "Go" button is pressed:

Courtesy Figure 53

Time Based

There are two general types of time-based triggering or synchronization. In one approach, all elements, including human performers, are synchronized by some connection to a master clock; cues (events) are then fired at specific, pre-programmed times. Alternatively, two or more types of time-based **linear media** (audio recording, video clip, or such) can be synchronized to each other (or to a separate master clock) on an ongoing basis.

A time-based system is less forgiving to human performers. If, for example, an audio clip rolls at one minute into the show and the explosion effects were triggered at 14 minutes and 35 seconds into every performance, the performer would be out of luck if his performance varied much. The performer becomes simply another slave to the master clock, synchronizing himself to the system, instead of the other way around. However, (to generalize) time-based systems have a major advantage over event-based systems: Once assembled and programmed, they are easy to run automatically and reliably, eliminating many variables. Such a system may sound limiting, but scores of theme-park shows, halftime spectaculars, and related productions have successfully run this way for many years.

Typically, the master clock comes from some sort of pre-prepared media that is already constrained and fixed in time, such as a video segment or audio sound track. There are many ways to synchronize all the other show elements to this master time clock; for example, some sort of "time code" can be sent from the master device to all the controlled devices. Alternatively, all the systems can be told to go at exactly the same time, and then each system runs "wild," with no continuing re-synchronization (this is feasible only if each local device has a very accurate internal clock, and is is not likely to drift out of time). In either case, the master clock determines what is going to happen and when.

Here are some examples of time based systems:

- The show starts at 3 p.m. Eastern Standard Time.
- A light cue is triggered at 10 seconds after 3 p.m.
- Two minutes, 22 seconds, and 12 frames into the video, a strobe is fired.
- Twenty-minute long audio and video clips are continuously synchronized together.

Time Line

Time-based systems are most typically represented in computer systems by "timelines," with the events or cue media displayed horizontally. The current point in time is traditionally indicated by a horizontally scrolling vertical bar:

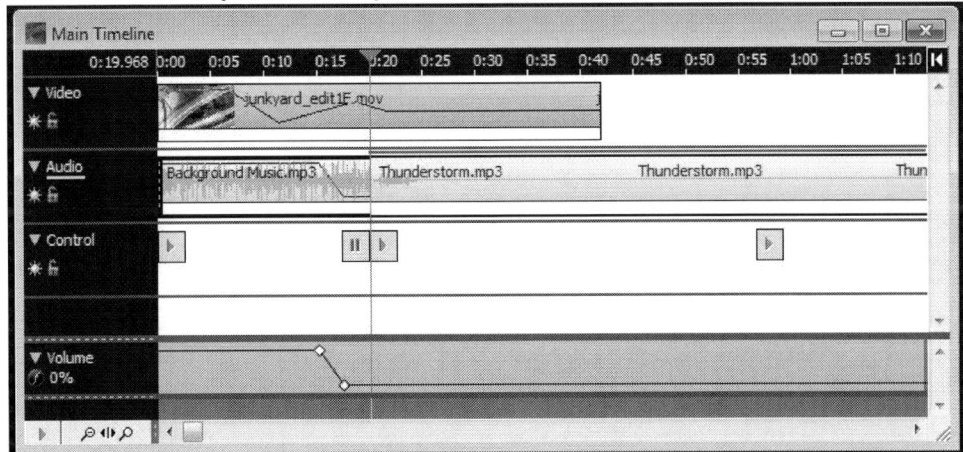

Hybrid Systems

Most entertainment control systems are hybrids of event- and time-based systems. Systems that are primarily event based can have built-in time-based sequences; conversely, a time-based system can stop and wait for an event trigger before continuing. In our example above, if complex explosion effects were built into a five-second-long time-based sequence, the system would then be hybrid. Most of the show would be event based, with the stage manager calling cues off various actions. When the actor yells, "Run for your life!", the stage manager commands a time-based sequence, consisting of the lighting and sound cues for the explosion sequence. The actors synchronize their movements to the time-based sequence—which is relatively easy, since the sequence is short and runs exactly the same way every night. At the conclusion of the time-based sequence (itself an event), the stage manager returns to a normal, event-based operating mode and continues to call cues based on the script, an actor's position on the stage, or other events.

An additional type of hybrid synchronization is "musical time," where systems are synchronized to musical songs, measures, or beats. Events are synchronized in this way by music score-reading stage managers in opera, musicals, and other types of productions, who call specific cues not on text or time, but instead on a particular beat of music.

OPERATIONAL MODES

Control systems often have multiple operational "modes." A system in a retail environment, for example, may have "day" and "night" modes, with day mode used when the store is open and night mode used when the store is closed. In day mode, all systems are operational and shows are performed; in night mode, show cues may be suppressed (or alternate shows could be run) and unused systems can be disabled. Other examples of modes might be "show enabled" and "show disabled"; these could be useful in a themed attraction, where you want to disable all the effects with a single button press when the attraction crew goes on break. Alternatively, a system may have a "maintenance" mode, where additional information and control is given to a maintenance technician who is working on a system.

Musical Time

Dr. David B. Smith and I implemented "musical time synchronization" using the musical data exported by RMS's Sinfonia® and software I wrote in Medialon Manager™. With this system, we take the current live musical performance and, using beat predictions generated by the Sinfonia system, compare the live performance to a "normal" performance. Real-time variance from normal can then determined, and technical elements such as video, audio, and so on, can be varied to sync up with the live musical performance.

COMMANDS/DATA

There are two general ways of controlling devices, and I've named them here "Preset/Command" and "Repetitive/Continuous."

Preset/Command

In preset or command-based control, the target (controlled) device needs some intelligence. It either needs to contain a previously recorded preset that can be recalled when the command is received, or it needs to understand enough to be able to interpret or act on a command immediately. For example, a command may be something like "Go Cue 66," which would recall a preset numbered 66 in a lighting console. Alternatively, commands can also take on a more detailed, non-preset-based form; for example, "Move to 43 feet, with an acceleration time of one second and a deceleration time of two seconds." To a control system, however, both of these approaches look pretty similar: a relatively short message triggering a potentially complex action. This structure requires a relatively significant amount of processing power in the target device, which is readily available on higher-level controllers (e.g., lighting consoles). For lower-level devices, which may not have as much processing horsepower, a "continuous" control approach may be taken, with the control data sent repetitively.

Repetitive/Continuous Data

"Dumb" target devices require less local intelligence. They don't understand anything about their larger context—they simply react to the data they are receiving *right now*. In a conventional dimming system, for example, there may be many dimmers, each of which needs to set individual lighting levels, while a lighting control console provides the global cue storage and interface for the system operator. Since there could be hundreds or thousands of dimmers in the system, it has historically been cost prohibitive (and overly complex) to engineer each dimmer to provide the ability to store, recall, and process cues. Instead, each dimmer is designed to simply set its connected light to the desired level, based on the data it is presented at that instant.[1] If the dimmer needs to fade from a level of 50% to 60%, then the target level is transmitted at 50%, then 51%, then 52% and so on up to 60%, with the dimmer immediately changing its level based on each piece of new incoming, repetitively-sent data. Alternatively, if the dimmer is sitting "off" at a level of zero, then the zero data value is simply sent repeatedly over and over and over.

---

1. You could look at this message as a "command," but for our purposes, commands are sent only once, not over and over and over again.

There are advantages to this approach: Controlled devices can be made very simply and cheaply and, if control data is lost or corrupted, the corrupt data will simply be overwritten by the next update. There are also drawbacks, of course. A lot of communications bandwidth and processing power is wasted in sending out data on a repetitive/continuous basis since, even if nothing is changing, data has to be sent out over and over and over again.

DATA RELATIONSHIPS

Absolute and **relative** are two more control concepts important to our field. In the *American Heritage Dictionary*, *absolute* is defined as "Relating to measurements or units of measurement derived from fundamental units of length, mass, and time."[2] Conversely, *relative* is defined as "Dependent on or interconnected with something else; not absolute."[3]

Relative

In a relative system (also called "incremental," especially in the automation industry), we only know where something is *in relation to the things before and after it*. A signal sent to a lighting console containing the message "next cue go" would be relative. The light board knows its *current* cue, so on receipt of this relative control message, it simply advances to the next cue in its list. However, if one control command in the middle of a series of commands is lost, the target system doesn't know about the error, so it can get out of whack and end up running behind the other elements to which it is connected. For example, let's say there is a series of cues such as 11, 12, 13, and 14, and these are triggered using the relative "next cue go" command sent for each cue change. We're sitting in cue 11, and then we want to go to cue 12, so the controller sends out a "next cue go" and the target devices advances its cue. Then some VIP backstage kicks out a plug from the back of the console, and no one notices. The "next cue go" for Cue 13 is sent, but never arrives. The unplugged cable is noticed and restored in time for cue 14, but, when this "go" is sent, the receiving device, which is still sitting in cue 12, now advances to cue 13 when it should be in 14. The system is now messed up until someone corrects it.

---

2. "absolute." *The American Heritage® Dictionary of the English Language*, Fourth Edition. Houghton Mifflin Company, 2004. 14 Mar. 2007. <Dictionary.com http://dictionary.reference.com/browse/absolute>.

3. "relative." *The American Heritage® Dictionary of the English Language*, Fourth Edition. Houghton Mifflin Company, 2004. 14 Mar. 2007. <Dictionary.com http://dictionary.reference.com/browse/relative>.

Data integrity is extremely important in relative systems because if data is lost, the controlled system will be in error. If a relative system loses power, when power comes back up the system must first re-initialize, find a predefined starting point, and then figure out where to go from there. But relative messages do have their place and are widely used—they are easy to implement and take up less data space because they carry less information.

Absolute

A control message of "cue 13 go" is absolute—the control data contains all the information necessary to place the cue in the show. If one absolute message in a series is lost, the system would simply be off track until the next absolute cue is received; well-designed absolute systems will eventually realign themselves. Let's look at our cue series example from above: Cues 11 and 12 are received and then 13 is lost. When cue 14 is sent, the receiving device jumps from 12 to 14. Cue 13 was lost, but restarting with cue 14, the system can now operate correctly.

Absolute systems are generally more robust and can recover from data corruption, and, therefore, are generally preferable, all else being equal. To generalize, they are more complicated (and therefore expensive) than relative systems, but in these days of ample computing horsepower and inexpensive memory, this is of decreasing concern.

Absolute vs. Relative Example #1

These concepts can be hard to grasp so, to illustrate the difference between absolute and relative, let's imagine a bi-coastal live video event, staged simultaneously in New York and Los Angeles. Sam, the scenic designer, has designed a series of ten three-foot square platforms laid out on six-foot centers, which line up with overhead lighting trusses. A scene shop in Toronto has constructed two identical sets of platforms and they have been delivered to both New York and Los Angeles. A lighting company has installed all the trussing. We have two stage carpenters, Allison and Roger, each installing an identical set in two similar spaces—Allison is in New York and Roger is in Los Angeles.

The designer has specified that the first platform should be six feet from the edge of the stage, so Roger measures six feet from the edge of the stage and then lays in the first platform.

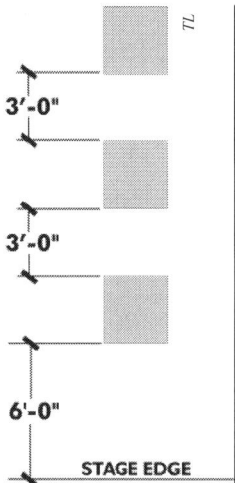

Since the platforms were each specified to be three feet square, and they are supposed to be on six foot centers, Roger simply measures three feet from the edge of the first platform to determine where the edge of the next should be, puts the next platform in place, and then does the same for each of the remaining platforms. He wraps up the job quickly and heads out, because the surf is up.

Allison measures six feet from the edge of the stage, and then lays in the first platform. However, instead of measuring from the edge of the first platform to determine where the edge of the second platform should go, as Roger did, she goes back to the edge of the stage and measures 12 feet (six feet to the first platform plus six feet for the edge of platform two). She then measures from the edge of the stage in order to place platform three at 18 feet, and so on. It takes a little longer, but Allison finishes up and heads home.

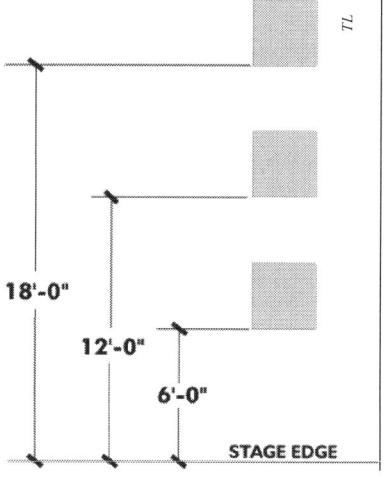

The set designer shows up in LA and goes ballistic. Each platform in the series is increasingly misaligned with the overhead pipes. The set designer's assistant in New York, however, checks Allison's work and everything is fine. What happened?

It turns out that the scene shop made a mistake in construction (English/metric unit error), and the platforms are actually three feet and three inches square. With Roger's relative measurements—from the edge of one platform to place the edge of the next—he never noticed that each platform after the first was off by three inches. The error accumulated, as a result the tenth platform was off by nearly 30 inches. Allison, however, used absolute measurements, referring back to the edge of the stage as a reference each time. Even though the distance between the platforms is three inches shorter than intended due to the size error, the platforms line up exactly with the overhead pipes as intended.

Absolute vs. Relative Example #2

Let's look at another example, with two ways to ensure that a series of cues run exactly the same way every performance. First, let's look at an approach where we program into our system the *relative* time each cue should wait after the previous cue before triggering:

Cue	Action	Wait Time Until Next Cue
1	Houselights Out	00:00:05
2	Music 1 Starts	00:00:04
3	Stage Lights Up	00:00:07
4	Video Start	00:00:44
5	Sound FX 1	00:00:10
6	Sound FX 2	00:01:03
7	Video Pause	00:00:02
8	Music 1 Cuts Out	00:00:45
9	Music 2 Starts	00:00:10
10	Lights Out	

This approach works fine, and is easy to program. However, the nature of show business is change; what happens if, after you have this programmed, the director wants to move Q3 four seconds later? You would simply change the wait time for cue 3 from 00:00:04 to 00:00:08. The problem is that now, since we increased a single wait time by four seconds, *all subsequent* cues will be four seconds late. To fix this, we now have go and subtract that four seconds off the wait time for cue 4 from 00:00:07 down to 00:00:03. Here's an excerpt of the update, with the changes shown in **bold**:

Cue	Action	Wait Time Until Next Cue
1	Houselights Out	00:00:05
2	Music 1 Starts	**00:00:08**
3	Stage Lights Up	**00:00:03**
4	Video Start	00:00:44

A more flexible solution might be to use a system that times out the cues using an absolute technique; here's the same initial cue sequence from above, but with the times shown as absolute trigger times, rather than relative wait times:

Time	Cue	Action
00:00:00	1	Houselights Out
00:00:05	2	Music 1 Starts
00:00:09	3	Stage Lights Up
00:00:16	4	Video Start
00:01:00	5	Sound FX 1
00:01:10	6	Sound FX 2
00:02:13	7	Video Pause
00:02:15	8	Music 1 Cuts Out
00:03:00	9	Music 2 Starts
00:03:10	10	Lights Out

Now, when the director wants to move Q3 four seconds, we simply change the trigger time for cue 3 from 00:00:09 to 00:00:13. Because all the cues in the list have *absolute* times, no other cue's trigger times will be affected, and we don't need to change anything:

Time	Cue	Action
00:00:00	1	Houselights Out
00:00:05	2	Music 1 Starts
00:00:13	3	Stage Lights Up
00:00:16	4	Video Start

With this absolute timed cue approach, any change we make affects only that cue, and no others in the sequence.

FEEDBACK

A key aspect of system architecture is the inclusion (or not) of "feedback." "Open-loop" systems do not use feedback; "closed-loop" systems do.

Open-Loop

Let's say your boss sends you an e-mail asking you to purchase a copy of this book and send it to a client. However, unbeknownst to you, the e-mail gets stuck in your spam filter and you never get it. Months later, the boss runs into the client on the street, and asks how she liked the book. "What book?," she says. Until the encounter on the street, this was an "open-loop" system—a request was made, but no confirmation of the requested action was issued.

Open Loop Communication

Open-loop systems are simple, fast, and are as reliable as the systems underlying the communications. If you got ten e-mails a day to send out books, and the requester trusted both the communications medium (e-mail) and the target device (you), and an occasional potential failure would not have serious consequences, then open-loop system is fine. However, if reliability is key, all else being equal, a closed-loop system will be more effective.

Closed-Loop

After the incident described above, your boss asks you to confirm shipment each time you send out a book. The boss sends you an e-mail, you ship it, and you copy the shipment tracking information back to the boss. You have now added feedback to this system, closing part of the loop.

In a closed-loop system, it's up to the sender (the boss in this case) to track the replies and see that each request is confirmed. However, with this structure, errors such as lost commands (e-mail stuck in the spam folder) can be detected and cor-

Closed Loop Communication

rected (the boss calls and says, "did you get my e-mail?"). Closed-loop systems are generally a bit more complex than open-loop systems, but this is less and less of an issue these days, since network protocols often close the loop for "free."

In our industry, you will likely see both closed and open loop control systems; critical functions are equipped with feedback to close the loop, and less important functions are left open-loop. For example, scenic automation systems are mostly run closed-loop for safety reasons. Lighting control systems, on the other hand, are mostly (as of this writing, anyway) run open-loop to keep things simple, fast, and responsive. Keep in mind, too, that loops can be closed in many ways. A human stage manager could call a "go" command to a lighting operator, but the lighting op doesn't need to say, "complete" back over the headset, since the stage manager closes the loop visually—she can see the lights changing on stage.

CONTROL STRUCTURES

There are two basic structures for control systems: centralized and distributed. We should note here that these terms often refer to either the physical arrangement of the system or its control hierarchy (or both).

Centralized Systems

In a centralized system, control functions are run from a single machine or system, typically in a central location (i.e., an equipment closet or a control booth). For example, a single lighting console may control all the dimmers on a show. This approach can be cheaper because less control hardware is required on the scale of an individual system. However, on a show scale, with centralized control, difficultly can arise during the technical rehearsal and programming process, since many departments will want to make changes simultaneously, thus creating a bottleneck. In addition, centralized, unified control sys-

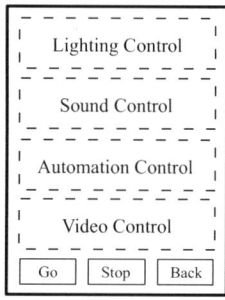

tems can end up being lowest common denominator solutions—functional for all departments but optimized for none. But even on a show scale, centralized control can be very cost effective for things like stand-alone animatronic displays.

Distributed Systems

A distributed system is generally made up of intelligent "subsystems," each optimized for its task and connected together into a larger system. Each subsystem decides, or is told by another system, when and how to execute its tasks. For example, a lighting console might control a number of moving lights, but the moving lights have built in processors that handle all the low level physical control details involved with physically panning and tilting the light, rotating its color wheel, etc.

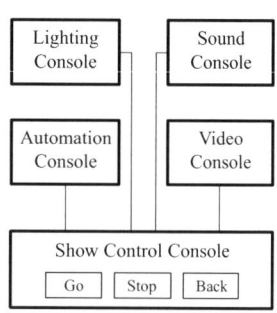

On the scale of a show, a distributed approach is often desirable, since a sound console's interface and operation can be optimized for sound, a machinery console for machinery, and so on. Such systems can still be linked to a single master controller, but the low-level control work is still distributed throughout the system. With the distributed control load, if one controller stops working, the show could continue on with the other elements. For this reason, all else being equal, I generally prefer distributed systems.

SYSTEM CONTROL ARCHITECTURES

For a system to work, all the connected devices have to agree on a management scheme or protocol to dictate how the devices communicate with one other. In this section, we will introduce several basic control architectures—some systems could, of course, be hybrids of the various types or incorporate various aspects of multiple architectures.

Master-Slave/Command-Response

In a "master-slave"[4] system, one device has direct, unilateral control over another. The target device can be fairly simple, or "dumb," since it only does what it is told to do. With this structure, a "command-response" approach is typically used, where the slave devices speak only when spoken to. If one controlled device needs to get data to another controlled device, the master must ask the first device for its data, retrieve it, and then deliver it to the second. If a target device has data that may change, the master device has to repeatedly and at a regular interval ask for this data from the slave; this process is known as "polling."

For example, lets take a look at a show control system that needs to know when a room in a theme park attraction is occupied. The sensor isn't able to tell the master what to do or what its state is; it just responds when asked with the current state either: "The room is occupied" or "The room is empty." The only way for the show controller to know that someone has entered the room is to repeatedly ask the sensor, "What is your state *right now*?, What is your state *right now*?, What is your state *right now*?" When empty, the sensor responds, "empty", "empty", "empty", and then after someone enters the room, to the next poll it will respond, "occupied."

Master-slave systems and command-response protocols are easy to implement and generally reliable, but are not particularly efficient or responsive. Data can be received from a slave only once per polling interval and, if there is no change in the target device condition, many polling cycles simply end up generating redundant data.

---

4. Master-slave systems are also sometimes referred to as "primary-secondary" systems.

Chapter 10: Entertainment Control Basics • **81**

Interrupt Driven

Interrupt-driven protocols are refinements of the command/response protocol. In these systems, whenever slaves need to send data, they send an "interrupt" signal to the master, using either a dedicated hardware line or a data control code. Upon receiving an interrupt, the master can deal with the node needing attention. This approach generally makes interrupt protocols faster than straight polling protocols, since the master needs to service nodes only when necessary. Internal computer busses often work in an interrupt-driven way.

Let's take another look at our show control system trying to determine if the room is occupied. With an interrupt-driven system, each occupancy sensor would have the capability to send a message to the show controller saying, "my data state changed." When the show controller receives this information, it can send out a poll. This could dramatically reduce the communications load.

Peer-to-Peer

The control intelligence in many systems is distributed throughout the system, and intelligent controllers or "peers" are connected using a network. In such a **peer-to-peer** system, each peer device has equal access to—or even control over—other devices in a system. This control structure allows for powerful and sophisticated systems, but also can be more complicated to design and troubleshoot, since contention issues—who gets control over the system at any given time—must be resolved by the system designers.

With our show control example from above, each sensor could simply send, as a peer, its data directly to the show controller whenever someone enters a room. This would be the most efficient and responsive approach; additional polls could be issued at regular intervals just to make sure any lost or corrupted messages are accounted for.

Client-Server

The client-server model, which has developed along with the rise of the computer network, is a system where a **client** can make requests of a **server**. The best known example of a client-server system is probably the interaction you have when loading a Web page. Your browser acts as a client for the remote server, which "serves" you up the Web pages you request. So, if you go to `http://www.controlgeek.net/`, the computer at my Web hosting company will serve up a file (i.e., the home page HTML file) to your browser (the client), which will then display it. Many different clients can be accessing the same server simultaneously.

Producer-Consumer

In the "producer-consumer" model, one device (the "producer") creates data of some sort and sends it out on a network for use by other devices (the "consumers"). For instance, a remote microphone input device may produce level data in dB, which is "consumed" by other connected systems, such as a remote meter display at the audio console. This approach can maximize network efficiency, potentially using less overhead than the other architectures.

PHYSICAL TOPOLOGIES

The way in which a system is physically connected, or laid out, is its "topology." The three most common system topologies are the star, the bus, and the ring, each named for its schematic appearance.

Star Topology

Each device in a star-topology system is effectively "home run" by a central device. The central connecting device can "repeat" or "broadcast" each incoming control message, or alternatively, it can route or "switch" particular messages to specific devices.

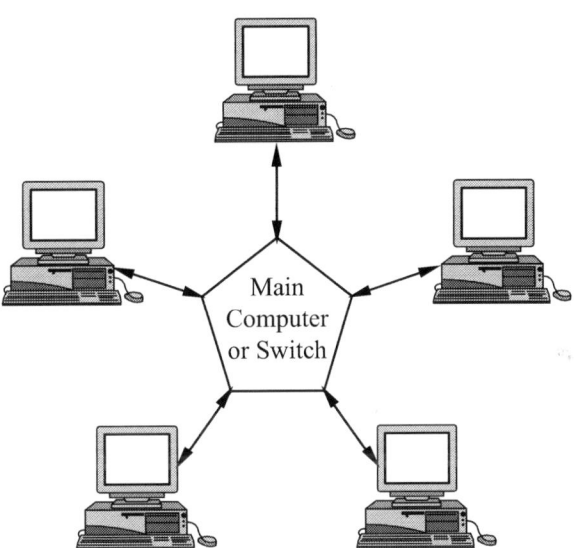

Star-topology systems are probably the most widely used, and are ideal for any application in which simultaneous multiple pathways between various devices are desired (most Ethernet networks use this topology; see Chapter 16). In addition, if a single device in a star topology system fails, other devices will typically not be affected. On the other hand, if the central star device fails, all communications will fail. However, high-reliability versions of these devices are available, and spares should be kept on hand for critical applications.

Chapter 10: Entertainment Control Basics • 83

Bus Topology

The bus topology is a simple scheme in which all devices are connected to a single system cable "bus."

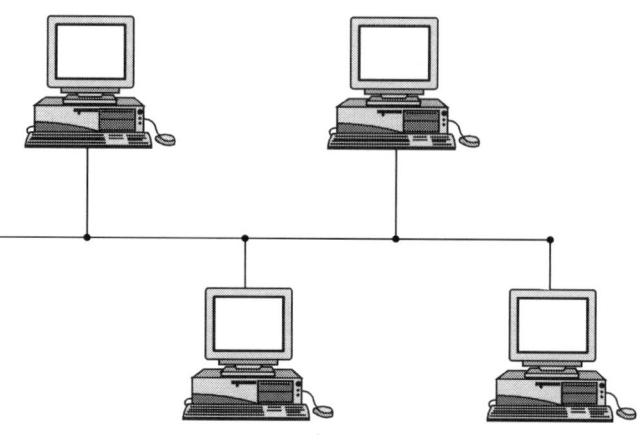

For electrical reasons, this is the topology used in some unidirectional control standards—a "daisy chain" is created through all the devices, with the final device terminating the line (this is how DMX works; see Chapter 19). A variation of bus topology is the "tree," where several independent buses are connected together via a "backbone," schematically giving the topology a tree-like appearance. Bus topology can also be found inside personal computers connecting internal expansion and peripheral devices. Bus topology systems are easy to wire, but suffer some distance limitations if the signal is not amplified or repeated as needed. Buses also have a disadvantage in that a problem on any one device can take down the entire system.

Ring Topology

In the ring topology, all devices are connected with a loop of system cable; system data is passed from one device to another around the ring sequentially. Ring topologies, which are often used with token-passing protocols, can cover greater distances than some other topologies, since data is repeated and amplified as it is passed around the ring. Like other topologies, however, the ring also has a disadvantage in that the failure of a single device, or a failure between devices, can bring down the entire

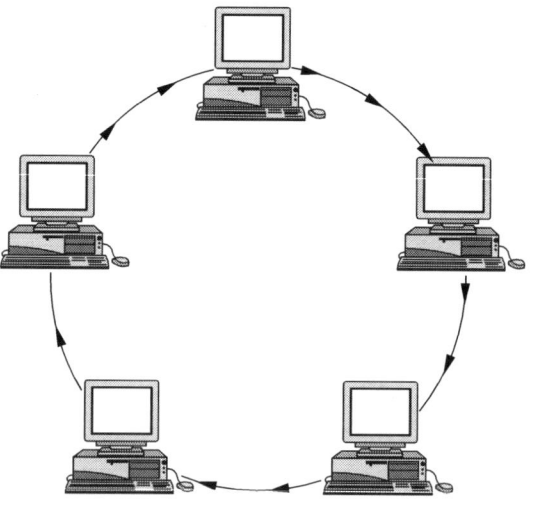

system. For this reason, critical ring systems often employ complete backup rings, which are active at all times. Another disadvantage is that the ring structure makes it difficult to add stations to the system after initial installation, since the entire system must be shut down in order to do so. Ring topologies are rarely seen in entertainment systems.

Combinations

Of course, it is also possible to combine various topologies—you could have a ring of stars, a bus of rings, and so on, or a "free" topology, where the system allows any sort of connection.

VARIABLES

Variables are a common concept in computer programming and are finding their way into more and more entertainment controllers and applications. A variables is essentially a computer memory location into which information can be written, or from which information can be recalled. For example, a variable named "Current Cue" could be used to store the number of the last executed cue on stage; each time a new cue is executed, the variable is overwritten with the new value. Variables can be filled with the result of mathematical or logical operations (see below) and come in several basic "types", such as integer, floating point, and string.

Integer variables can be used only to store "whole numbers," which can be thought of as numbers with no fractional or decimal components. But what if you want to deal with a fractional number? Floating point storage is the answer. The floating point approach can be thought of (to simplify) as a sort of binary implementation of scientific notation. In this binary representation, a number of bits are allocated to store the coefficient, the sign (+ or –), the exponent, and so on. In this way, complex fractional numbers can be stored and manipulated in a binary representation on a computer. Strings are simply strings of alpha-numeric *characters*. The words on this page were stored as strings in the computer on which I wrote them.

Depending on the operating system and programming language, you may have to convert one type to another in order to use it. For example, if you enter numbers on a screen as a string, you may be storing the *characters representing the number* (more on this later in the "Character Encoding" section, on page 126). To use these numbers in a mathematical operation, you would have to convert them from a string into an integer or floating point number. Finally, keep in mind that the amount of memory allocated to a variable determines how large or how accurately a number can be stored, or, with a string, how many characters can be stored.

LOGICAL OPERATORS

Logic has many applications throughout the world, from critical thinking to entertainment control systems. In a logical operation, we typically take information from some number of inputs, process that information through one or more logical operators, and create an output state.

If-Then

While If-Then isn't really a logical operator, it is very important for control applications. If-Then says that *if* the input condition equals some value (such as "true," "seven," or "red"), *then* the output condition will now equal some predetermined value. This operator is very useful in entertainment systems. For instance, *if* the performer is in place, *then* execute light cue 1, or *if* the time is one minute and thirty five seconds, *then* fade out the sound cue.

Or

Or is the simplest of the logical operators and, simply, is a rule stating that if *any* of the inputs are true (otherwise known as "on" or "1"), then the output will be true; otherwise, the output will be false (otherwise known as "off" or "0"). Let's imagine a system with two inputs, A and B, and apply the Or operator to it. If either input A *or* input B is true, then the output will be true. If *neither* input A *or* input B is true, then the output will be false. Logical operators are generally easier to understand through the use of a "truth table," which shows all the possible conditions of a set of inputs and their output value based on the operator. Below is the truth table for the Or operator (it shows only two inputs; more would, of course, be possible).

Input A	Input B	Output
False/0	False/0	False/0
False/0	True/1	True/1
True/1	False/0	True/1
True/1	True/1	True/1

The standard Or operator is also known as the "inclusive" Or, because there is a variation known as the "exclusive" Or, where only one input will cause the output to be true (if all the inputs are true, the output will be false).

Below is a two-input truth table for the exclusive Or operator.

Input A	Input B	Output
False/0	False/0	False/0
False/0	True/1	True/1
True/1	False/0	True/1
True/1	True/1	False/0

And

With an And operator, all of the inputs must be true in order for the output to be true. Let's use the And operator as above with our two input (A and B) system. If input A *and* input B are both true, then the output will be true. Any other condition will cause the output to be false. Below is the And operator expressed in a truth table:

Input A	Input B	Output
False/0	False/0	False/0
False/0	True/1	False/0
True/1	False/0	False/0
True/1	True/1	True/1

Not

A Not operator "inverts" the state of any input. If the input is true, the output of the Not operator will be false. If the input is false, the output is true. This Not, or "negating" condition, is often represented by a horizontal bar over a label, such as $\bar{A}$, or referred to as an "Inverting" input. Below is a truth table for the Not logical operator:

Input	Output
True/1	False/0
False/0	True/1

Logic Combinations

Complex systems can be easily built by combinations of simple logical operators; for example, we could build a complex statement as follows:

IF button A is pressed *AND* button B is pressed (indicating the performer is safely in position),
AND the fire system is *NOT* activated *AND* the wind speed is *NOT* above 15 m.p.h.,
AND the day of the week is Monday *OR* Tuesday *OR* Wednesday,
AND the technical director has recently pressed button C,
THEN fire the explosion effects.

Chapter 11: **Electrical Control System Basics**

Now that we've covered some general control concepts, we can move on to the basics of control systems. The vast majority of entertainment control systems are operated and controlled electrically, so electrical systems are the focus of this chapter (keep in mind, however, that there are many other types of control systems—pneumatic, optical, etc.). In this book, I assume that you already know the difference between voltage and current, and so forth; if you need more information on those areas, please consult the bibliography on my website.[1]

SENSORS AND SWITCHES

Inputs to a control system can come either from the output of another system (remember from Chapter 1 that outputs always connect to inputs) or from connected **sensors** and switches,[2] which detect some condition in the environment and report that condition's status to the control system. A sensor or switch could be a button pressed on a control console by a human operator, a photo-electric beam crossed by an actor, a proximity switch that senses when a metal machine part is near, or a sensor measuring the air temperature. There are many types of sensors and switches, and because they're common in a variety of entertainment control systems, we'll cover some basic types here.

Operator Controls

This category can include anything from an industrial control actuated by a theme park employee to a "go" button on a lighting console.

Courtesy Automationdirect.com

---

1. http://www.controlgeek.net/
2. I have not been able to find any precise definitions describing the difference between sensor and switch. So, for the purposes of this book, a switch involves something mechanically operated, while a sensor perceives conditions in its environment in other ways. So a mechanical limit would be a "switch," but a temperature probe would be a "sensor."

Touch Screens or HMI

In industrial automation, the term **human-machine interface** (HMI) generally means a **touch screen** control interface. These interfaces are very useful because they can be easy to use, if the screen layout is well defined, and are very flexible in that changes can simply be uploaded to the touch screen unit or computer.

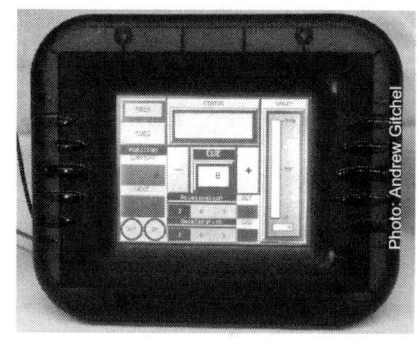

Wireless Remote Controls

Wireless (typically radio or infra-red) remote controls are another form of operator control that are useful in many types of entertainment control systems. Handheld units for our market come from the security or industrial control industries, where actuation reliability is taken seriously. They can be useful where a client or performer wants to trigger an effect discreetly, control a private entrance, and so on or, of course, in any situation where wires are not desirable. For reliability reasons, I always advise using a wired interface if at all possible.

Limit Switches

A limit switch is typically actuated by a machine part that physically contacts the switch, indicating that a "limit" has been reached. Limit switches offer absolute positional control and come in a variety of designs, ranging from rotary-cam limit switches to simple mechanical switches mounted on a track or piece of machinery. Both are actuated when the machinery system hits a "limit," or desired target position.

Courtesy Automationdirect.com

Encoders

Encoder is a very general term, but in entertainment control systems, we typically use the term to refer to a device that takes a position, and senses and "encodes" that position into some kind of data readable by a control system. Encoders come in a wide variety of types, but most common are optical rotary-shaft encoders, which take a rotating motion and generate a digital signal to send to a control system, and linear encod-

90 • Part 2: Entertainment Control

ers, which are laid out in a linear track. The rotary shaft encoder encloses a rotating disk, which interrupts a beam of light to generate electrical pulses. The control system can then decode this signal to determine the tracked device's position or the encoder can be a user interface, interpreted by the system.

Encoders come in both absolute and relative (also known as incremental) varieties (See "Data Relationships," on page 74, for an explanation of the differences between absolute and relative). A typical absolute optical encoder uses a glass disk (see photo for an example) through which a number of light beams read out a discrete digital signal describing where in its rotation the encoder's shaft is, since each rotary position of the shaft (to the resolution of the encoder) generates a dis-

tinct combination of on and off states. Even if the power to an absolute encoder is lost, or the system suffers data corruption, the digital value representing the encoder's position is retained by the encoder itself and can be easily re-read. However, to get a useful signal with no repeated positional data using an absolute encoder, you must gear the entire movement of the measured system down to a limited number of revolutions (typically one) by the encoder.

A relative optical encoder, on the other hand, simply generates a pulse train when its shaft is rotated (see photo for an example). A typical encoder contains a simple disk with a number of lines, through which a light beam[3] is passed. If the encoder's shaft is stationary, we get no signal. If we rotate it 10° clockwise, some number of pulses are generated on the output. So we now know that in relation to the starting position, the shaft (and, therefore, the attached system) has

moved by that number of pulses; this data can be translated into a positional value. These types of encoders must be "initialized" to a known start position on power up and use external counters to maintain a positional value. Most relative encoders in our industry are quadrature encoders which, when in motion, generate two pulse streams out of phase with each other. By looking at which pulse train makes a low-to-high (or high-to-low) transition first, a controller can determine in which direction the encoder is moving.

---

3. Or two beams in a quadrature encoder.

Photo-Electric Sensors

Photo-electric sensors typically send an infra-red beam of light directly or indirectly (via a reflector) to a receiver, which then sends out a control signal when the beam is either interrupted or made, depending on the application. "Light operate" sensors turn on their outputs when light is detected; "dark operate" sensors turn on when the light beam is broken. "Retroreflective" sensors utilize a beam that bounces off a reflector, and are so named because the reflector gives back light at the same angle from which it arrived.[4] "Thru-beam" photo sensors are those that send a beam from a transmitter to an active receiver. These sensors are useful in entertainment when you want a performer or audience member to activate something by simply breaking an invisible beam.

Proximity Switches

Using no moving parts, proximity switches are able to detect the presence or absence of a nearby piece of metal or other material. They are generally very reliable and durable and find a wide variety of applications in simple machinery systems.

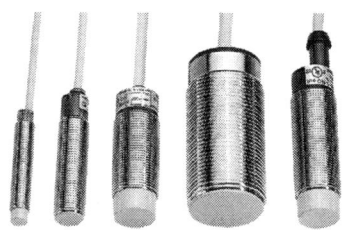

Courtesy Automation direct.com

Motion Detectors

While most of the sensors we've discussed so far have come from the world of industrial control and factory automation, motion detectors were developed for the world of security and alarm systems. These devices typically send an infra-red light, microwave radiowave, or ultra-sonic audio signal into a space; if someone moves through that space, they disturb the field and the motion can be detected from the change in reflections. Systems are also available that monitor a video signal and look for visual change. Whatever the technology, these sensors can be useful for a variety of entertainment applications—from haunted

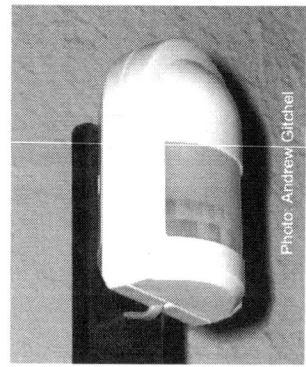

---

4. Retroreflectivity is the reason stop signs reflect car headlights back to the driver from any angle.

houses to museums. However, you should keep in mind that these sensors are typically designed with reliable actuation, not necessarily quick response time, in mind.

Radio Frequency IDentification (RFID)

Radio Frequency IDentification (RFID) technologies are increasingly affordable and are now being used in entertainment technology applications. Typically, RFID "tags" contain a unique ID number that can be recalled by a remote radio system. Some tags actually get their power from the radio system and do not need batteries. The applications for such systems are endless: a performer might carry the tags so they could be identified onstage for wireless microphone delay time location; cars in a dark ride could be identified and tracked; and so forth.

Other Sensors

We've only scratched the surface here of what's available. Sensors exist for nearly any physical characteristic you can imagine: temperature, pressure, velocity, force, strain, position, flow, and so on.

CONTACT CLOSURES

One of the simplest (and most limited) ways to interface two things electrically is to use a **contact closure**, otherwise known as a dry contact closure (DCC) or general purpose interface (GPI). This type of interface is a lowest-common denominator way of connecting two systems, since only two states can be communicated—on or off, 1 or 0. However, contact closures are quite useful for simple interfaces, such as a proximity switch (with "on" indicating that the sensed item is in range), a go button (with "on" telling the system to go), a thermostat, pneumatic valve, or other similar device.

A contact closure interface may contain physical, conductive contacts (like those shown in the photo). Alternatively, the on-off functionality may be provided by a solid-state (no moving parts) switch such as a transistor or a solid-state relay. In

What Makes A Contact "Dry"?

According to several engineers I contacted some years ago, "dry" contacts are those capable of working reliably with very low currents, and "wet" contacts need higher voltages and currents to work properly. Wet contacts are no longer commonly found and were generally made of inexpensive base metals, such as copper, brass, or tin, which oxidize easily, unlike precious (and more expensive) metals, such as silver and gold. The oxides which form on base metals are often nonconductive, so, to keep the corrosion from affecting performance, telephone engineers "wetted the circuit" by running a constant DC voltage across the terminal to ensure that the AC signal switching would work reliably.

any case, the contact closure typically switches (or accepts as an input from a switch) a small voltage or current to actuate another system or device.

However, even though contact closure interfacing is conceptually simple, it can be difficult to get two devices interfaced properly because there are so many variations and no widespread standards.

CONTACT NOMENCLATURE

An important characteristic of sensors, switches, and contact closure interfaces is the contact arrangement. These arrangements can range from simple, mechanical relay contacts in single or multiple poles to sophisticated transistor outputs.

Normally Open or Normally Closed

A switch or contact closure is typically either **normally open** (NO) or **normally closed** (NC). In an NO contact, the flow of current is interrupted when the switch is in its "off" state; current flows when the switch is in its "on" state. A standard light switch is normally open—in a "normal" state, the switch is off, the contacts are open, no current is flowing, and the light in your room is off. When you flip the light switch, you turn it "on," the contacts close, current flows, and the light comes on.

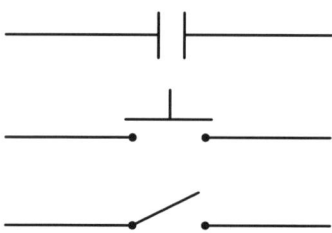

Three different ways of indicating a normally open contact

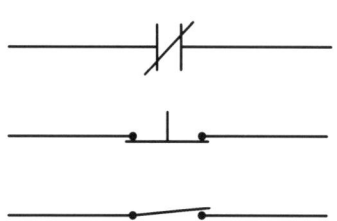

Three different ways of indicating a normally closed contact

An NC contact is the opposite of an NO contact. In its "normal" state, current *is* flowing; when the switch is actuated, the flow of current is interrupted. Why would you ever want a switch like this? This type of switch is very useful in control system applications where you want current to flow until a condition is met, such as pressing an "emergency stop," or "e-stop," button (see "Emergency Stop Systems," on page 112). In this NC application, you want the system to work normally *until* the button is pressed. In other words, you want current to flow in the "normal" situation and the flow to be broken in the emergency, or switch-actuated, state.

Contact Arrangements

Contacts on switches or sensors come in a wide variety of designs, and can offer multiple, independent control circuits that are actuated simultaneously. In industry, these contact arrangements are often referred to by their "form" schedule or by a more recognizable nomenclature, such as SPDT.

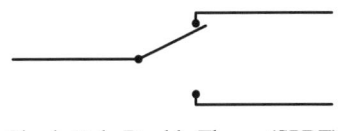
Single Pole Double Throw (SPDT)

SPDT simply means "single pole double throw," and could be represented schematically as shown here. In such an SPDT ("form C") switch, one wire or circuit can be routed to *either* of two outputs, hence the nomenclature: single pole (one wire) double throw (two output positions).

With a DPDT switch contact, two independent wires can be routed to two separate sets of independent contacts. While the switch contacts are electrically isolated, they are mechanically linked together so that they change states at the same time. This nomenclature can represent nearly any type of switch contact; for instance, a 4PST switch would have four wires on separate circuits, and only two possibilities—on or off—for all four wires simultaneously.

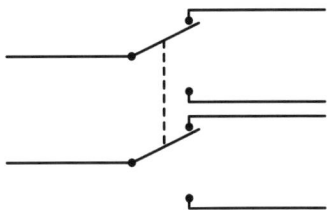
Double Pole Double Throw (DPDT)

Contact Options

In addition to the contact arrangement, there are a number of other options to consider when selecting a contact closure device. For instance, you need to consider whether the contacts should be "momentary," returning to a "normal" position after the actuating force (e.g., an operator pushing the button or the machinery moves away) is removed, or "maintained," where the contacts stays in the actuated state until some other action resets it. There are dozens of other considerations too numerous to detail here, but keep in mind that the seemingly simple act of selecting something like an operator control can involve quite a bit of work.

Relays/Contactors

While computers and controllers can typically produce small electrical currents for control, we often want to run larger loads. A simple electro-mechanical device that will do this is a **relay**, or a **contactor**. Relays generally refer to devices that carry smaller currents and contactors generally are devices that carry large currents. Both work in the same way: A small control current is used to energize an electromagnet, which, in turn, physically pulls a set of contacts together, closing the circuit.

These contacts can be designed to handle very large amounts of current and, therefore, allow large, high-power devices to be operated from small control signals. Solid-state relays using transistors, **silicon-controlled rectifiers** (SCR), **Insulated Gate Bipolar Transistors** (IGBT), and other devices are also available and offer the same functionality with no moving parts. Care must be taken to suppress the inductive spikes caused by electromagnetic coils found in control electronics devices. **Snubber**s are typically used for this purpose.

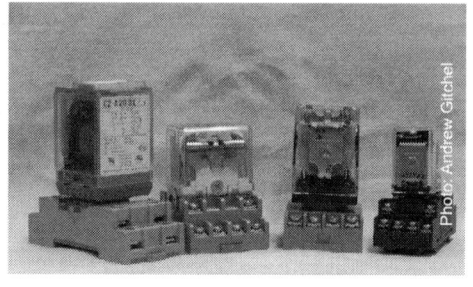

Sourcing/Sinking Transistor Interfaces

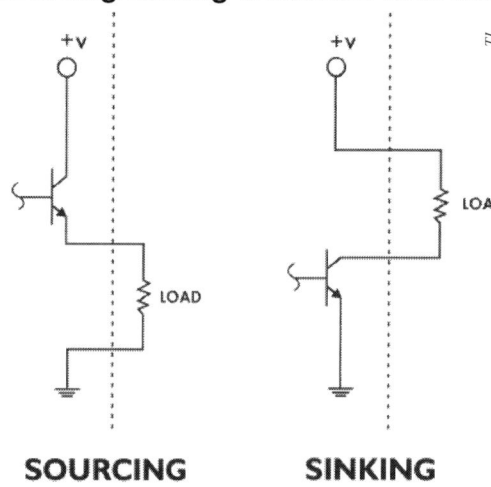

SOURCING   **SINKING**

Many contact closures or general purpose interfaces actually use transistors for interfacing, instead of physical contacts. There are two configurations in which transistors are used for this purpose: **sourcing** and **sinking**. When a sourcing interface turns on, current flows through the transistor, out to the load, and then back into an electrical common, often shared with the transistor. With a sinking interface, current flows from the power supply, through a load, into the transistor, and then to an internal common. When interfacing to sourcing or sinking interfaces, care must be taken to get polarity and other characteristics configured correctly.

Open Collector Transistor Interfaces

To fully describe an "open collector" transistor interface would require that we get into electronics a little more in depth than we will cover here, but the basic concept can be important to the system designer. In an open collector interface, the "collector" pin is presented to the outside world, and the system designer must provide an external power supply through the transistor terminals and is also responsible for ensuring that the system does not blow up the transistor.

This output type is very flexible and allows the transistor to work with a wide variety of voltages, but it requires some lower-level engineering by the system designer and, frequently, additional devices or components to complete the interface.

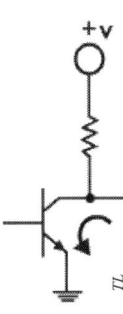

ISOLATION

When you connect electrical devices or systems, all sorts of nasty things can happen. For instance, the connection could create a ground loop, where unwanted current flows from one point of a system to another through the ground connections; this can cause data corruption in control systems or hum in sound systems. Another possibility is damage. What happens if high-voltage, high-power electricity (such as found in a dimmer rack) connects to low-voltage electronics (such as that found in a control console)? Fire and smoke, typically! The easiest way to overcome these problems is to electrically disconnect the two devices or, in other words, provide electrical **isolation**. The isolation devices are usually inexpensive and expendable—if a significant fault occurs, the small, inexpensive isolation device may burn up, sacrificing itself to protect the whole system. Isolation comes in two general types: optical or galvanic.

Optical Isolation

In an **optical isolator**, a DC input control signal turns on a light source (typically an LED). A separate and electrically isolated output circuit, incorporating a photo-transistor, turns on when the light indicating the input status is illuminated, and off when it is not. In this way, the data is changed from an on/off (1/0) DC electrical signal to an on/off optical signal, and back to an on/off DC signal. This approach provides a 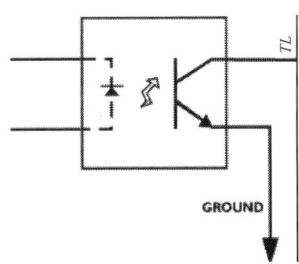 complete electrical disconnection between input and output, but typically only works for on/off (not proportional analog) applications. Optical isolators can block electrical ground loops and some surges and other anomalies (within the electrical limits of the optical isolating device), and are inexpensive insurance to help protect against faults and failures.

Galvanic Isolation

According to the *American Heritage Dictionary*, *galvanic* means, "of or relating to direct-current electricity . . .,"[5] so galvanic isolation means to provide direct current (DC) isolation. The most common device for this purpose in our industry is the

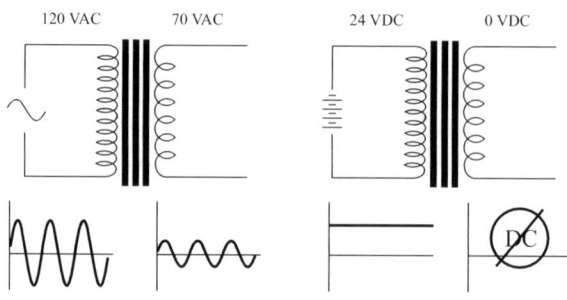

transformer. Transformers generally are used to "transform" voltage and current based on the ratio of their windings. An AC signal presented at the transformer's primary (input) winding generates a varying magnetic field, which then induces corresponding voltage and current onto the secondary (output). Transformers provide galvanic isolation because they pass AC signals, but block DC signals.[6] However, for this reason, transformers cannot pass typical DC on/off control signals unless the signals are modulated into some sort of AC signal (as is typically done in Ethernet transceivers).

Relay Isolation

An additional form of galvanic isolation is the relay (see "Relays/Contactors," on page 95). In a relay (see schematic), the control coil is completely separated electrically from the actuating contacts.

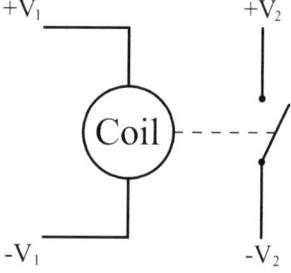

---

5. "galvanic." *The American Heritage® Dictionary of the English Language*, Fourth Edition. Houghton Mifflin Company, 2004. 14 Mar. 2007. <Dictionary.com http://dictionary.reference.com/browse/galvanic>

6. For reasons outside the scope of this book.

Chapter 12: Numbering Systems

Before we move on to the details of data communications, we have to cover one more set of basic concepts—numbering systems. Please be patient while working through this section—the ideas presented here are not really that complex, but many people have difficulty with them since they may not have studied these concepts since grade school. Understanding number systems, however, is critical to understanding computers and networks, and networks form the foundation for many entertainment control systems.

BASE 10 (DECIMAL) NOTATION

Humans generally deal with quantities in **base 10**, where numerals represent quantities of ones, tens, hundreds, thousands, and so on.[1] Numbers in base 10 are represented with the "Arabic" digits 0 through 9; in base 10, the symbols "235" represent a quantity of two hundred and thirty-five units. Each position in the numeral 235 has a certain "weight." The least-significant (right most) digit (5) has a weight of 1, or 10^0; the most-significant (left most) digit (2) has a weight of 100, or 10^2.

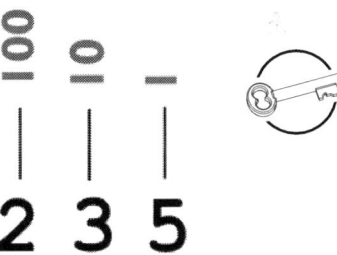

The numeral 235 breaks down as follows (right to left, or least to most significant digit):

Symbol	Weight	Quantity	Total
5	$10^0 = 1$	5×1	5
3	$10^1 = 10$	3×10	30
2	$10^2 = 100$	2×100	200
			235

---

1. Why base 10? Ask evolution and count your fingers.

BASE 2 (BINARY) NOTATION

While humans deal intuitively with base 10, digital machines operate natively in a **binary**, or **base two**, universe. At the lowest level, they deal with only two states: on or off. Correspondingly, when base 2 is represented using Arabic numerals, only the first two digits are used: 0 and 1.

In a digital system, each digit of a binary number is called a "bit," which is short for binary digit. A bit is simply an on or off state (1 or 0), referenced to a specific point in time. Since a single bit can represent only two possible quantities (one or none), bits are generally grouped into "words." The most common word size is a group of eight bits called a **byte** or, more accurately, an **octet**.[2] A part of a byte (usually four bits) is sometimes called a **nibble**.

The bits in an octet are numbered 0–7 (right to left), with the bit position numbers corresponding to the digit's weight. An octet can represent 256 different quantities (0–255). As in base 10, the least-significant (right most) binary digit represents units: in base 2, this digit can denote only 0 or 1. Weights in binary numbers are powers of 2, so the next digit to the left, or the next most-significant digit, represents the quantity of twos, the next the quantity of fours, the next the quantity of eights, and so on.

```
128  64  32  16   8   4   2   1
 |   |   |   |   |   |   |   |
 1   1   1   0   1   0   1   1
```

The quantity of 235 units has a binary equivalent of 11101011, which breaks down into base 10 as follows (right to left, or least to most significant digit):

Symbol	Weight	Quantity	Total
1	$2^0 = 1$	1×1	1
1	$2^1 = 2$	1×2	2
0	$2^2 = 4$	0×4	0
1	$2^3 = 8$	1×8	8

---

2. The term "byte" is sometimes (and somewhat incorrectly) used to refer to a word containing any number of bits; the term "octet" is more specific, and refers only to eight-bit words. For this reason, you will find it used in many network standards.

Symbol	Weight	Quantity	Total
0	$2^4 = 16$	0×16	0
1	$2^5 = 32$	1×32	32
1	$2^6 = 64$	1×64	64
1	$2^7 = 128$	1×128	128
			235

BASE 16 (HEXADECIMAL) NOTATION

Since humans work primarily in base 10, and computers use base 2, you may be wondering why we are covering **base 16**, or **hexadecimal** ("hex"). The reason is that hex is very useful as an alternative (and much easier) way to deal with eight-bit binary numbers, because each binary nibble (four bits) can range in value from 0 to 15 decimal (one digit in hex) and, therefore, an octet can easily be represented by two hex digits.

Since there are only ten symbols in the Arabic number system, the digits 0–9 are used to represent hex values 0–9, and the letters A–F are used to represent hex values 10–15:

Decimal Digit	Hex Digit
0	0
1	1
2	2
3	3
4	4
5	5
6	6
7	7
8	8
9	9
10	A
11	B
12	C
13	D
14	E
15	F

In hex, the least-significant digit represents (as is the case with all the other bases) the quantity of ones (16^0). The next most significant digit represents counts of 16 (16^1), and so on.

So the decimal quantity 235 is EB in hex, and breaks down as follows:

Symbol	Weight	Quantity	Total
B	$16^0 = 1$	11×1	11
E	$16^1 = 16$	14×16	224
			235

To further clarify, here are a few examples of some hex numbers and their decimal equivalents:

Decimal	Hex
000	00
001	01
015	0F
016	10
127	7F
255	FF
256	100
512	200
4095	FFF

As shown above, 01 and 10 could be decimal, hex, or even binary. In this book, numbers are assumed to be base 10, unless some context is given (such as in a table of binary values), or I have appended a $_{subscript}$ notation after the number to indicate the base. For example, a subscript of $_2$ indicates a binary number, while $_{16}$ indicates hex. 01 binary would be shown as 01_2, while 01 hex would be shown as 01_{16}.

102 • Part 2: Entertainment Control

In programming, you may also see a notation like 0x preceding a hex number, or you may see an upper- or lower-case "h" (h or H) appended to the number. But they all mean the same thing: the number is hexadecimal.

BINARY-CODED DECIMAL NOTATION

In **binary-coded decimal** (BCD), binary numbers are used to represent base-10 *digits* (not values). For instance, the BCD representation of the decimal number 82 is 1000 0010, where the nibble 1000 represents 8 (the original base 10 tens digit) and the second nibble 0010 represents 2 (the units digit). (The higher possible binary values in each four-bit nibble are not used, since they represent more than ten states.) This notation is not very efficient, since it leaves some bits unused, but it is very useful in control system applications where you want a few control lines to represent many human-readable (base 10) combinations. For example, BCD switches might be used to set the ID number on a controlled unit. A base 10 number is presented on the front of the switch to the user, while on the back a group of on/off, digital control lines are switched to indicate the quantity in binary to the control system.

MATH WITH BINARY AND HEX

Keep in mind that binary and hex numbers are still just numbers, therefore, the same mathematical rules and operations that apply to decimal numbers also apply to numbers of other bases.

For example, what is the result if we add these two numbers?

```
00001111
      +1
       ?
```

Remember that this binary number is simply an indication of quantities, and so we "carry the 1," and so on. So what is the result of this operation? It's simply:

```
00001111
      +1
00010000
```

Here's a subtraction example:

```
00001111
     -10
────────
00001101
```

Of course, hex works the same way:

```
FE
+1
──
FF
```

```
 FF
 +1
───
100
```

BITWISE OPERATION

In our field, we are primarily working to connect computers, and since computers work in binary, we will look at one additional operation that works only on binary numbers: **bitwise** operations. These operations are a little different from standard logical operations such as And, Or, or Not in that they operate on the number on a *bit-by-bit* basis, rather than by comparing the entire number.

To explain, here are a few bitwise operations on two binary octets:

```
11111111
00001111  Bitwise AND
────────
00001111
```

```
10101010
00001111  Bitwise AND
────────
00001010
```

```
11111111
11110000  Bitwise AND
────────
11110000
```

```
      10101010
      11110000   Bitwise OR
      11111010
```

Computers can execute bitwise operations very quickly, so are often used in very low-level operations. They are also the basis of Network Masks (page 194).

CONVERTING NUMBER BASES

It's easy to convert numbers from one format to another—many computer operating systems have built-in calculators that make this easy, or you can buy a "programmers" calculator. However, it's also important to understand how conversion between bases is done, so let's work through several examples.

Binary to Hex

Creating hex numbers from binary is quite simple. First determine the value represented by the least-significant nibble, and write the hex symbol denoting that value (0–F_{16}). Next, determine the value of the most-significant nibble and write that hex symbol to the left of the character you just wrote. For example, let's figure out the hex equivalent of the binary number `11101011`. First break the octet into two nibbles:

Least-significant nibble: `1011`

Most-significant nibble: `1110`

Working through the bits of the least-significant nibble from right to left (least significant bit to most significant bit), `1011` has a value of 1 ones, 1 two, 0 fours, and 1 eight; 1 + 2 + 0 + 8 = 11 (decimal). The symbol representing 11 in hex is `B`. Now let's analyze the most-significant nibble, `1110`. It has a value of 8 + 4 + 2 + 0 = 14, or `E` in hex. So the hex representation of `11101011` is EB_{16}—our old friend 235.

Hex to Binary

What do we do if we want to convert hexadecimal numbers into binary? You can easily work through the process detailed above in reverse. However, for purposes of understanding, let's work through a conversion to base 10 first. Let's try the hex number $D4_{16}$:

Symbol	Weight	Quantity	Total
4	$16^0 = 1$	4×1	4
D	$16^1 = 16$	13×16	208
			212

Decimal to Binary

Now, lets convert the decimal number 212 into binary. First, let's determine if an eight-bit word can represent a number as big as 212. What is the biggest binary number we can represent with eight bits (`11111111`)? Math tells us that it is $2^n - 1$, where n is the number of bits.[3]

But let's look at that in a more intuitive way:

Symbol	Weight	Quantity	Total
1	$2^0 = 1$	1×1	1
1	$2^1 = 2$	1×2	2
1	$2^2 = 4$	1×4	4
1	$2^3 = 8$	1×8	8
1	$2^4 = 16$	1×16	16
1	$2^5 = 32$	1×32	32
1	$2^6 = 64$	1×64	64
1	$2^7 = 128$	1×128	128
			255

The table shows that an octet can represent decimal numbers up to 255, which is bigger than the 212 we're trying to convert. So, to make the conversion, we'll simply subtract out the biggest number possible, and proceed until there is nothing left from the subtraction.

---

3. We subtract one because 0 is a valid value. For example, 2^8 would be 256, so an octet can represent 256 possible values. But since 0 is a valid value, the highest decimal value that can be represented by eight bits is 256 – 1, or 255.

106 • *Part 2: Entertainment Control*

First, let's determine the quantity represented by the most significant binary digit in our decimal number 212. In this case, we're dealing with an octet, so the most significant digit has a weight of 128, and we can either have one or zero 128s. Since 128 (2^7) is less than 212, we'll subtract, and then progress through each binary digit.

$$\begin{array}{r} 212 \\ -128 \ (2^7) \\ \hline 84 \end{array}$$

We now know that the most significant bit will be a one, but we don't know the value of the other bits. So this gives us `1???????`, and we can move on.

$$\begin{array}{r} 84 \\ -64 \ (2^6) \\ \hline 20 \end{array}$$

This gives us `11??????`. Now let's try the next most significant digit.

$$\begin{array}{r} 20 \\ -32 \ (2^5) \\ \hline ?? \end{array}$$

Twenty, our remainder, is smaller than 32. So what quantity of 32s can we use to represent it? Zero. So this gives us `110?????`, and now we can try the next most significant digit.

$$\begin{array}{r} 20 \\ -16 \ (2^4) \\ \hline 4 \end{array}$$

We can subtract a quantity of one 16, giving us a remainder of four. So we now have `1101????`, and now we can move onto the next most significant digit.

$$\begin{array}{r} 4 \\ -8 \ (2^3) \\ \hline ? \end{array}$$

Chapter 12: Numbering Systems • 107

Eight is bigger than four, so we get zero eights, and we now have `11010???`.

$$\begin{array}{r} 4 \\ -4 \quad (2^2) \\ \hline 0 \end{array}$$

Subtracting 1 fours digit, we get `110101?`. Since we have zero left over, we don't need to do any more subtraction, and we can fill in the rest of the number with zeroes, giving us `11010100`.

So, we converted our hexadecimal number `D4` into the decimal number 212, and that decimal number into the binary `11010100`.

SAMPLE NUMBERS IN DIFFERENT FORMATS

Just to review, the table below shows some examples of the same number represented in different number schemes; a complete conversion list of numbers between 0 and 255 is included in the "Appendix: Decimal/Hex/Binary/ASCII Table," on page 427.

Decimal	*Binary*	*Hex*	*Binary Coded Decimal*
231	11100111	E7	0010 0011 0001
137	10001001	89	0001 0011 0111
033	00100001	21	0000 0011 0011
085	01010101	55	0000 1000 0101
240	11110000	F0	0010 0100 0000
175	10101111	AF	0001 0111 0101
219	11011011	DB	0010 0001 1001
255	11111111	FF	0010 0101 0101
020	00010100	14	0000 0010 0000
070	01000110	46	0000 0111 0000

STORING NUMBERS IN COMPUTERS

As we discussed, at their lowest level, computers work with binary numbers. This is because the `1/0`, on/off nature of binary closely matches the on/off states of memory and the processing chips that manipulate that memory. The fact that computers use these numbers, however, has some implications for their use.

Binary numbers, by their nature, do not have fractional components, and how many bits available to store a number represents the "precision" available to the system. For example, 8 bits offer 256 integers (0–255); 16 bits offer 65,536; 32 bits offer 4,294,967,296; and so on. A variable (page 85) of 32 bits is often referred to as **single precision** or just "single"; 64 bits is called **double precision** or just double.

This precision can have a critical impact on the accuracy of calculations, because if the binary memory space overflows, serious errors can occur. In addition, keep in mind that if you want to have negative numbers as well as positive, the operating system has to save room in the binary memory space to store sign information as well.

Chapter 13: System Design Principles

Before moving on to the details of data communications, let's take a "big picture" view of design considerations for entertainment control and show control systems. Everyone designs things differently; what follows is a list of the general principles and objectives I use when designing any kind of system:

1. Ensure safety.
2. The show must go on.
3. Simpler is always better.
4. Strive for elegance.
5. Complexity is inevitable; convolution is not.
6. Leave room for unanticipated changes.
7. Ensure security.

Of course, these are highly generalized principles and, as we know, there are always exceptions to any rule (except Principle 1: Ensure safety).

PRINCIPLE 1: ENSURE SAFETY

"Safety is no accident" is the cliché, but this one is true. Safety can only exist if it's not an afterthought—it must be considered for every situation and action, from the top of a system, process, or design to the bottom, from the beginning to the end. Safety must always override any other consideration: If you can't afford the resources (time, money, and so on) to do the project safely, you can't afford to do it at all.

There are a variety of safety standards[1] that cover many issues related to entertainment control, and you should always check with the **authority having jurisdiction** to see which apply to your situation. Beyond that, there are a few principles that I follow in my designs. Use these if you agree but, of course, I make no claim that following these principles will make anything safe! Safety is up to you—the system designer.

---

1. The NEC, the Life Safety Code, etc.

Fail-Safe Design

Fail-safe design means that a system is designed to fail in a safe way. For example, relays or contactors can be wired using any variety of contacts, but in case of a control system power loss, it's important that the relay (which is now uncontrolled) switches the connected device to a safe state (typically de-energized). For example, on some trucks and trains, the brakes are controlled using air pressure, but the air pressure actually *releases* the brakes, rather than applying them. That way, if a brake hose becomes detached or develops a leak, the system will fail into a safe condition, applying the brakes.

Single Failure-Proof Safety

Another good design principle is **single failure-proof design**, where a system is designed so that if any single (anticipated) failure occurs, the system will not fail in a way that creates more danger. The classic example of this principle in the entertainment industry is the lighting safety cable.[2] If the lighting fixture falls for any reason, the safety cable keeps it from hurting anyone. Of course, the light may end up pointing in the wrong place, and you may need to get out a ladder to get the light down, but the goal here is safety, not performance. Single failure-proof design can also help to increase reliability through redundancy.

Emergency Stop Systems

Emergency stop, or E-Stop systems, are critical for any dangerous entertainment control system that could injure someone. An E-Stop generally works by *disconnecting* power to or from a drive at very low levels in a system. E-stop systems should be very simple and independent of computerized or electronic elements.

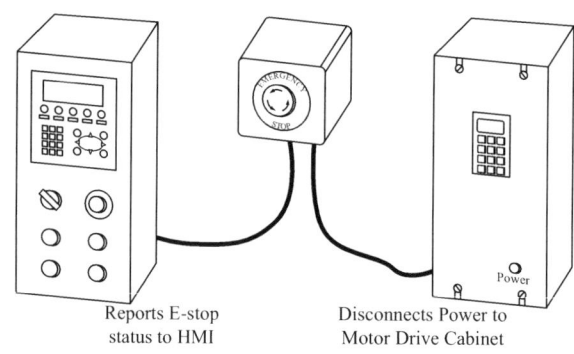

Reports E-stop status to HMI   Disconnects Power to Motor Drive Cabinet

Chapter 9 of NFPA 79, the *Electrical Standard for Industrial Machinery*[3], is one of the key standards covering E-Stop in the US, and it outlines the type of control

---

2. For those not familiar, the safety cable is a strong, rated wire rope cable looped around the hanging position (typically a pipe) and through the "yoke" of the unit, or onto a specially made hook built into the unit.

3. You can read the 2012 edition of this standard online for free at: http://www.nfpa.org/aboutthecodes/AboutTheCodes.asp?DocNum=79

components allowed in the E-Stop circuit, and how the system should react once an E-Stop command has been issued by a human. It also mandates that E-Stops must override all operations, and that the system remain in the emergency state until the E-Stop is reset at the location where it was initiated. Once reset, the system must not restart the machinery, but only allow the system to be restarted.

Large, dangerous effects may not stop smoothly, and catastrophic damage to a system (possibly causing further unsafe conditions) could occur due to large inertial loads or other factors. All these issues must be considered when designing the entire system (mechanical and control); a control engineer needs to work with the mechanical designers to create a system that can be stopped safely.

In addition, many control systems have a "controlled stop" mode for less serious cases (e.g., a performer walks on stage a bit too early, but is not yet in imminent danger), and, of course, the E-stop is always available for emergencies (e.g., the performer is in the wrong place and is about to get hit by a moving platform).

Humans in the Loop

In my opinion, humans should *always* be in the control loop of any entertainment effect or system that could hurt someone. In factory automation and robotics, they build big fences around dangerous robots and add interlocks so that if someone opens the fence gate, the machine stops. This approach is useful in some entertainment applications (e.g., under a stage), but there are many effects, particularly on a stage, that can't be fenced in without ruining the show. Computerized control systems can apply tremendous processing power to get highly precise effect timing and cuing; but any dangerous effect must never be allowed to operate without some positive action from a human somewhere—on or offstage. **Enabling** systems are often used to incorporate this human feedback into control systems.

Effect-Enabling Concepts

Enabling systems allow humans to "authorize" or "enable" dangerous effects. For example, a computer could continuously monitor a number of safety sensors and, if all the sensed conditions are safe at the appropriate time in a show, open a timing "window" that enables the execution of a dangerous effect. When this window opens (typically for a limited amount of time), the effect is enabled and, in many cases, a button lights up on a user interface to indicate this state. The operator can then fire the effect at the right moment in the show after confirming that conditions are safe. The actual job of the operator in this case is to *NOT* fire the effect if things are not safe for the performer. Another control structure (also depicted) would be for the operator to enable the computer system (typically by pressing a button on a console) to fire the effect at a specific time in coordination with other systems.

Still another possibility is that the computer could open a timing window, an operator would enable the effect, and the *performer* would actually fire it by stepping on a switch or some other means. Through the use of enabling systems (and appropriate system design, of course), a single operator (often called the "technical director" in theme parks) can control entire sophisticated attractions and shows full of potentially dangerous effects.

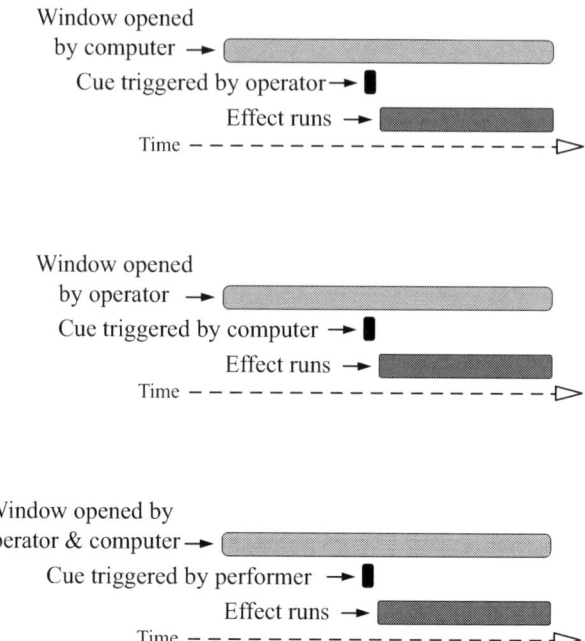

Operator Issues

In applications such as theme parks, where shows may run dozens of times a day, system designers also have to take operator boredom or complacency into consideration. Well-designed enabling systems typically look for some sort of *transition* (button pressed, button released, or button pressed and released) from the operator's control, not just a steady state. This keeps a bored operator from placing a rock on the enable button in order to take a break. If the enable is not seen by the system at the appropriate time, the effect can be programmed to fail in a safe manner or go into some sort of "graceful abort" mode. A related device is the "deadman's switch," which stops an effect if the operator becomes incapacitated; for example, on a subway, the train will stop if the motorman drops dead (since he's no longer able to push down on the operating control).

There are many common-sense techniques to counteract operator fatigue or complacency. For example, you could place part of the operator's control or indicating system *inside* or near the dangerous effect. Another tactic is to use two controls spread across some distance, to make sure that the operator is facing the appropriate direction and is not distracted while pressing the execute button.

Failure Mode and Effects Analysis (FMEA)

Failure Mode and Effects Analysis (FMEA) is the formalization of a safety technique that designers often use intuitively. The basic FMEA process anticipates possible failures and then prioritizes them in terms of severity, likely frequency, and how the failures can be detected. Risk Priority Numbers are then assigned to the identified failures, and actions can be taken by the designers to try to eliminate, or at least mitigate the risk.[4]

Peer Review for Safety

No one can see every eventuality, and anyone can make mistakes. The best way to deal with this is to have a qualified peer review your work and this should, of course, start when the project is still in the design phase, when things can be easily changed.

PRINCIPLE 2: THE SHOW MUST GO ON

This axiom is second nature to those of us in the entertainment industry, and while the rest of the world is getting to be more like our industry all the time, some system designers still fail to take this principle into account. Entertainment systems must be designed to be flexible and quickly and easily understood, diagnosed, repaired, or even bypassed. An operator must be able to run the system manually if necessary. Murphy usually strikes about three minutes before show time, so it's critical that backup systems are in place and are easy for the operator to use, even when under enormous pressure. The amount of redundancy, of course, is always subject to budget and other resource considerations.

Redundancy

One of the best ways to ensure that the show goes on is to make sure there is redundancy in your system. Computers are relatively inexpensive, so if you're running a sound effects system, it's quite easy and economical to have another machine run in parallel with the main machine as a backup. If the main machine hard drive dies in the middle of the show, you can simply switch over to the backup. How smoothly this transition goes typically depends on how the system is designed, and in networked systems, how carefully the implications of switching over have been analyzed. If you can execute the transition seamlessly, without losing your place in the show, this redundant approach is typically called a "hot" backup system.

---

4. Alan Hendrickson's book *Mechanical Design for the Stage* is a great resource on developing a FMEA.

Redundancy can also be implemented through the use of multiple power supplies built into a system, or through the use of Uninterruptable Power Supplies (UPS), which offer a backup battery to run the system for some period of time.

Ensuring Maximum Computer Reliability

While of course reliability is a concern with every kind of device in a control system, we are increasingly dependent on computers, so we should take some special precautions. Statements about which operating system "never crashes" are pointless and inevitably lead to ridiculous, drawn out flame wars on Internet forums and mailing lists. Every system is capable of crashing (and if it's not the operating system, a hardware failure is always possible), so it's best to plan your system design with that in mind. Following are some general good-practice principles that I apply to any computer running a system on a show.

Disconnect Your Show Network from the Internet

In addition to security issues (see "Principle 7: Ensure Security," on page 118), connecting your show network to the Internet gives all kinds of software processes on your machine the opportunity to update themselves, register software, report on your usage, and so on, which may disrupt events. If your show network is physically disconnected from the Internet, these processes are far less likely to pop up in the middle of your show. If you do need to connect your show network to the Internet to, for example, get updated virus definitions for your machines, then try to do it only when needed and then disconnect your network again. Alternatively, you need to learn a lot about security, routers, etc., that is beyond the scope of this book.

Install Only What You Need, Remove What You Don't Need

Computers are ever-cheaper, so you should dedicate your machines for show purposes only—they should contain nothing but the software you absolutely need for the show. Disable any automatic operating system features that you don't absolutely need—these often run in the background, consuming computer resources, causing seemingly erratic performance problems, and potentially opening security vulnerabilities. For similar reasons, do not install games or other "fun" software on the show machine—bring your own laptop for that purpose.

If you are taking an older machine and converting it to show purposes, it's best to strip it down and do a clean install of the operating system and your show software, and then update the show software to the latest version, and so on. When that is complete, lock the configuration for the run of the show. Unless a major bug is causing problems, wait before installing that new software version—do that between shows when you can thoroughly test it.

Manage Automatic Virus Checking

It used to be that if your network was off the Internet, and your operators were following good practices, there should be no way for a virus to "infect" your machine in the first place. But in recent years, the threats have transitioned into things like USB thumb drive-borne "trojans" and other malware, so the need for virus checking is, unfortunately, increasing. But virus checkers often operate in an automatic mode, and you really don't want them deciding that it's a good time to thoroughly check out your hard disk while in the middle of the most demanding cue of the show. So it's best to carefully manage the settings on your virus checking software, and be very careful about what kind of removable media you allow to be connected to your show computers.

Store Your Show Files Locally

It's great to have a network available to backup your show files or to load media, but in my experience, it's generally a bad idea to actually pull your show files from a drive over a network. This increases network use, which opens you up to potential failures if the network has a problem during the show. Use a local drive instead.

BACK UP YOUR DATA

I am fanatical about backing up. It's always surprising to me how even experienced people can be cavalier about backup, when your data is your work! If you lose a day's changes to your show, you have lost a day's work. I typically keep large data files (sound, video, and so on) backed up on multiple drives in different locations, and I save a new version of the show file itself each day, or maybe even multiple versions within a single day. This way, if I totally screw up my file, I can always go back and get yesterday's file, or even last hour's file. Take one copy of your data offsite (USB drives are great for this)—what if the sprinklers go off in your building? With the ever plummeting price of disk storage, RAID (redundant array of inexpensive/independent drives/disks), which can (if administered properly) seamlessly deal with hard drive failures, are also increasingly affordable. And keep in mind that data doesn't have to be lost by some mysterious cause; it can just as likely be lost when someone spills coffee on the machine.

PRINCIPLE 3: SIMPLER IS ALWAYS BETTER

I have never seen a situation where a more complicated solution is better than a simpler system that achieves the same result. Simpler systems are easier to design, install, program, and troubleshoot (and all that generally adds up to less expensive). No one will ever be impressed by a complex system that no one can troubleshoot, no matter how many blinking lights it has. If it's possible to cut something from the system without compromising performance or decreasing flexibility, I'd say get rid of it.

PRINCIPLE 4: STRIVE FOR ELEGANCE

An elegant solution uses a minimum of resources to best accomplish a task.

PRINCIPLE 5: COMPLEXITY IS INEVITABLE, CONVOLUTION IS NOT

Big systems are inevitably complicated, but there is *never* a reason to have a convoluted system—one more complicated than necessary. Convoluted systems are generally the result of poor design, planning, implementation, or documentation, all of which lead to an unsightly mess that no one can ever figure out. Everyone I've ever known who has created convoluted systems as a means of "job security" has been eventually fired when they were not able to fix the system—sooner or later even they couldn't decipher their own convolution. To paraphrase Einstein, "Everything should be made as simple as possible, but no simpler."

PRINCIPLE 6: MAKE IT SCALABLE, AND LEAVE ROOM FOR UNANTICIPATED CHANGES

The cost of a piece of cable is extremely low relative to the costs of engineering the cable and the labor involved in running and terminating the cable. Always run spare cables, order more than you think you need, and buy spare parts. Do this even if you cannot imagine any possible way the system could need expansion—someone will soon figure that out for you. I've never heard anyone complain that spare capacity was in place, but I have heard plenty of complaints about not having room for expansion. While it's often difficult to convince bean-counters, it's much cheaper to put in room for expansion now rather than later.

PRINCIPLE 7: ENSURE SECURITY

These days, with so much literally "riding on" our systems, we have to make sure that they are secure. Security applies to all types of devices in a control system, but especially, of course, to computers and networks. Following is a basic list that I use for any computer-based, networked control system (which is just about all of them these days).

Keep Your Show Network Off the Internet

It's generally a lot tougher for an attacker to get into your network if it's not connected to the Internet. If you need to connect your network to the Internet for a software update (and your IP address scheme can accommodate it—see "Internet Protocol (IP)," on page 183), be sure to disconnect it again before show time. If you must keep your network connected to the Internet, learn everything you can about firewalls and use one. Also consult a talented IT security expert.

Use Security Features on Wireless Networks
There are so many portable wireless network devices out there now, it's impossible to know who might try to access your network just for fun. Be sure to encrypt or hide any wireless networks you might have (see "Security Issues," on page 178, for more information).

Shut Down Wireless Networks When Not Needed
It's great to use a wireless tablet to move around the venue and adjust things during technical rehearsals. However, if you are not using wireless during the show, disconnect or power down your wireless access points (see "IEEE 802.11 "Wi-Fi"," on page 175, for more information).

Use Passwords on Critical Machines and Systems
This simple, but often overlooked, step will stop people from casually trying out a control screen they happen to encounter.

Control Physical Access to Critical Infrastructure
Simply locking your critical network equipment up can make it very difficult for outsiders to get access to your network (if it is to be an inside job, he or she will likely have access to keys anyway).

Run Only Things Needed for the Show on Any Show Machine
That fun website your operator might like to visit may also contain all kinds of viruses, spyware, and so on. Tell the operator to bring in a laptop if they want to surf the Net or play video games.

SYSTEM TROUBLESHOOTING

Troubleshooting is often a daunting challenge, but it needn't be. In fact, I find it fun (as long as no one's breathing down my neck). Because every system is different, there are no real standard ways to troubleshoot anything, but I've developed a basic approach which seems to work in a variety of situations. I've included it here in case you might find it useful.

Don't Panic!
First, Don't Panic! In my experience, the vast majority of system problems are simple power, configuration, or connection problems. Entertainment control equipment is mostly very reliable, and connections, cabling, and power are often the cause of many problems.

Before Starting
Before starting to troubleshoot, be sure that:

- You have a signal flow diagram.
- You *understand* the signal flow through the system.
- The system is *capable of working*.
- You can conceptually break down the system into functional parts.
- You have a known test signal source (audio player, signal generator, data source, etc.)
- You know how to use test systems, software, and equipment such as oscilloscopes and multi-meters, Ethernet testers, DMX testers, and so on. These systems allow you to look inside complex systems, and without them, there is little you can do to understand what's really going on in computerized, networked systems.

Cure the Problem and Not the Symptom
Try to cure the problem and not the symptom. There are times when taking a shortcut to cure a symptom rather than a problem is a necessary course of action, such as five minutes before show time. But I've found such shortcuts will come back to bite you eventually, so go after the problem whenever possible.

A Troubleshooting Process
After taking care of everything listed above, following is the process I use for troubleshooting:

1: Verify the Tester
- Verify the test signal source!
- Test your test equipment. (Is your meter or tester in the right mode? Is it broken?)

2: Quantify the Problem
- Is no signal at all coming out of the system?
- Is a distorted signal coming through?
- Is the signal coming through somehow changed?
- Is the system working as anticipated?
- What part of the system is not working as anticipated?

3: Check the Obvious

- Is each device in the system turned on and powered up? Ensure that all the power lights and other indicators are showing correct operation.
- Are all the connectors connected properly?
- Are the output and input indicator lights or meters for each device in the system indicating correct signal function?
- Will the system work as designed? Did it ever work? If it's a new system that is not functioning, it's possible that there was a design flaw and the system cannot work. If it did work previously, simplify the system as much as possible—bypass any unnecessary equipment or features.

4: Determine "Verified," "Unverified," and "Suspect" devices

"Verified" devices are those you can determine to be working; "Unverified" are those you have not yet tested. "Suspect" devices are those you have tested, but don't seem to be working as expected.

5: Go Through the System Until You Have Verified Every Device

Substitute known good (tested!) parts or components for suspect ones. Work logically, systematically, and carefully through the system until every device, component, or connection has been verified. This should allow you to find the problem, or at least lead you closer to the answer. Also, it's often best to work through these steps by yourself, or with a very small group. Other people may have great ideas, but they may add confusion when you are trying to reduce the problem to a single variable.

6: If You're Stuck, Clear Your Head

If you have the system design in your head and understand it, but you're still stumped, take a break. Walk around the block, go have lunch. I can't tell you the number of solutions that have come to me as soon as I walked away from the problem and thought about something else. Clearing your head leaves room for a new solution, or at least a new troubleshooting direction.

7: Consult Others

If you're still stuck, consult others. Often, if you have been looking at something for too long, you can no longer see the obvious. If you're still stuck, consult an expert.

Part 3: **Data Communication and Networking**

In parts 1 and 2, we covered an overview of entertainment control disciplines and introduced the basic building block concepts of entertainment control. Now we move on to data communications (datacom) and networking, the fundamental building blocks of all computerized entertainment control systems. We start with fundamentals of data transmission, and then move in to data communications standards and networking.

I've broken down data communications into two types: point-to-point (covered in Chapter 15, on page 145) and network (covered in detail starting with Chapter 16, on page 157). The distinctions between the two can get blurry but, generally, point-to-point interfaces are meant for connecting two (or a few) devices together in a point-to-point fashion; networks are more generic connection methods used to connect many stations.

If these topics are new to you, please don't be intimidated—many others in entertainment are in the same situation—these technologies have not historically been covered in traditional entertainment technology training. But if you haven't yet worked on a job site where Ethernet is ubiquitous, rest assured you will encounter networking technologies soon, as more control systems—lighting, sound, video, scenery, pyrotechnics, show control—are networked every day.

Chapter 14: **Data Communication**

Data communication is simply moving some sort of data from one system to another and, yet, this simple concept offers incredible power in practice, and lies at the core of all the amazing things we see in shows. To get started in our exploration of this field, let's first cover some general concepts that are relevant to all datacom and networking systems.

AN INTRODUCTION TO COMMUNICATIONS LAYERING

Many of the older standards commonly used in the entertainment industry (DMX, MIDI, etc.) specify the details of everything needed to establish communications: the physical connections (voltages, connectors, interfaces, etc.), data transmission details (number of bits in a word, error detection scheme, routing structure, etc.), and the commands ("go"). Control standards developed in the era of the network, however, generally only specify the commands themselves and leave the details of transmission to independent, underlying technologies. Key to this approach is a concept called "layering." More details on actual layering schemes can be found in the Chapter 16, but a brief conceptual overview is useful here.

In a layered communications system, the communication tasks are broken down into component parts. For example, let's imagine a hypothetical (simplified) three-layer system, with the layers called "Application," "Transport," and "Physical." The Application layer contains the system processes that interface with the user; Transport ensures that the message gets from one place to another, while keeping it intact and in order; and Physical specifies the interface type, the type of connectors, and other similar details.

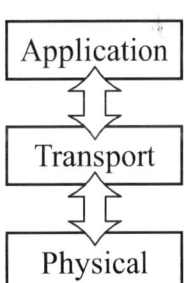

With this layered control approach, the tasks of each layer are completely compartmentalized and independent, since each layer only needs to know how to talk with the layer above it and the layer below. For example, the Application layer really doesn't care about the details of how its messages get from one machine to another; it simply has to receive its data from the user and then present it to the Transport layer (and vice-versa, of course). The transport layer doesn't know anything about the actual connectors, and it doesn't make any difference if the signal is travelling over a wired or wireless link or a fiber-optic cable. It simply needs to know how to talk to the Physical layer, which takes care of all those details.

This layered approach (along with tremendous amounts of inexpensive computer horsepower, memory, and network bandwidth, of course) has dramatically reshaped the way we build control standards. Protocols developed today allow us to easily build on other, standardized approaches, saving us time and simplifying the entire process. And, as we move forward in this chapter, keep asking yourself with each new topic: how and where does this topic fit into the layered approach?

CHARACTER ENCODING

To allow systems to communicate, we must first agree on the way that machines model and represent the physical world, and make sure that all devices communicating with one another use an agreed upon approach. Many standards for encoding specific types of data (e.g., "go," or "dimmer at 50%") are covered later in Part 4, "Standards and Protocols Used in Entertainment," on page 239, but there are three interrelated, open data character encoding standards prevalent in the world of data communications that we should cover here: the **American Standard Code for Information Interchange** (ASCII), **Unicode**, and the Universal Character Set's (UCS) widely-used **UTF-8** variant.

ASCII was standardized in the early 1960s[1], and is basically a grown-up version of a communication game you may have played as a child: substituting numbers for letters of the alphabet in order to send coded messages. For example, to send the text "Ethernet" in ASCII, the following hex numbers would be used:

```
45   74   68   65   72   6E   65   74
E    t    h    e    r    n    e    t
```

ASCII was, and UTF-8 now is, one of the most widely used standards in computing and networking, and many other standards reference or use them. UTF-8 and Unicode are backwards compatible with ASCII, and in basic control systems we're likely to be using pretty simple ASCII characters, so in this book, I will refer to these character encoding standards together simply as ASCII. A listing of the basic ASCII character set, along with decimal, hex, and binary equivalents, is in the "Appendix: Decimal/Hex/Binary/ASCII Table," on page 427.

---

1. ASCII's roots actually reach back to 1874, when Emile Baudot patented a five-bit communication code for semi-automated telegraph transmission.

DATA RATE

A digital data link carries a binary stream of 1s and 0s. The rate of transmission is

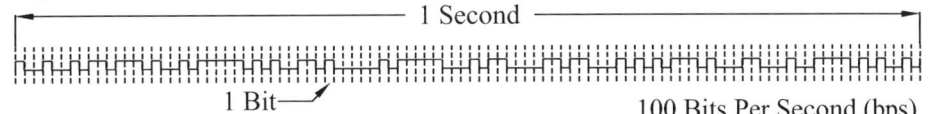 

known as the **data rate**, which is measured in **bits per second** (bps or bit/s). Data rate measurements use the International System of Units (SI) prefixes, so it's very common to see something like kbit/s, meaning 1,000 bits per second, or Mbit/s, which is 1 million bits per second.[2]

In the world of datacom, you may also encounter the term **baud**[3] rate," and the distinctions between baud and bit/s can be confusing. Baud actually refers to the number of symbols or "signalling units" sent over a data link per second, and since, with some sort of encoding, a signaling unit could in some way actually encode more than one bit, the baud rate of a communication link could be different than the link's bit/s. To avoid any confusion, bit/s will be used throughout this book.

BANDWIDTH

Whatever the communications medium, there is always some limit as to how much data a single communications connection, or "channel", can handle; this capacity is known as the channel's **bandwidth**. Bandwidth is measured by how much information can be sent in a given period of time (bit/s), and a "high bandwidth" connection can carry more than a "low bandwidth" link.

MULTIPLEXING

One way to get more information over a single communications channel is to **multiplex** the data. There are two forms of multiplexing in common use: frequency-division multiplexing and time division.

Time Division

In time-division multiplexing, multiple communications channels are each chopped into slices, which are sent sequentially over a single link, with each incoming channel taking turns on the communications channel. DMX (see Chap-

---

2. A similar (but not as widely used) unit is the number of bytes per second, which is notated with an upper-case B, as in B/s, kB/s, or MB/s.
3. Named after Maurice Emile Baudot, who invented a five-bit digital teletype code in *1874*!

ter 19, on page 241) is an example of time-division multiplexing. At the receiving end, a demultiplexer breaks up the multiplexed signal into discrete channels. In other words, each channel occupies the data link for some fraction of the total time available. For instance, four 25,000-bit/s channels could be sent over a single 100,000-bit/s link, assuming that each channel occupied the multiplexed link a quarter of the time.[4] Obviously, the multiplexed data link must have a faster information-transfer rate than the sum of all the individual channels.

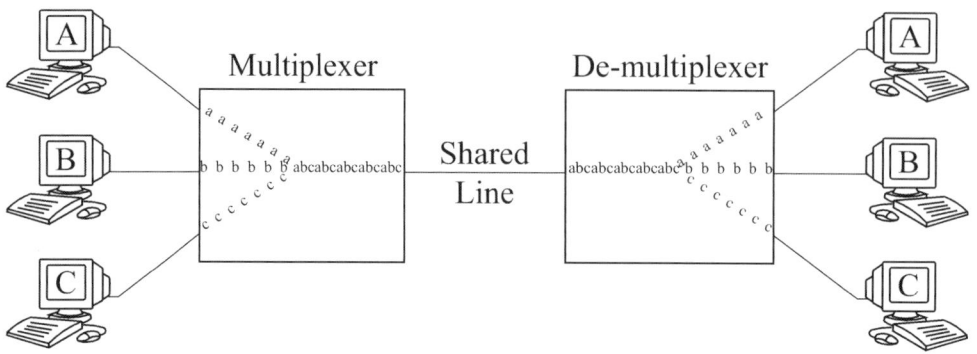

Frequency Division

In frequency-division multiplexing, multiple channels of information, each modulated at a different frequency, are sent over a single communications link. An example is (analog) cable television, where hundreds of channels are sent over a single coaxial cable from the central office to your home. Each channel is modulated at a different frequency, so the link is frequency-multiplexed.

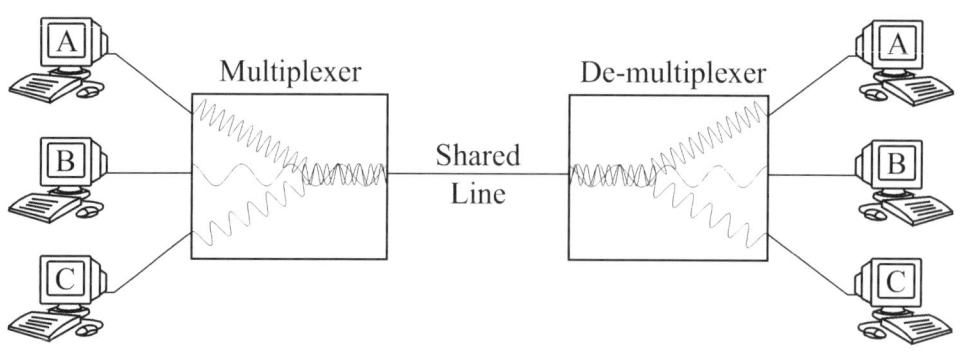

---

4. Not counting communications overhead, of course.

COMMUNICATIONS MODE

A communications link can be designed to operate in one or more modes, transmitting data in one direction only, one direction at a time, or both directions simultaneously.

Simplex

A **simplex** communications link allows a transmitter to send information in one direction and one direction only. The receiver can only listen and not reply. A "cue light" (see "Cues," on page 68) is an example of a simplex communication: The stage manager who turns the cue light on can send a message (cue light on or off) to the cue light, but the operator has no way to respond.

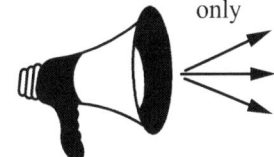

Half Duplex

In a **half-duplex** link, two parties can speak to each other over the same link, but only one party can speak at a time. A pair of hand-held "walkie-talkies" works in half-duplex mode: You can talk to someone else on the radio, but you can't hear while you're talking. Conversely, you can't talk while you are listening.

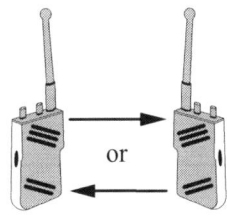

Full Duplex

Full duplex means that two parties can each talk at the same time. A telephone uses a full-duplex link: While you are speaking, you can hear the other party, and vice-versa.

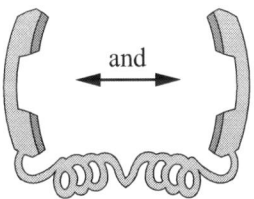

ERROR DETECTION

No communications link is perfect; there is always some possibility of error occurring in the transmission—whether caused by noisy lighting dimmers or a loose connector. Errors can detected in a number of ways; covered here are three common data **error detection** schemes, each of which adds some data to the

Chapter 14: Data Communication • **129**

information traveling over the link, and offers the receiver a mechanism to determine if the data was corrupted during transmission (or storage).

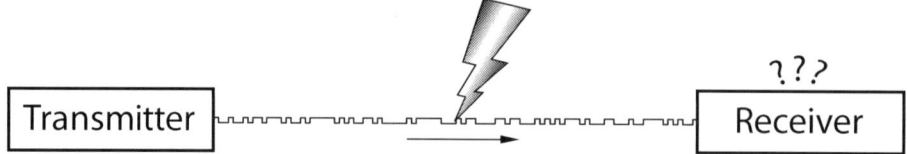

Parity Checking

In the simplest form of error detection, **parity checking**, the transmitter adds a bit to each binary word, making the arithmetic total of the word's bits either even or odd, depending on the parity type (which is determined by the system designer and configured into the system). The parity-containing word is sent to the receiver, which does the same calculation as the transmitter (sums the bits), and checks to see if the sum is even or odd. If the sum is not what the receiver has been configured for, it can ask for retransmission, or discard the word; if the sum is correct, it assumes that the data was received intact. (Obviously, both receiver and transmitter must be set for the same operational mode for this approach to work properly.)

This technique will be easier to understand if we look at an example. Suppose we transmit an octet of 10111001, and the transmitter and receiver are set for even parity. The transmitter notes that the arithmetic sum of the bits in the octet 10111001 is five, an odd number. Since the transmitter is set for even parity, it adds a parity bit with a value of 1 to this octet, bringing the number of "1 bits" to six, an even number: 101110011. If the data travels over the communications link and remains intact, the receiver adds up the bits in the message, determines that the total is even, and accepts the octet as valid. If the transmitter is set for odd parity, it adds a parity bit of 0 to this same octet, making the sum odd: 101110010. Note that neither the transmitter nor the receiver is concerned with the binary weights or meanings of the bits; it simply counts the number of 1s. Simple parity checking adds one bit of overhead to every transmitted word (12% for an octet).

Parity is simple to implement, but does have a cost: Parity checking can detect only an odd number of bit errors. To see why, assume that the octet shown above is transmitted using odd parity and a single bit gets corrupted along the way. Here, one bit (which is *italicized*) is corrupted: 100110010. The receiver catches this single bit error because the sum of the received octet is four (an even total), and since the receiver is configured for odd parity, it was expecting an odd total. Here's the same octet with two bit errors: 100111010. When the receiver counts the bits in the corrupted octet, it gets an odd sum, 5, so it assumes that the data was correctly transmitted, which is not the case.

Even with these drawbacks, parity is still useful because it is so simple to implement and catches at least 50% of errors, and it is available for use in many point-to-point connections. If a system designer has reason to believe that data corruption is highly unlikely to occur, and if the consequences of a receiver acting incorrectly on corrupted data are not severe, than parity error detection is useful.

Checksum

Checksum error detection uses what's called a "redundancy check." The transmitting system determines a "checksum" by adding a block of data, as shown:

```
     10010100
     11110111
     01101000
    ─────────
    111110011 Checksum
```

The data and the checksum value are then sent to the receiver. Upon receipt, the receiver adds up the value of the block (not including the transmitted checksum) and compares the checksum it has calculated with the checksum transmitted along with the block. If the transmitted values agree, the receiver assumes that the transmission was good. If the checksum does not agree, then the receiver can alert the transmitter or ignore the corrupt data.

While a checksum approach is more effective than parity schemes, checksums still can't catch all errors. For instance, if these same three data octets somehow arrived at the receiver out of order, the checksum would still come out the same. Also, if there were two single-bit errors in the same bit in two octets, the checksum would remain the same:

```
     10010110  Error
     11110101  Error
     01101000
    ─────────
    111110011  Same checksum,
               but error not detected
```

Another drawback of the simple arithmetic checksum approach is that the redundant data word can end up larger than any of the data words (as in our examples above, where the data was all eight bits and the resulting checksum was nine bits long). If a transmission system used eight-bit blocks, a nine-bit checksum would either have to be transmitted over several octets, or bits would have to be discarded, compromising the checksum's accuracy.

Cyclic Redundancy Check

The **cyclic redundancy check** (CRC) approach is an extremely effective error-detection method, with near 100% accuracy in many applications. While the concept of CRC error detection is fairly simple, the theories explaining its efficacy involve complex mathematical proofs well beyond the scope of this book.

In the CRC approach, the data to be checked is treated as a block of bits instead of a series of digital words to be summed. This data block is typically divided using a specially designed polynomial, and the result of the division is the checksum, which is transmitted along with the data for verification by the receiver. Because of the properties of the specially designed polynomial, the remainder for any given block is one of a huge number of possible check values, and the probability that both pure and corrupted data would generate the same checksum is extremely low. CRC is the error detection approach used in Ethernet.

FLOW CONTROL

In a data link, the transmitter might not be able to blindly send data forever, because the receiver might have limited memory capacity and could "overflow," causing data to be lost. For this reason, in some applications, the receiver needs to be able to signal the transmitter when to stop and start transmission—a process known as **flow control**.

There are two types of flow control: hardware and software. In hardware flow control, additional control lines are run between the receiver and transmitter; when the receiver's buffer starts filling up, it signals the transmitter to halt transmission. Once the receiver is ready to accept data again, it signals the transmitter to restart. In software flow control, specially designated characters act as start and stop signals. One common approach to software flow control, known as XON/XOFF, uses the ASCII control character DC3 (13_{16}) to instruct the transmitter to stop sending and DC1 (11_{16}) to instruct it to start again. For reasons of data link efficiency, however, hardware flow control is usually preferable.

Flow control is less commonly implemented now than in the past, because as memory has gotten cheaper, equipment is able to have large data input "buffers," which store excess data until the receiver's processor can deal with it. In addition, data rates are now generally fast in comparison to the amount of data we typically send, and slow in comparison to typical processing horsepower.

ELECTRICITY FOR DATA TRANSMISSION

The most common physical communications method is the transmission of electricity over a wire, with the state of the voltage or current representing (and communicating) the state of the data signal. This approach is cheap, reliable, well understood, and easy to install and troubleshoot.

Transmission Schemes

There are a couple basic schemes typically used in electrical data interconnection. A **point-to-point connection** is generally one where two devices are directly connected to each other. In "multipoint" or **multidrop** communications, multiple stations share a common connection. In such a multitransmitter link, "contention" issues (who gets to speak when) must be resolved. In addition, there are two primary ways used to transmit the electrical signals: voltage loop and current loop.

Voltage-Loop and Current-Loop Interfaces

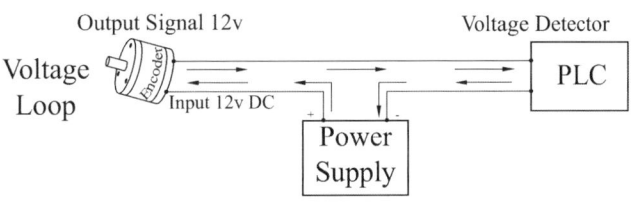

To indicate a particular on or off (1 or 0) state to a receiver, a transmitter uses either the presence or absence of voltage or the flow of current. **Voltage-loop** interfaces—where a certain voltage indicates a particular state—are common and well defined. In order to ensure successful data transmission (and to avoid equipment damage), the voltage, polarity, and other characteristics of a voltage loop interface must be agreed upon, and there are a number of internationally recognized voltage-loop standards. Voltage loops are the most common way to send data electrically over a wire in our industry (for example, see the "TIA/EIA Serial Standards" section, on page 147, and Ethernet in Chapter 16, on page 157).

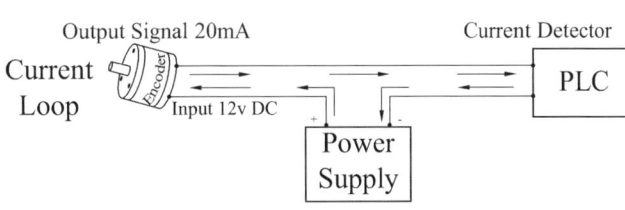

Current-loop interfaces have not been as formally standardized, although there is an informal standard for digital communications: a flowing current of 20 milliamperes (mA) equals a logical 1; the absence of current equals a logical 0. In addition, 4–20 mA analog current loops have often been used to indicate proportional values in the industrial control. Since current can't exist without voltage (and vice versa), voltage is of course present on current loops, but the precise voltage is only important in that it allows the proper current to be delivered. Current loops have some advantages over voltage loops: They interface easily with opto-isolated equipment and are highly immune to **Electro-Magnetic Interference** (EMI), since the currents induced by EMI are very low relative to the signaling levels. MIDI (see Chapter 22, on page 281) is an example of a current-loop standard.

Single-Ended/Unbalanced or Differential/Balanced Transmission

There are two primary voltage-loop transmission designs: single-ended or differential. In **single-ended**, or unbalanced, transmission, all lines in the communications link, both data and control, share a common ground, to which all signals are referenced. If EMI induces voltage on a line, this noise is simply summed into the data stream and can confuse the circuitry in the receiver.

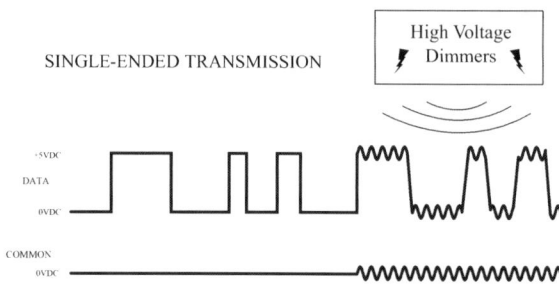

In **differential**, or balanced, transmission, the signal is sent in opposite polarity over a pair of wires, and the receiver looks at the *difference* in potential (hence differential) between the two wires, irrespective of ground. Differential transmission lines are also known (particularly in audio) as balanced.

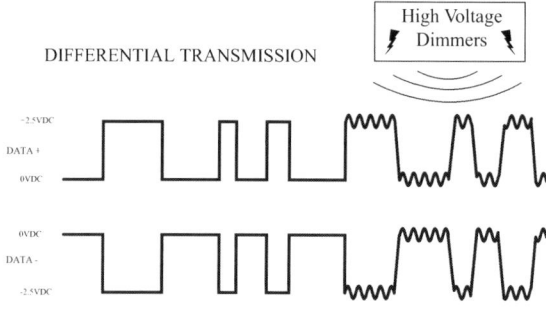

Differential transmission is more robust than single-ended, but costs a little more and requires more transmitting and receiving components. However, in our industry, where large EMI sources (dimmer racks, transformers) may be present and we may be transmitting over long distances at high data rates, differential is almost always the preferred method.

Wire and Cable Types

So far, we've talked about how electrical signals are transmitted; now we'll move on to discuss the media over which those signals travel: **wire** (a single conductor) and **cable** (an assembly of wires). Selecting an appropriate cable is a seemingly simple task, but is something that system designers often spend a considerable amount of time doing, especially when dealing with the transmission of unfamiliar signals. Cable cost is a factor, but a bigger factor—particularly in a permanent installation—is labor, and installing or running the wrong cable can lead to system problems and costly mistakes leading to re-pulls. Cable manufacturers design their products with specific applications in mind and offer application support; use this information to choose a cable carefully. Entire books have been written on the subject of wire and cable (see the bibliography on my website), but we only cover the basics here. When thinking about wire and cable, keep in mind that we always need an electrical "circuit," so we always need a loop, and that generally means you need a pair of wires in order to have a communications circuit.

Unshielded, Untwisted Cable

The simplest type of cable consists of a pair of inexpensive, unshielded, untwisted wires, such as that used in a doorbell or table lamp. However, since this type of cable is very susceptible to EMI, and since communication signals are relatively weak, this type of cable should generally not be used for data transmission.

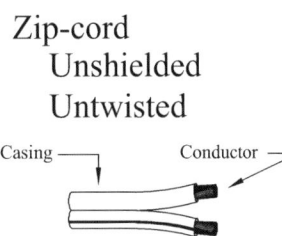

Zip-cord Unshielded Untwisted

Unshielded, Twisted Pair (UTP) Cable

Twisted pair cable,[5] made of two pieces of wire twisted tightly together, works especially well with differential, balanced transmission. Twisted pair cable is susceptible to "common-mode" EMI-induced voltage, but since the noise presented on both conductors is very similar (since the wires are occupying as close as possible to the same space), maximum cancellation can be accomplished by the differential receiver. This type of twisted pair cable, without a shield, is called **unshielded, twisted pair** (UTP). UTP is cheap, effective, and easy to install, and is used in most networks.[6]

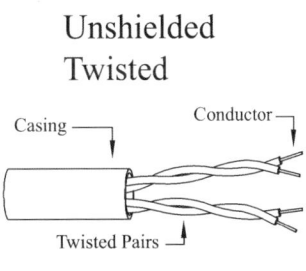

Unshielded Twisted

---

5. Twisted pair cable was actually patented by Alexander Graham Bell in 1881.
6. At least in North America—STP is used more widely in Europe.

Shielded, Twisted Pair (STP)

Adding a shield to a cable—such as metal foil or a braid of tiny conductors—can offer additional resistance to noise, since the shield (if grounded) can conduct some EMI-induced voltages away. This type of cable is known as **shielded, twisted pair** (STP).

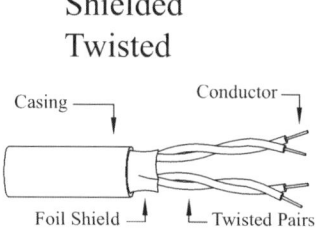
Shielded Twisted

Coax Cable

Coaxial cable, or "coax," consists of a center conductor surrounded by a "dielectric" insulator and a coaxial shield, and is very robust. Capable of very high bandwidth operation, it has been the choice for everything from cable TV systems to older local area networks. Because it is difficult to manufacture, coax is expensive, and the coaxial structure also makes the cable difficult to terminate. While coax was used extensively in early networks, it has now mostly been replaced by UTP.

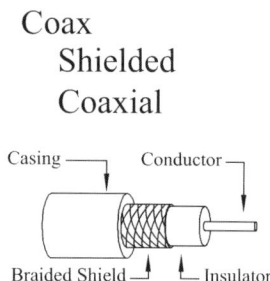
Coax Shielded Coaxial

TIA/EIA Category Cables

While you may encounter almost any type of cable in entertainment control, there is one group of widely used standard cable types, and it's worth introducing them here. Their specifications are grouped into "categories" in a standard called "ANSI/TIA/EIA-568-B" by the Telecommunications Industry Association (TIA) and the Electronic Industries Alliance (EIA). The most common type is referred to as **Cat 5 cable**, even though this type was superseded by Category 5e. Before Cat 5e came to rule the networking world, there were several other types, and we will cover them briefly here.

Category 1

Category 1 was defined as an unshielded, twisted pair cable made for "plain old telephone service" or "POTS." The cable impedance is not standardized, and this type is now primarily found in legacy applications or in applications where the data rate is irrelevant.

Category 2

Category 2, another legacy type, was defined as UTP cable at any impedance up to 4 Mbit/s.

Category 3
Category 3 cable contains four pairs of wire, is either UTP or STP, has a standardized impedance of 100 Ω, and data rates of up to 10 Mbit/s.

Category 4
Category 4, another obsolete type, was four twisted pairs of either UTP or STP, had an impedance of 100 Ω, and typical data rates of up to 20 Mbit/s.

Category 5
Category 5 (which has been superseded by Category 5e, see below) is probably the most widely used cable type in the world, run in countless networking installations as the most common cable type used for Ethernet. "Cat 5" cable, which also consists of four twisted pairs, comes in both UTP and STP varieties, has an impedance of 100 Ω, and is most commonly used at speeds of up to 100 Mbit/s.

Category 5e
The Category 5e specifications made the Cat 5 specifications obsolete in 2001. Cat 5e was designed to better handle crosstalk, and accommodate speeds of up to 1,000 Mbit/s (or 1 Gbit/s). As of this writing, this cable is the most widely used for the bulk of basic networking needs at network speeds.

Category 6/6e
Category 6 and 6e, designed to handle 1 Gbit/s, is used for ultra high-performance applications, although it is not (yet, anyway) really necessary for many control applications.

Installation of Cat Cables
To make connections work at incredibly high bandwidths, there are installation standards for category cables. Under the standards, all hardware used must be rated for the appropriate category, the maximum length of wire that can be untwisted is specified, as is the bend radius, and many other criteria. Installing a critical category-based cable system is not something that a casual user should attempt; there are many contractors who have experience with it and have the test equipment needed to certify the installation.

RJ45 (8P8C) Connectors

In addition to a huge variety of cable types, you are likely to encounter a wide variety of connectors in entertainment control. One connector you are very likely to see now is the connector shown in the photo, which is actually an **8P8C** (eight positions, eight conductors), but is more widely (although incorrectly) referred to as an **RJ45**, available both in plug and jack form.

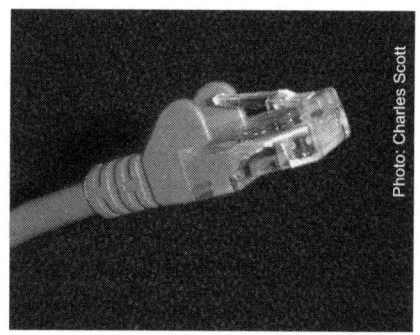

RJ stands for "registered jack," and the RJ series connector, developed by the telecommunications industry, specified both the connector *and* the wiring configuration. The 45 variation was originally designed as an 8-position connector carrying (in some applications) only a single pair of wires, but it looks very similar to the 8P8C connector specified for Ethernet. The actual telecom RJ45 connectors and configuration are now obsolete, but the 8P8Cs are almost universally referred to as RJ45s. Because of this, I will refer to these connectors as RJ45 (8P8C) throughout this book.

RJ45 (8P8C) connectors are now used to carry everything from low- or high-speed data to plain old telephone signals. The RJ45 (8P8C) has eight pins and can carry four pairs of wire, and is a slightly bigger version of the typical "modular" phone connectors (RJ11) used for home telephones. RJ45 (8P8C) connectors are made of easily-shattered plastic, and are not well

suited to the backstage environment. However, they are extremely inexpensive and can be easily installed in a few seconds with "crimping" tools. Some available varieties of these connectors have small ramps on either side of the release tab, protecting them when a cable is pulled through a tangle. In addition, the industrialized Ethercon® connector from Neutrik (see photo) was designed with our industry in mind.

TIA/EIA-568B Cabling Standard

The TIA/EIA-568B standard specifies a "structured" layout of network wiring in a building. The standard is very complex, but basically it takes a typical multistory office building and breaks its cable infrastructure into "vertical" and "horizontal" runs. Vertical cables are high-bandwidth "backbones," which connect through network distribution systems to a floor's individual "horizontal" lines (limited to runs totaling 100 meters or less). The horizontal cables are sometimes run through air-handling plenums, so a type of cable for this application is often called "plenum" cable. The vertical cables may run through cable risers in buildings, and because they run vertically, "riser" cables must have a high fire resistance rating. Further details of the 568 standard are outside the scope of this book, but there is one important aspect you may encounter: the pin designations in the standard.

T568A and T568B for RJ45 (8P8C) Connectors

The wire pair and color code pin out on RJ45 (8P8C) connectors is perhaps the best-known aspect of the 568 cabling standard, and also is perhaps the most confusing. The 568 standard designated two pin outs for RJ45 (8P8C) connectors: one confusingly called **T568A**, and the other called **T568B**. On the RJ45 (8P8C) plug, the pins lay out as shown in the graphic.

Cat 5e cable is the most widely used cable with this sort of connector, and has four tightly twisted pairs of wires. To accommodate fast network speeds, the physical configuration of the wire pairs is critical, and this explains the seemingly strange pin out specified in T568A and B.

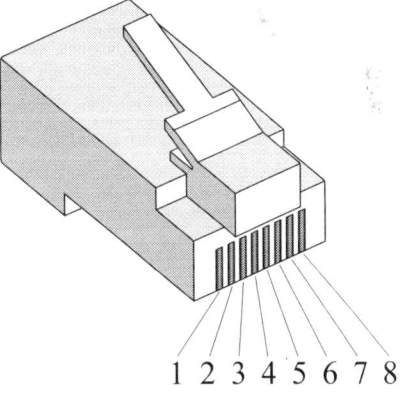

The first pair of wires is in the center of this family of connectors (to give some backwards compatibility to pair 1 on RJ11 and similar connections). The second pair splits across the first, and then both wires of pairs 3 and 4 are adjacent and out toward the edges of the connector (they could not continue the splitting scheme further due to crosstalk and other issues). The only real difference between the T568A and T568B schemes is that two pairs are swapped; so, if you are wiring a facility, it really doesn't matter (functionally) whether you use A or B, as long as you use the same standard on both ends of every cable.

However, even though the A pinout is most widely referenced in the standard (and is required by some US government agencies), the vast majority of real-world installations use the T568B pin out, and this is the one recommended by PLASA in its networking documents:

Pin	T568B Pair	Wire	T568B Color
1	2	Tip	White with orange stripe
2	2	Ring	Orange solid
3	3	Tip	White with green stripe
4	1	Ring	Blue solid
5	1	Tip	White with blue stripe
6	3	Ring	Green solid
7	4	Tip	White with brown stripe
8	4	Ring	Brown solid

Here is the standard for the far less common "A" pin out:

Pin	T568A Pair	Wire	T568A Color
1	3	Tip	White with green stripe
2	3	Ring	Green solid
3	2	Tip	White with orange stripe
4	1	Ring	Blue solid
5	1	Tip	White with blue stripe
6	2	Ring	Orange solid
7	4	Tip	White with brown stripe
8	4	Ring	Brown solid

TRANSMISSION/MODULATION METHODS

While it's possible to send data directly across the communications link, it is often desirable to "modulate" the data onto some sort of other signal. This "carrier" signal actually transports the data on the transmission link, and a link may carry multiple channels simultaneously. The receiver then "demodulates" the signal, recovering the original ones and zeros of the data stream. A device capable of modulating and demodulating information is known as a "modem," short for modulator/demodulator. There are many transmission and modulation schemes, but here we cover a few that you may encounter in our industry.

Non-Return-to-Zero

Non-return-to-zero (NRZ) is a simple scheme that uses two different voltages (or other states) to represent the 1 or 0 bits. For instance, zero volts could represent a 0 bit, and +5 VDC could represent a 1 bit.[7] If a long string of 1s or 0s is sent using NRZ encoding, the signal can remain at one voltage level for a period of time. This means that the sending and receiving systems must be synchronized using some other method, and that the signal will have some direct current (DC) component, which can cause problems for some transmission devices (such as transformers, which cannot pass DC). To solve this and other problems (beyond our scope here), schemes such as Manchester encoding have been developed.

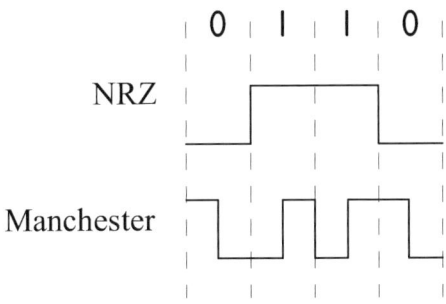

Manchester

Manchester encoding solves the DC problem by guaranteeing that the voltage will swing back through zero within a given period of time, and there is a signal transition in the middle of every bit period. There is no single official standard for Manchester; however, for many devices, if that middle transition goes low to high, it represents a binary 1; a high to low transition indicates a binary 0. These transitions allow a receiver to synchronize to a transmitter, and can also act as an additional error check—if there is no transition, the receiver will know something went wrong. Manchester encoding is the core of some implementations of Ethernet.

Frequency Shift Keying

In frequency shift keying (FSK), digital data is typically used to modulate the frequency of some sort of analog carrier. For a bit with a value of 1, one particular frequency is transmitted; for a 0 bit, a different frequency is transmitted. In entertainment control systems, this modulation method might be used to record digital data as an audio signal.

LIGHT FOR DATA TRANSMISSION

Instead of sending electrons over wire to send data, it is possible to send a beam of light either through the air or over a piece of glass or plastic "fiber optic." With this approach, light is turned on or off to represent the data bits being transmitted.

---

7. More common in data communications interfaces is a scheme where a negative voltage represents a 1, and a positive voltage represents 0.

Fiber Optics

While sending light through the air is often useful (as in infrared remote controllers), it is typically more useful in control system applications to send light through a glass or plastic cable known as "fiber optic." Light travels down a length of **fiber-optic cable** by bouncing back and forth off the boundary edge between the center glass or plastic and the air, or between the center and a jacket known as the cladding, or buffer.

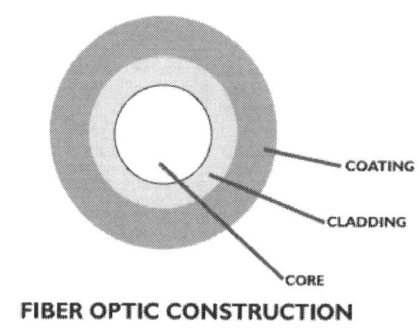

FIBER OPTIC CONSTRUCTION

MULTI-MODE LIGHT TRANSMISSION

Fiber comes in two types: multi-mode and single-mode. Multi-mode allows multiple "modes", or pathways, for the light to bounce. Single-mode allows fewer possible pathways for the light beam, which leads to fewer bounces, which in turn leads to less loss. These characteristics allow extremely high bandwidth, but single-mode fiber is much more difficult to terminate, needs more accurate end alignment, and is therefore more expensive to use.

Because fiber transmits light, fiber-optic data links are completely immune to all forms of electrical interference. Potential bandwidth in fiber is extremely high, but this high bandwidth and noise immunity comes at a price: compared to wire, fiber-optic cable is more expensive and more difficult to terminate. So, while fiber has become more common in our industry, it's still usually installed only when high bandwidth or extreme noise immunity is required. Common fiber connectors used in our industry include the types LC and ST (pictured in the photo).

RADIO FOR DATA TRANSMISSION

"Radio modems," which accept digital data and retransmit it over radio waves, are commonly available and can be used to connect a variety of computers. Radio modems using "spread spectrum[8]" technologies take a single data stream and spread it across many possible radio transmission paths. This approach makes it much more robust and more resistant to interference. And since backstage is typically a hostile environment for RF signals (with wireless mics, hand-held radios, multiple RF noise sources, etc.), this type of transmission works well in entertainment data applications.

The most common radio data transmission we are likely to encounter is IEEE 802.11 "Wi-Fi," which is covered in the "IEEE 802.11 "Wi-Fi"" section, on page 175. Also, there are some commercial approaches to transmitting DMX over a spread spectrum signal (see "Wireless DMX Transmission," on page 250). However, because of the potential for radio interference, RF links are rarely as robust as wire or fiber-optic links, so I still advise using wire or fiber if at all possible.

---

8. Patented in 1942 by the actress Hedy Lamarr and the composer George Antheil (best known for his *Ballet Mécanique*, which involved many forms of synchronization).

Chapter 15: **Point-to-Point Interfaces**

Now that we've covered control concepts and the fundamentals of data transmission, we can finally get into the details of some data communications standards and start talking about ways to actually interface machines. I've broken down data communications into two types: point-to-point (covered in this chapter) and network (covered in detail starting with Chapter 16, on page 157). The distinctions between the two can get blurry, but generally point-to-point interfaces are meant for connecting two (or a few) devices together in a point-to-point fashion; networks are more generic connection methods used to connect two through a near-infinite number of stations.

PARALLEL INTERFACES

The simplest way to connect two digital devices is through a group of wires—one for each bit transmitted across the interface. An additional wire, known as a strobe line, indicates to the receiver when the transmitter is ready to have its data read. When data is sent using this technique, transmission is said to be in **parallel**. The figure shows the ASCII text "Penn" being transmitted across an eight-bit parallel interface (note that the text appears backwards because time is indicated across the x axis, and the "P" is sent first).

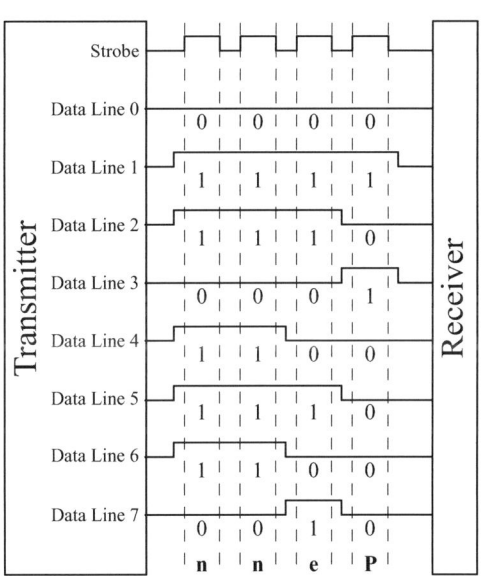

Parallel communications are fast and efficient. However, parallel transmission requires a large number of wires, which is expensive when data must be sent over significant distances. For this reason, parallel interfaces were used primarily for short-haul communications, such as connecting a computer and printer. Even for this application, however, they are now rare and have largely been replaced by high-speed networks such as the "Universal Serial Bus (USB)," on page 154.

The 25-pin female "D" connector was a parallel interface commonly used in original IBM PCs, but is rarely seen in PCs any more. It is useful, however, to take a look at an example illustration of a parallel port (see figure).

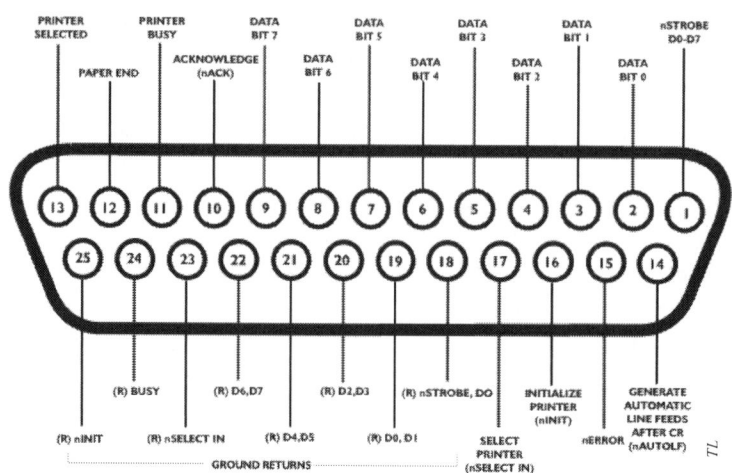

SERIAL INTERFACES

Instead of sending all the bits simultaneously in parallel over a group of wires, it is also possible to send bits "serially," one after the other, over a single communications line. **Serial communications** are ideal for long-haul applications or for any application where a large number of wires is not practical. While parallel interfaces have a strobe line to indicate when the receiver should read the data, serial interfaces require instead that the transmitter and receiver agree on a timing or synchronization scheme.

Synchronous Interfaces

The simplest form of serial communication is the synchronous serial link, where both the receiver and the transmitter clocks are locked together and precisely synchronized. The figure below shows the synchronous transmission of the ASCII text "`Teller`."

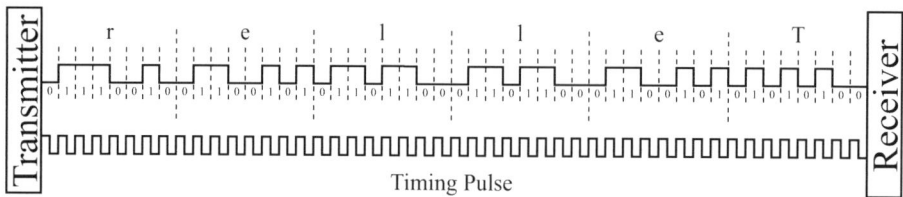

The easiest way to lock the clocks of the transmitter and receiver together is to run an additional synchronization line from the transmitter to the receiver, delivering clock pulses (as depicted in the figure). Doing so, however, adds more wiring, which defeats one of the primary advantages of serial communications. A synchronous link is indeed possible without external clock lines: If the data is sent according to a scheme (beyond our scope here) agreed on by both the transmitter and the receiver, the receiver can actually derive its bit-timing clock from the incoming data stream itself. However, both ends of the link must have a certain amount of sophistication, which can be costly and complicated to implement. Another solution is to run the link "asynchronously."

Asynchronous Interfaces

In asynchronous communications, the receiver's clock essentially starts and stops (i.e., "resyncs") at the beginning of each and every transmitted digital word. To enable the receiver to detect the beginning of a word, the transmitter adds a "start" bit (0) to the beginning of each data word and a "stop" bit (1) to the end of each word. The diagram below depicts a typical asynchronous link sending the text "Dawkins," encoded using ASCII.

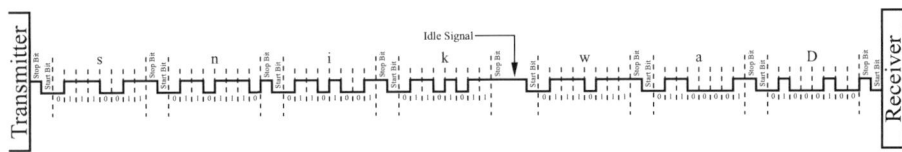

The receiver can synchronize itself to this asynchronous signal because the start/stop-bit structure *guarantees* a transition from 1 to 0 at the beginning of every word, even if a data word contains all 1s or all 0s. This is true because at the end of each word, the stop bit returns the line to the "mark," or 1, condition—the state in which most asynchronous serial links idle.

Of course, this scheme works only if both the transmitter and the receiver are set for exactly the same parameters: bit/s, parity, start-bit enabled, the correct number of stop bits. A mismatch can result in corrupted data. The start- and stop-bit scheme also allows the link to resync itself if the data is corrupted, since the receiver doesn't accept data unless a stop bit (typically a binary 1) validates the word. If bits get corrupted or dropped, a properly configured asynchronous link will eventually resync itself, but how long it takes depends on the placement of the bits and the corruption. In general, though, more stop bits mean faster resynchronization.

TIA/EIA SERIAL STANDARDS

Some of the most widely used computer interfacing standards are the Telecommunications Industry Association (TIA) and Electronics Industries Alliance (EIA) standards for serial communication. As with most successful open standards, these standards describe only a "lowest-common denominator" functionality, and in this case that means they specify only the physical layer: the electrical connections between equipment. They do not dictate the data rate, number of bits in a data word, method of error detection, how many start/stop bits are to be used, or other parameters critical to successful communication. As a result, the interfaces are extremely flexible; with this flexibility, however, comes confusion and there can be a wide variation in the way the standards are implemented. So, even though two pieces of gear may have "standard" serial ports, that in no way means that the machines will be able to communicate—at least not without some interfacing or programming work and possibly even conversion hardware.

DTE and DCE

This group of serial standards was designed (years before the advent of the PC) to connect **data terminal equipment** (DTE), such as "dumb" computer terminals, with **data circuit-terminating equipment** (DCE), such

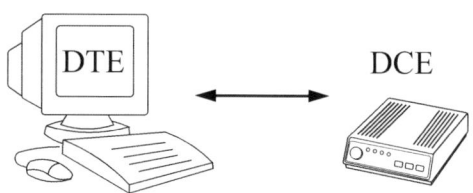

as **modems**. These acronyms, while seemingly designed to be as confusing as possible, are important to understand and remember, because the standards enable DTE devices to communicate with DCE devices, and vice versa.

RS-232 (EIA-232, TIA-232)

The "232" standard is by far the most prevalent point-to-point serial standard. It was originally known as **RS-232**, but the official name eventually changed to TIA-232, or EIA-232, as the TIA and EIA now maintain it. In this book, I call it RS-232, since that's how the vast majority of people in the industry refer to it (same with the other RS standards below).

Whatever it's called, RS-232 is a single-ended, voltage-loop interface, and was designed for relatively low data rates. In 232 (and the others detailed below), the cable length is not given as a specific number, but instead is supplied in a chart, which depends on characteristics of the cable and the data rate. A good working "rule of thumb" maximum length for 232 is 50 feet or less, although much farther is possible using a good quality, appropriately specified cable.

In RS-232, the voltage polarities are the opposite of what you might expect. A logical 1, called a "mark" value, is any negative voltage between –3V and –25V with respect to ground. Logical 0, called a "space," is any positive voltage between +3V and +25V. When the line is idle, it is held in the mark, or negative voltage state. Just to confuse things even more, control lines are inverse in sense to data lines: A control line true condition is any voltage from +3V to +25V, and control false is any voltage from –3V to –25V.

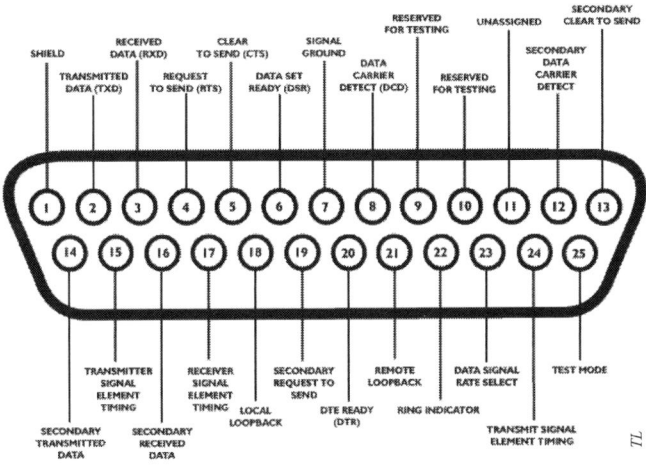

There are 25 lines designated in RS-232, and a D-shaped, 25-pin connector was originally in widespread use, but it is now mostly obsolete. According to the specification, DTE uses male connectors and DCE uses female. You may still encounter this older implementation, so I'm including the control signals of a 25-pin connector here. This connector enables synchronous communications (through the timing lines) and a wide variety of flow-control options.

For asynchronous connections, a subset of the 25 lines is generally sufficient. This subset was implemented by IBM in its AT computers as a 9-pin connector, and is the dominant serial connector implementation (see figure); these nine pins (or even just three) form the heart of RS-232 for asynchronous applications. Transmitted Data (TD or TX) is the line over which the DTE sends data to the DCE; Received Data (RD or RX) is the line used by the DCE to send data back to the DTE. Signal Ground is the common point for all the data and control lines in this single-ended link.

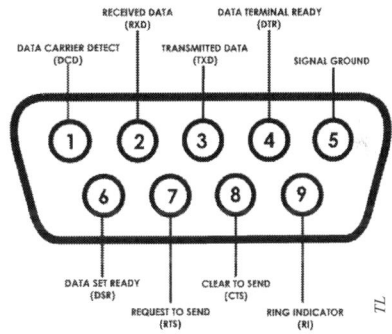

For flow control (when needed), "DCE Ready" indicates to the DTE that the DCE is powered up and ready; "DTE Ready" indicates to the DCE that the DTE is ready. Request To Send (RTS) is used by the DTE to indicate to the DCE that it would like to send some data; Clear To Send (CTS) is the DCE's response.[1] With

---

1. I know all this DTE/DCE information is confusing, but it's necessary to understand. Sadly, this is not the last time you will encounter these acronyms!

larger buffers and high-speed processors, many 232 implementations simply omit all these flow control lines, and just use the TD, RD, and signal ground lines.

RS-232 was provided on PCs for many years and, therefore, there have been literally millions of these interfaces out there. For this reason, RS-232 is widely used to connect sensors, as a control port for some broadcast video control equipment, and for short haul, relatively slow computer-to-computer connections. However, on newer PCs, 232 is a special option rather than standard equipment, and is for the most part being replaced by Ethernet (see Chapter 16, on page 157) and USB—see "Universal Serial Bus (USB)," on page 154.

RS-422 (EIA-422, TIA-422)

Of course, a major drawback of RS-232 is that it uses single-ended (unbalanced) signalling. This is particularly problematic for applications in our industry, since the link is incapable of resisting EMI for all but the least harsh environments.

RS-422 offers us a major improvement over 232, since it uses differential (balanced) signalling (page 134), with two balanced wires for each signal. RS-422 has a typical working length of about 4,000 feet (as usual, limited by data rate, cable type, etc.) and can broadcast to as many as ten receivers. In 422, a binary 0 is between +2V and +6V; binary 1 is between –2V and –6V. These lower voltages, coupled with differential's improved noise immunity and common-mode rejection, give 422 specified typical data rates of up to 10 Mbit/s. Connectors are not standardized in 422. RS-422 is the foundation of Sony 9-Pin (see "Sony 9-Pin Protocol," on page 350), and is used for devices such as sensors where more distance is required than RS-232 can provide.

RS-485 (EIA-485, TIA-485)

RS-485 shares many of the same specifications as RS-422, with one key difference: RS-485 is a "multidrop" standard, which allows 32 transmitters and 32 receivers. RS-485 uses lower voltages than any of the other general-purpose serial standards: voltages between +1.5V and +6V are 0 (space), and from –1.5V to –6V are binary 1 (mark). RS-485 has the same approximate maximum data rate as RS-422: 10 Mbit/s. As in 422, connectors are not standardized in 485, but RS-485 is the foundation for one of the most widely used standards in entertainment control: DMX512-A (see Chapter 19, on page 241). In addition, it is often used for connecting sensors and other industrial devices.

Comparison of TIA/EIA Serial Interfaces

Each of the general purpose EIA recommended standards is suited to different applications, and each has strengths and weaknesses. Factors determining use include cost, noise immunity, and data transmission rate. The following table summarizes the features of each standard.

	RS-232	RS-422	RS-485
Mode	Single-ended	Differential	Differential
Drivers	1	1	32
Receivers	1	10	32
Typical maximum link distance	50	4,000	4,000
Mark (1) voltage	–3V to –25V	–2V to –6V	–1.5V to –6V
Space (0) voltage	+3V to +25V	+2V to +6V	+1.5V to +6V

PRACTICAL SERIAL CONNECTIONS

Serial interfacing can be easy or complex, depending on the situation. Here is an introduction to a few miscellaneous issues related to serial connections.

Using RJ-45 (8P8C)

One of the reasons that Cat 5e cable and RJ-45 (8P8C) connectors are so ubiquitous is that they are so versatile and inexpensive. One application for them is to carry standard serial data, and adaptors are available for use between 9-pin "D" connectors and RJ-45 (8P8C). There is no official standard for the pin out, however.

Connecting DTE to DTE (Null Modem)

Although this set of serial standards was designed for connecting DTE devices to DCE devices, the line between those two categories is now fuzzy. For example, we often want to connect two computers together, and computers typically have DTE ports. If you plug a "straight through" cable from a DTE to DTE, the connection won't work, because the transmit line on one system will be connected to the transmit line on the other, and the same goes for the receive lines. For devices to communicate properly, the data transmit and receive lines and, possibly, some other control lines have

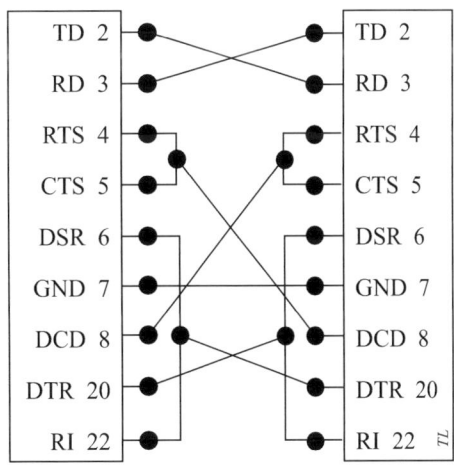

to be swapped. An adaptor that does this is called a crossover or **null modem**

Chapter 15: Point-to-Point Interfaces • 151

adaptor. The figure is a schematic of a typical 25-pin null modem cable, implementing all the flow control lines.

Nine-pin null modem cables don't have (and therefore don't swap) as many auxiliary control lines; bare minimum null modems swap only transmit and receive lines, and connect the signal ground:

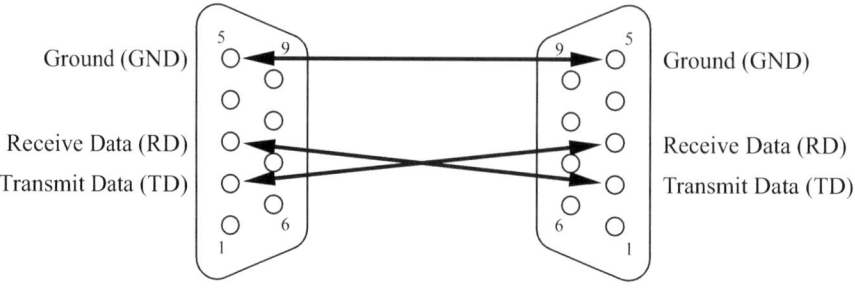

SERIAL CONNECTION EXAMPLE

To bring all this together, let's take a look at a practical example. Let's say we want to control a DVD player from a computer:

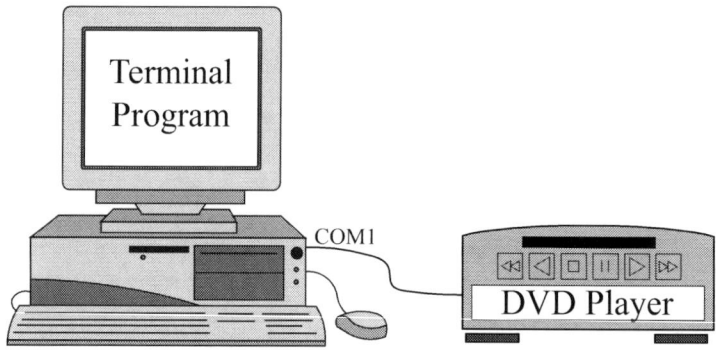

To communicate using a serial interface, you need some sort of software to convert the messages you want to send (i.e., ASCII text) into the binary format understandable by the chip that actually connects to the physical communication lines. Many kinds of control software can generate these messages; however, when you are first trying to control a piece of gear new to you, it's often advantageous to connect the device manually, test the connection, and verify the commands to ensure that everything is working correctly. Once you have all the settings in order and the control strings verified, you can easily implement those known, working command strings and port settings into the control system. A very useful program for this purpose is a **terminal emulator**. These programs emulate old style hardware "dumb" computer terminals; hence the name. One such program is Tera Term,[2] written by a group of Japanese programmers; we will include some information here from Tera Term as an example.

When getting started with a program such as Tera Term, you first typically configure the serial connection to match the settings of the equipment you want to control. In this case, the DVD player is configured to communicate using "9600/8/n/1." So, with this information, we configure the system as shown in the screen capture: we set the hardware port to COM1 (in PCs these are called "Com" ports); we set the bits per second (shown as baud rate) to 9600; the number of data bits in a data word to 8, no parity (error detection) scheme, 1 stop bit, and no flow control.

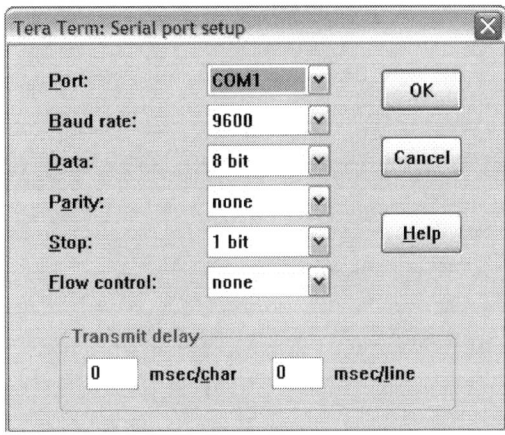

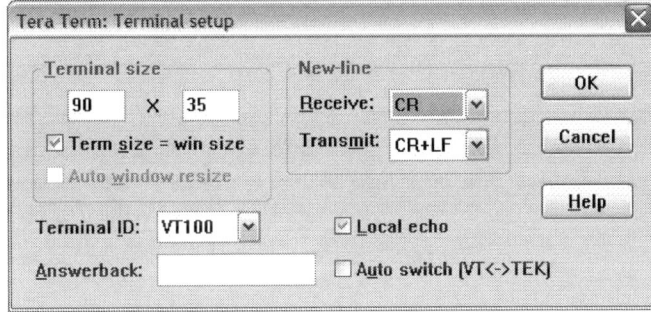

In the old days of dumb terminals and mainframes, you wouldn't have been able to see the characters you had typed unless they had been "echoed" back from the remote system in some way. So, most of these terminal emulation programs default to a mode where you won't be able to see what you type. If you want to see characters as you type them, you need to turn on "local echo," which will print the characters on your screen. In addition, many serially controlled devices execute only the action contained in the command on receipt of a "carriage return" ($0D_{16}$ in ASCII) or "line feed" ($0A_{16}$) character. Many terminal programs send these automatically when you press the Enter key.

---

2. Available at http://ttssh2.sourceforge.jp/index.html.en

Here's text captured from an actual conversation between Tera Term and a Pioneer DVD player (with local echo on, formatted into a table for ease of understanding, see "Pioneer LDP/DVD Control Protocol," on page 347, for more info).

Message to DVD	Response from DVD	Description (included here for explanation—not captured)
ch2se		Tells the DVD to search to Chapter 2
	R	Sent on completion of search
pl		"Play" command
	R	
st		"Still" command
	R	
play		Invalid command typed for example
	E04	Error response from DVD
pl		"Play" command
	R	
?c		Chapter Query
	02	Current Chapter Number from DVD
?f		Frame Query
	0000380	Current Frame Number from DVD
?f		Frame Query
	0000446	Current Frame Number from DVD (increase because DVD is playing)
?f		Frame Query
	0000497	

HIGH-SPEED SERIAL POINT-TO-POINT INTERCONNECTS

While the EIA serial standards are still used, they are for many applications now replaced by very fast, point-to-point connection methods. While these interfaces are consumer oriented and not often used directly for critical entertainment control applications, they are used very often in our market for peripheral connections and other similar applications, so I'm including a brief overview here.

Universal Serial Bus (USB)

The **Universal Serial Bus** (USB) was originally standardized in 1995 by a consortium of computing manufacturers as a "plug and play" and "hot-patchable" serial standard; as of this writing, the current version of USB is USB 3.0, which was released in 2008 and runs at speeds up to 5 Gbit/s. USB was designed primarily for connection between a computer and a few desktop peripherals such as keyboards, speakers, video capture boards, and even Ethernet adaptors or standard EIA ports. It was not really designed to network computers, but you may encounter it as a connection between computers and peripherals in the show environment. It is also widely used in the world of physical computing.

USB supplies 5-volt power on two conductors to power and/or charge peripherals, with the remaining lines carrying the data. USB Hubs can connect multiple devices to a single port. USB can provide both "isochronous" (bandwidth-guaranteed) or asynchronous (nonbandwidth-guaranteed) service, and has a built-in CRC check.

USB connectors are designated as either Type A or B, with Type A connectors typically found on computers, hubs, or other devices that supply power, and Type B connectors on peripheral devices, which consume power. Standard cables connect Type A to Type B (see photo), and cable length at full speed is limited to a few meters. The picture shows the full size connectors which have four wires; there are also mini and micro connectors with fire wires. Here's the standard pin out:

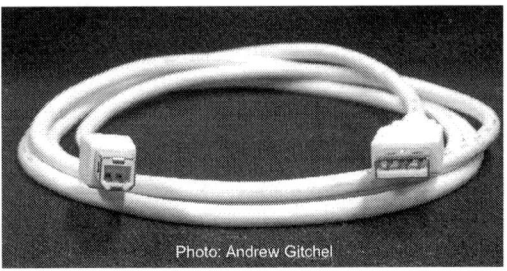
Photo: Andrew Gitchel

Pin	Name	Cable color	Description
1	VCC	Red	+5 V
2	D–	White	Data –
3	D+	Green	Data +
4	GND	Black	Ground

The micro/mini five pin connection scheme is:

Pin	Name	Cable color	Description
1	VCC	Red	+5 V
2	D–	White	Data –
3	D+	Green	Data +
4	ID	None	Host/Slave Indicator
5	GND	Black	Signal ground

In this scheme, pin 4 is connected to the signal ground on the host end, and disconnected at the slave.

FireWire™ (IEEE 1394)

IEEE 1394 was originally developed in the mid-1980s by Apple Computer as **FireWire™**, a high-performance, real-time, isochronous transport system for connecting devices or connecting peripherals to computers. FireWire became an IEEE standard—1394—in 1995. The standard is used primarily for high-speed digital interconnect (such as connecting a sound interface to a computer), and can carry a huge number of high-quality digital audio and video channels simultaneously. 1394 originally worked at speeds of 100, 200, 400, and 800 Mbit/s, and can run under various implementations (in 1394B) up to 3.2 Gbit/s. As many as 63 nodes 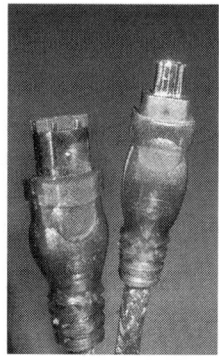 can be connected to up to 1,024 buses, giving a total of about 64,000 possible nodes, and the standard can be used for simple networking. 1394 can address as much as 16 *petabytes* of memory space, is hot pluggable, and plug and play. Cables and connectors originally had either four or six conductors; the 4-pin connector doesn't carry any power, while the 6-conductor version does; other connectors are on the market as well. Data is transmitted on two shielded twisted-pairs, with one pair for data and the other for synchronization. Six-conductor cables have an additional pair for power and ground. Unfortunately for entertainment applications, cables are limited to a few meters; however, fiber-based cable solutions to this limitation are available. Firewire was widely used for media transmission (although rarely for control beyond simple control of camcorders, etc.), and as of this writing, USB (page 154) is displacing Firewire for many applications.

Bluetooth

Bluetooth is a network primarily designed for the short-range connection of peripheral devices, such as a wireless headset to a cell phone. It's not well suited for the rigors of show use, but you might find it useful to connect simple components to a show computer. It comes in three classes: Class 1 runs at 100 mW, for a range about 100m; Class 2 operates at 2.5 mW, for a range of about 10m; and Class 3 runs at 1 mW, for range of about 1m. It operates in the 2.4 GHz band, and data rates (as of this writing) range up to 3 Mbit/s. Typically, two devices must be "paired" in Bluetooth in order to communicate, and encryption is available.

MOVING ON

Now that we've covered point to point connections, let's move on to the core of many entertainment control systems: networks.

Chapter 16: Networking Basics

Way back on page 3, we defined a network as a shared physical and computing infrastructure that, through a number of communications links, allows connected devices to communicate. This simple concept has become the common backbone of hugely sophisticated systems, allowing an incredible amount of data to be sent across our venues and around the world.

OPEN SYSTEMS INTERCONNECT (OSI) LAYERING SCHEME

As we introduced in "An Introduction to Communications Layering," on page 125, communications systems can be broken down into modular functional building blocks known as "layers," with each layer only having to know how to interface and communicate with the layer immediately above and directly below it in the stack. "Upper" layers are closer to the (human) user of the system, while "lower" layers are closer to the machines, cabling, and so on.

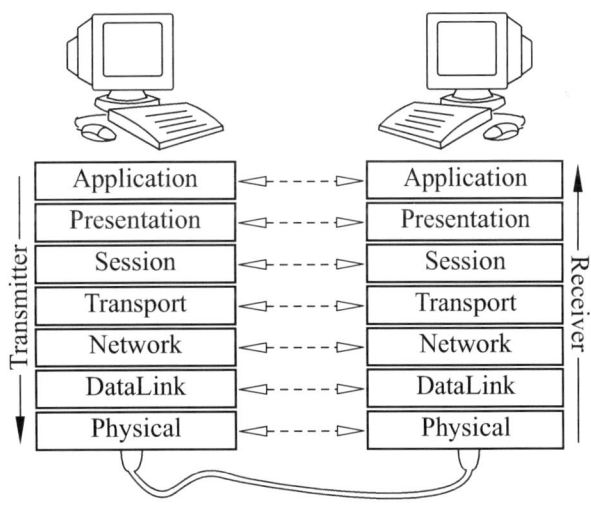

The most well-known layering model is the Open Systems Interconnection (OSI) model. Work on OSI started in 1977 when the International Organization for Standardization (ISO) began developing a standard way to connect network nodes together. The ISO 7498 standard, finalized in 1984, consists of seven discrete layers, each performing one part of the communications and networking task, building on the services offered by the next lowest layer. The OSI layer approach is complex and represents an idealized model; many network systems (such as "TCP, UDP, and IP," on page 181) combine or leave out layers altogether and do not strictly follow the model. However, OSI is a helpful tool for understanding the functions of networks, and you may hear its terminology used in descriptions of network hardware and design, so a brief description is included here.

Through OSI, a layer in one device can effectively communicate with its counterpart layer in another device on the network, even though the only actual physical connection between the two devices is the physical layer. OSI is incredibly powerful but can also be incredibly confusing—it is often difficult to discern separate functions for some of the layers, especially the ones near the middle. Fortunately, we are generally able to leave this task to networking experts, but it is important to understand the concepts, so a brief introduction to each of the OSI layers is offered here.

7—Application

The application layer, the "highest" layer in OSI, is where the meaning of data resides, and this layer presents resources for applications programs run by users. The application layer includes resources for transferring files, messages, or even electronic mail.

6—Presentation

The presentation layer is responsible for presenting the raw data to the application layer; file and data transformation and translation are the responsibility of this layer, allowing different applications to access the same network. Encryption might be included in this layer as well.

5—Session

The session layer is responsible for managing and synchronizing the overall network conversation, or "session."

4—Transport

The transport layer is responsible for ensuring reliable end-to-end transfer of data; multiplexing is one of the transport layer's responsibilities, as is error handling.

3—Network

The network layer is the traffic control center for the network; it determines how a particular message can be routed and where it will be sent.

2—Data Link

The data link layer packages raw data for transport; it is responsible for error detection, octet framing, start and stop bits, and so on.

1—Physical

The physical layer is the "lowest" layer of OSI, and defines the nuts and bolts (or bits and volts) of the network, including bit timing, data rate, interface voltage, mark and space values, connectors, and so on. The physical layer has no intelligence about what kind of data is being sent; data is simply treated as raw bits.

PACKET SWITCHING

In a point-to-point connection (see Chapter 15), a continuous, physical data communications link exists between two (or a few) communicating devices, which simply send bits or bytes of data down the line to each other. In a network, however, at least some part of the connection is shared (one of the main benefits of having a network), so the data traffic must be managed, or "packaged," in some way. The most typical approach is to break the data up into **packets**, with each packet containing a small chunk of the larger data stream. Protocols and systems that manage the transmission of these packets add "header" information to each packet to tell the network how to switch[1] or route the data. The packet typically exists on Layer 3; when it is passed down to Layer 2 (e.g. Ethernet) and has some additional header and synchronization information added, it's often then called a **frame**.

This packetized nature of networks adds another level of complexity, since it's possible that packets arrive at the receiving node delayed, out of order, or corrupted. Higher-level protocols are required to handle these issues, but the benefits of flexibility and sophisticated cross-system interoperability far outweigh the drawbacks.

ENCAPSULATION

In a layered networking scheme, data packets get passed down to lower layers and is often **encapsulated** into another protocol. In the figure below, you can see that the original application data gets more and more information added on as it passes from upper layers to lower layers and is encapsulated (indicated in gray). The receiver reverses this process, extracting the data from successive layers, eventually presenting the application data back to the top layer process.

---

1. The "switching" idea may be confusing here, but keep in mind that the roots of this practice are in telecommunication "circuit switching," where physical circuits were switched (think of the old style operator switchboards).

Here is a graphical representation of encapsulation, with two protocols adding their "wrappers" to the initial application data:

```
                            Application
                               Data
                           Encapsulation
                              Process
              +------------+------------+------------+
              | Protocol 1 |Encapsulated| Protocol 1 |
              |  Preamble  |Application |  Postamble |
              |            |   Data     |            |
+-------------+------------+------------+------------+-------------+
| Protocol 2  | Protocol 1 |Encapsulated| Protocol 1 | Protocol 2  |
|  Preamble   |  Preamble  |Application |  Postamble |  Postamble  |
|             |            |   Data     |            |             |
+-------------+------------+------------+------------+-------------+
```

PACKET FORWARDING SCHEMES

Packets on a network can be delivered in different ways, depending on the application, and the network components can make a decision on a packet by packet basis whether to forward. (or not) a packet to a particular interface. The three basic delivery types we will cover are unicast, multicast, and broadcast. The different approaches can each have a place in a network, depending on what is needed, and networks can operate in different modes at different times, broadcasting data at some times, multicasting at others, and then unicasting as well.

Unicast

Unicast delivery simply means that packets from one sender are forwarded through a network to a single destination.

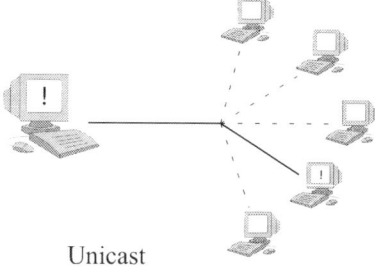

Unicast

Multicast

When data is forwarded from one transmitter to two or more receivers, this is called **multicasting**. Multicast communications on a network maximize efficiency by allowing the sender to send a particular packet of data only once, even though the data is delivered to multiple receivers.

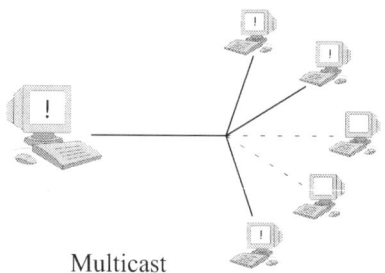

Multicast

160 • Part 3: Data Communication and Networking

Broadcast

Broadcast delivery is where data is sent to every single receiver on the system (or network segment, see the "Broadcast Domain" section, on page 213). This approach is very simple and effective. On the other hand, broadcasting is inefficient, since every packet uses network bandwidth and receivers that don't need the data still have to deal with it.

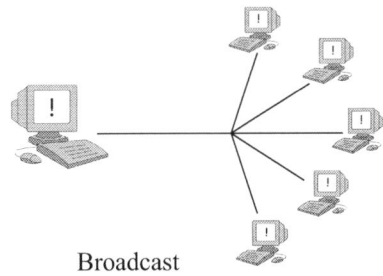

Broadcast

NETWORK TYPES

There are several general types of networks, each covering a specific application. As in so many other areas, the line between the network types can get blurry, but we can generally break networks down into two basic types: Local Area Network (LAN) and Wide Area Network (WAN).

Local Area Network (LAN)

A **Local Area Network** (LAN) covers a "small area," such as a single building or a small group of buildings, and is typically owned and maintained by one organization. Most show networks are LANs based on Ethernet.

Wide Area Network (WAN)

A **Wide Area Network** (WAN) covers long distances, a wide area, or a broad geographic area. Of course, the best known example of a WAN is the Internet (see below). WANs typically use a "common carrier" such as a phone company for some or all of their connections and, therefore, are rarely entirely owned and operated by a single organization. WANs are a separate area of specialty, mostly outside of the scope of this book, and are rarely used for live entertainment applications except in the largest applications (e.g., a theme park).

Internet

The **Internet** is basically a network of networks. To the user, the Internet appears to be one giant wide-area network; but, in fact, the user's network, through their internet service provider (ISP), is simply connected to many other private and public networks. The Internet Protocol (page 183) is the basis of this system.

Intranet

Many large corporate WANs or MANs (or pieces of them) are referred to as internal private **intranets** to differentiate them from the public Internet.

Chapter 16: Networking Basics • 161

ETHERNET

Full-duplex, switched **Ethernet** is the de facto networking standard for entertainment control. While there certainly are other networking standards in use in the larger market (especially in WAN applications), you're not likely to see them on a show, so we will be focussing here on Ethernet, which you will find even on small productions, carrying lighting control data, connecting video control equipment, linking show control computers to sensing systems, transporting multichannel digital audio and control data throughout entertainment facilities of all sizes, and even connecting components in scenic motion-control systems.

Ethernet was developed in the 1970s and early 1980s by Xerox, Intel, and Digital Equipment Corporation (DEC) to enable users at their "workstations," then a radical concept, to transfer files using a nonproprietary network. In 1985, Ethernet was first standardized by the Institute of Electrical and Electronic Engineers (IEEE), and the IEEE 802.3 Ethernet Working Group is still very actively developing extensions and new systems (and the 802.11 group works on wireless Ethernet technologies). IEEE 802 is such an unwieldy mouthful—and few people use the formal name anyway—so the network will be referred to in this book simply as Ethernet.

Ethernet is responsible only for transporting bits from one place or another, while higher-level protocols (such as TCP, IP, ARP, etc.; see Chapter 17) are responsible for packaging the data and making sure that the message is delivered reliably and appropriately. Ethernet breaks down into three general layers (from top to bottom): Logical Link Control (LLC), Media Access Control (MAC), and Physical (PHY). LLC and MAC can be thought of as occupying OSI Layer 2, DataLink, and the Ethernet PHY layer of course fits into OSI Layer 1, Physical.

OSI	Ethernet Layer
Upper Layers	Not Ethernet. HTTP, DHCP, ARP, FTP, TCP, UDP, IP, etc.
Layer 2: Data Link	Logical Link Control (LLC)
	Media Access Control (MAC)
Layer 1: Physical	Physical (PHY)

A brief introduction to the function of each Ethernet layer follows.

Logical Link Control (LLC)

The **LLC layer** receives data packaged by an upper-level protocol such as IP (and, therefore, TCP or UDP, and everything above those layers) and passes it on to the MAC layer for physical transmission onto the network (of course, data also flows in the other direction upon return).

Media Access Control (MAC)
The **MAC layer** connects the LLC layer (and, therefore, everything above it) and the shared Ethernet media in the PHY layer. Much of the MAC layer's functions are outside our scope, but there are three topics that are important to understand, so we will introduce CSMA/CD, the "MAC frame," and the frame check sequence.

CSMA/CD

Picture a group of people standing around in a circle at a cocktail party. Anyone in the group can speak, as long as they wait until no one else is speaking. This approach works fine as long as everyone is polite, no one talks over each other, and everyone waits their turn to speak. This is basically the approach modeled in Carrier Sense, Multiple Access  (CSMA). All the nodes (people) in the network have access (multiple access), and all nodes can hear everything and know whether or not someone else is speaking (carrier sense). With CSMA, any node can transmit its data frame to any other node, as long as it checks first to see if the network is busy. If the network is found to be busy, the node will wait until the link clears, and then transmit its data.

The problem comes when two people at the party start talking at the same time. Humans can hear this problem occurring and one or more of the talkers can "back off," but computers need a mechanism to handle this "data collision," and this is where the final part of the acronym comes in: "Collision Detection (CD).

With CSMA, the nodes should wait until the network is clear before transmitting, but in a network, collisions might occur because two nodes are too far apart (in time) to "hear" each other. It takes time for a data signal to propagate through a cable, and this speed is measured in "propagation velocity"—a percentage of the speed of light. A typical cable might have a propagation velocity of 66%, and this means the signal moves about a foot in 1.5 nanoseconds. That is extremely fast, but computer processors and networks work faster, and so it is possible that nodes sharing the same "collision domain" might start transmitting without realizing that another node is already talking.

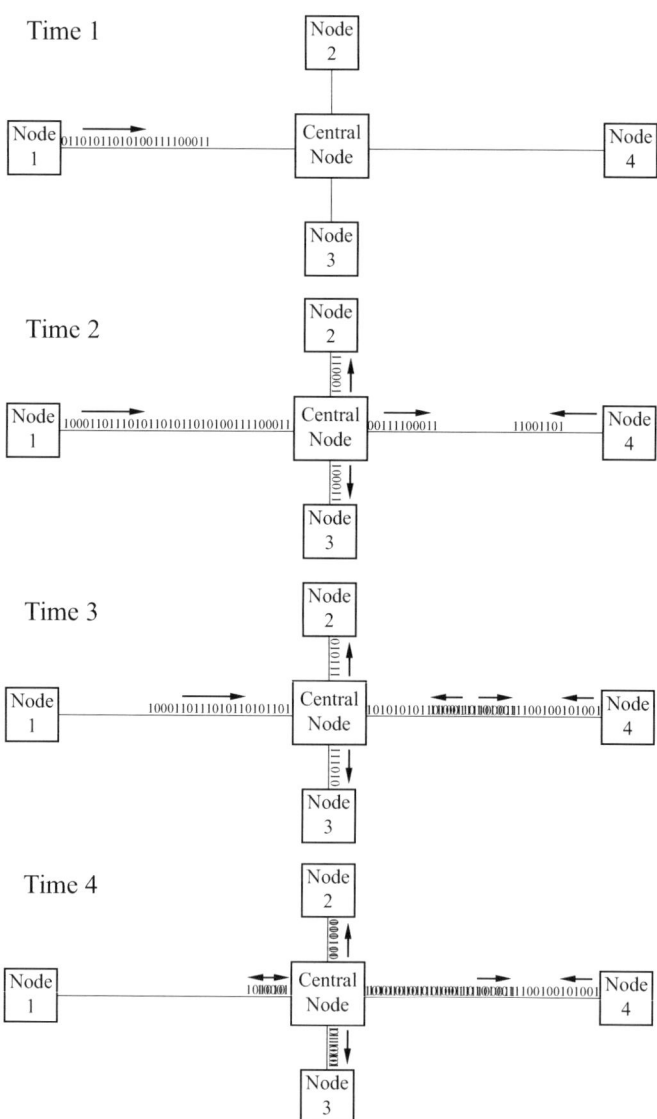

Let's take a look at an example, shown in the figure at right. At "Time 0" (some time before "Time 1" on the graphic), Node 1 had some data to send and checked to see if the network is busy. Finding no other traffic, it begins sending its data as shown in "Time 1." At the same time, however, Node 4 also has some data to send and it checks the network to see if it is in use. Since the message from Node 1 has not yet arrived, Node 4 determines that the network is free and it begins sending its data. At "Time 2," neither node yet knows that the other has started sending data, then at "Time 3," a data collision occurs, corrupting the

164 • *Part 3: Data Communication and Networking*

data. All the nodes are watching the network (collision detection) and will eventually see the problem, and activity on the network grinds to a halt.

Now, if each node waiting with data to transmit simply started sending when it saw the network clear, another data collision (likely the first of a series) would occur and cause data to back up, and this could eventually lead to the effective collapse of the network. One solution to this problem might be to assign a different delay retry time to each node, but the problem with this approach is that the nodes with the shorter backoff times could dominate the network. To address this, each node generates a *random* backoff delay, which ensures that a swarm of data collisions will not occur, and also ensures that each network node maintains *equal access* to the network.

The CSMA/CD approach is highly efficient, since the network is occupied only when data needs to be sent, and there is very little overhead involved in managing the control of and access to the network. However, access and response times in CSMA/CD networks are neither predictable nor deterministic, because of the potential for data collisions and the random time element. Additionally, with heavy network utilization, the number of collisions may increase and network performance will then decrease.

So, CSMA/CD networks are at their core nondeterministic, which was initially a big problem for us in entertainment: no one wants a light cue to go in .1 seconds today and 1 second tomorrow. However, fortunately for us in entertainment, this problem has been overcome by the larger IT industry, with full-duplex switched networks (described on page 170), which eliminate collisions and (typically) disable CSMA/CD, removing the random backoff timing element.[2]

> ### Collisions No More
> With the advent of inexpensive, full-duplex switches two terms that had long been important for the design of Ethernet systems became mostly irrelevant: network collision domains and network diameters. A connection between a node and a hub, or two nodes, is called a network "segment." Ethernet uses broadcast transmission, so any group of nodes capable of simultaneously receiving a frame forms a "collision domain," since any two (or more) of the connected nodes could cause a collision affecting all the nodes in that domain. The physical distance in meters from the two most distant nodes in a single collision domain forms the "network diameter." A physically larger collision domain means longer transport times for messages, which means a longer time window for potential collisions. More nodes in a collision domain mean more potential for collisions. However, in full-duplex systems, there are no collisions, since the collision domain is just two nodes, and the traffic is bi-directional on physically separated lines.

---

2. Except in wireless Ethernet, where, with a shared media, collisions are still issue.

MAC Address

Contained in the MAC frame is the "physical" address of the destination and sending nodes, which is a globally unique physical 48-bit **MAC address**, burned into an Ethernet Network Interface Card (NIC) at the factory, or otherwise stored in nonvolatile system memory. MAC address conflicts are avoided through the use of the Organizationally Unique Identifier (OUI), a 24-bit number used as part of the MAC address and purchased as a block by a Ethernet interface manufacturers from the Institute of Electrical and Electronics Engineers (IEEE). As long as the manufacturer ensures that they don't put the OUI into more than one product, this scheme creates globally unique addresses.

Frame Check Sequence

The **frame check sequence** occupies the final four octets of the MAC frame, and is a CRC check of all the octets in the frame, except the preamble, the start frame delimiter, and the FCS itself. This error detection is one of the benefits of Ethernet—every single frame includes a Cyclic Redundancy Check (CRC). This all makes for an Ethernet frame that lays out like this (simplified version for clarity):

Preamble	MAC Destination Address	MAC Source Address	Data Payload	Frame Check Sequence (CRC)

Physical Layer (PHY)

While Ethernet specifies a number of aspects of the connection, the **Physical layer** (PHY) details with the lowest levels of the system: the hardware. There are a wide variety of Ethernet types, each notated by a data rate in Mbit/s, signaling method and an indicator of physical media type. So for example, "100BASE-TX" means 100 Mbit/s data transmission rate and baseband transmission over Cat 5 (or above) cable.

Electrical Isolation

Ethernet uses a variety of highly complex data encoding, coupling, and transmission methods, but all (standards-compliant) versions of twisted pair Ethernet use transformers to couple to the cable, which results in galvanic isolation (page 98) to at least 1500V. Isolation, of course, is something we always want on shows.

ETHERNET IMPLEMENTATIONS

Computer and networking hardware advances at a blinding rate, but 100 and 1000BASE-T are the most widely used Ethernet types as of this writing, with the faster standards used for backbone or high-performance applications. One of the major strengths of Ethernet is backward-compatibility, and Ethernet interfaces use sophisticated "autonegotiation" processes to enable each link partner to determine what the other end is capable of, and to automatically adjust for the maximum data rate. So a device with a 1000BASE-T interface is typically still able to connect with devices using 100BASE-T or even 10BASE-T.

10BASE-T

10BASE-T has been one of the most widely used networks in the world, although you are more likely to see one of its faster cousins (detailed below). 10BASE-T Manchester encodes (page 141) its data and has a data rate of 10 Mbit/s. There were originally several wire and connector types for 10BASE-T, but you are now only likely to find Cat 5e UTP (unshielded, twisted pair) cable and RJ45 (8P8C) connectors in a star topology. Two pairs of the four available in Cat 5e are used, one for transmission and the other for reception. The T in the 10BASE-T acronym stands for twisted pair, and the typical maximum segment distance is 100 meters, or about 328 ft. (The length was not actually specified in the standard, only conditions for proper operation, but the 100m is an accepted rule of thumb, with most permanent installations limited to 90m to ensure that there is length available for the patch cables.)

10BASE-FL

10BASE-FL is less common than the faster variants detailed below, but allows 10 Mbit/s data to be sent over multimode fiber (see "Fiber Optics," on page 142) in a point-to-point manner for a distance of up to 2,000 meters. Two pieces of fiber-optic cable are used in each link, one for data in each direction. Manchester coding is employed, with a "high" state indicated by light on and a "low" state indicated by light off. 10BASE-FL was used in entertainment for special applications, such as long-haul links, or where extreme electrical noise immunity or lightning protection is needed.

Old Ethernet

Ethernet, of course, has advanced tremendously since its introduction. Two versions, in fact, are now totally obsolete, but I'll mention them here just for historical background. 10BASE5, or "ThickNet", was the original Ethernet, and was often called "Frozen Yellow Garden Hose" because the 13 mm diameter, 50Ω coaxial cable was huge, hard to work with, and yellow. 10BASE2, or "ThinNet," was a second-generation Ethernet. It used 5 mm diameter 50Ω coax and BNC connectors in a bus topology to send 10 Mbit/s Manchester-coded data (see "Manchester," on page 141) over network distances of 200 meters. Each node was connected using a "T" BNC connector, and the ends of a line had to be electrically terminated. With the robust BNC connector, 10BASE2 was originally adopted by many entertainment manufacturers.

100BASE-T "Fast Ethernet"

100BASE-T, introduced in 1995, is the general class name for "Fast Ethernet," which is a star-topology system running at a data rate of 100 Mbit/s. There are several variations of 100BASE-T; the most common for entertainment applications are 100BASE-TX and 100BASE-FX. As of this writing, 100BASE-T is one of the most popular forms of Ethernet, since it offers an obvious performance improvement over 10BASE-T and can be run over inexpensive Cat 5 or better UTP cable. 100BASE-T uses a 4B5B encoding method, which is beyond the scope of this book, and 100BASE-TX uses one pair of Cat 5e UTP or STP for transmission and another pair for reception. RJ45 (8P8C) connectors are used, and the maximum segment length is specified at 100 meters.

100BASE-FX

With 100BASE-FX, two multimode optical fibers are used, one for transmission and one for reception, with maximum segment lengths of 2,000m.

1000BASE-T—"Gigabit Ethernet"

1000BASE-T, or Gigabit Ethernet (also known as "Gig" or "GigE"), offers 1,000 Mbit/s over copper UTP, and was published by the IEEE in 1999. It uses all four pairs of Cat 5 or better cable using RJ45 (8P8C) connectors. This implementation was designed primarily for extremely high–bandwidth "backbone" applications, but many PCs come with 1000BASE-T interfaces, and these are often used in applications on shows where huge files (e.g., video) must be transferred. Of course, to take advantage of the higher speeds, gigabit switches and so forth must also be used. 1000BASE-T uses extremely sophisticated echo cancellation and pulse-amplitude modulation, which are outside the scope of this book.

1000BASE-X

1000BASE-SX is a multimode fiber gigabit standard, offering 1,000 Mbit/s over a standard distance of 220m. 1000BASE-LX is a longer range, single-mode fiber standard, offering transmission distances of several kilometers.

10GBASE

10 gigabit Ethernet is, as of this writing, finding some applications for very high bandwidth entertainment applications (very large numbers of HD streams, for example). There are many issues with running these network rates over copper, so fiber standards like 10GBASE-SR (short range, 33-400 meters) are more commonly used.

Higher Rates

Ethernet is under continuous development by IEEE, and higher rate standards are available and being used in high-performance backbone applications. However, it's not clear if any of these variants will be widely used in our market.

Ethernet Capacity Comparison

We now have an enormous amount of bandwidth available, but pushing to higher data rates comes with a cost—a possible increase in errors, dropped frames, touchier cable runs, etc. So going to gigabit isn't always necessary. For example, here's a comparison of capacity of three types of Ethernet transmission[3]:

	10 Mbit/s	100 Mbit/s	1000 Mbit/s
DMX Universes using Art-Net	40	400	4000+
Bi-Directional 48kHz/24bit audio channels using Audinate's Dante	~9	96	1024
30fps 640x480 IP Cameras	1	12	125

ETHERNET HARDWARE

You can't have an Ethernet system without hardware components; here's an overview:

Network Interface Card

The **Network Interface Card** (NIC) is the actual interface between a computer and a physical network. In many ways, the NIC is actually the node, although functionally, the whole device encompassing the NIC can also be thought of as the node. Historically, this was a peripheral card (hence the name) but it's likely now to be integrated right onto the motherboard.

Direct Connection Cables and Auto MDIX

For a LAN of only two nodes, it is possible to connect the two devices by simply switching the transmit and receive lines using a **crossover** or "direct connection" cable (these are similar to "null modem" cables for point-to-point serial applications—see page 151). However, such a cable may not be necessary—most Ether-

---

3. All numbers approximate. From the Art-Net3 protocol document, `http://www.audinate.com/index.php?option=com_content&view=article&id=99#How%20many%20audio%20channels%20does%20Dante%20support?`, and `http://www.video-insight.com/Support/Tools/IP-Bandwidth-Calculator.aspx`

net interfaces now can do automatic transmit/receive configuration using **Auto-MDIX** (MDI stands for automatic Medium Dependent Interconnection, the X stands for "crossover"); try it to see if it will work on your system, so that you don't need to keep track of a separate crossover cable. Alternatively, with switches now so inexpensive, it may be easier to simply connect the two devices to a switch.

Repeating Hubs

A **hub** is the central connection point for a "hub and spoke" (star) topology LAN. The repeating hub has a number of hardware "ports," and receives networking messages from any port, connects the transmit and receive lines correctly, and copies and repeats those messages to every other port on that hub.

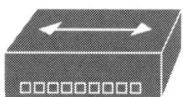

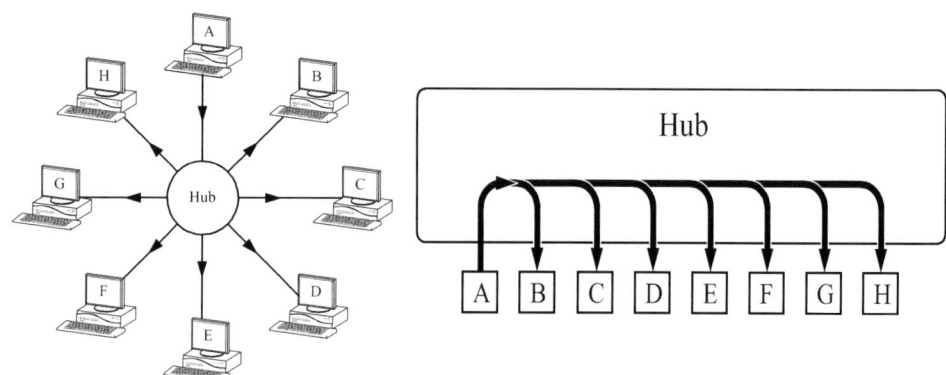

One port on the hub could also be run some distance to another hub, allowing networks to be expanded (more on this in Chapter 18). Generally, repeating hubs operate at OSI Layer 1, and do not contain any routing intelligence, which makes them very inefficient since all data is repeated to every port, regardless of whether or not the port needs the information. For most applications, switches offer some important benefits, and they are so inexpensive now that it's often hard to even find a hub for sale.

Switches

Switches (previously called "switching hubs") are sophisticated versions of multiport hubs, incorporating enough resident intelligence to know the OSI layer 2 MAC address of each connected device.

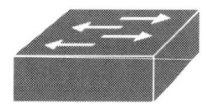

With that knowledge (either user configured or found automatically), the switch will forward to a node only the messages intended for that particular node, creating virtual "private channels."

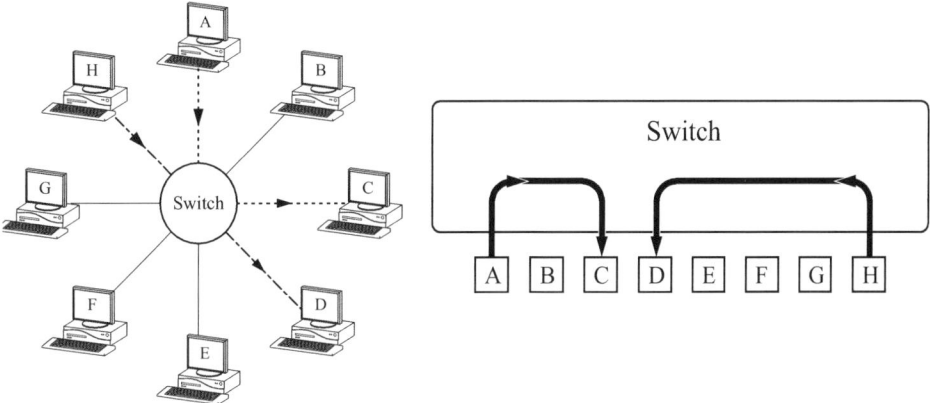

This approach dramatically improves the performance of a network, particularly when used with CSMA/CD, since the collision domain is smaller, and, therefore, the switch can reduce or eliminate collisions altogether on a network segment.

Full-duplex switches take the switching concept one step further, providing cir- circuitry to enable each node to transmit and receive simultaneously on separate pairs of wires. With this approach, collision detection and avoidance mechanisms can actually be disabled, since with only two connected nodes and full-duplex communication, there *can't be any collisions*.

The "switching fabric" of the highest-performance switches, sometimes called **line-rate** switches, can operate at full "wire speed" for all the connected ports' possible total bandwidth. These cost a bit more but are usually worth it for entertainment control applications.

In a network with an all full-duplex switching infrastructure (no "repeater" hubs), effective point-to-point communications to and from all network nodes simultaneously can be achieved, since each node sees only the traffic destined for it—many virtual communications channels can be active simultaneously. Full-duplex switches are now so inexpensive and solve so many problems for our industry that there really is rarely any reason to use anything else.

Managed and Multi-Layer Switches

Many show networks (as of this writing), are built from simple, **unmanaged switches**, which come configured from the factory for basic operation and run automatically. For more sophisticated network configuration and operation, like implementing Virtual LANs (VLANs, page 221), you need a **managed switch**, which is simply a switch which can be configured in some way by a user or network administrator. Some switches are configured via a command line interface (CLI), and a terminal emulation program (page 151); others use an IP address and serve up a Web-based interface:

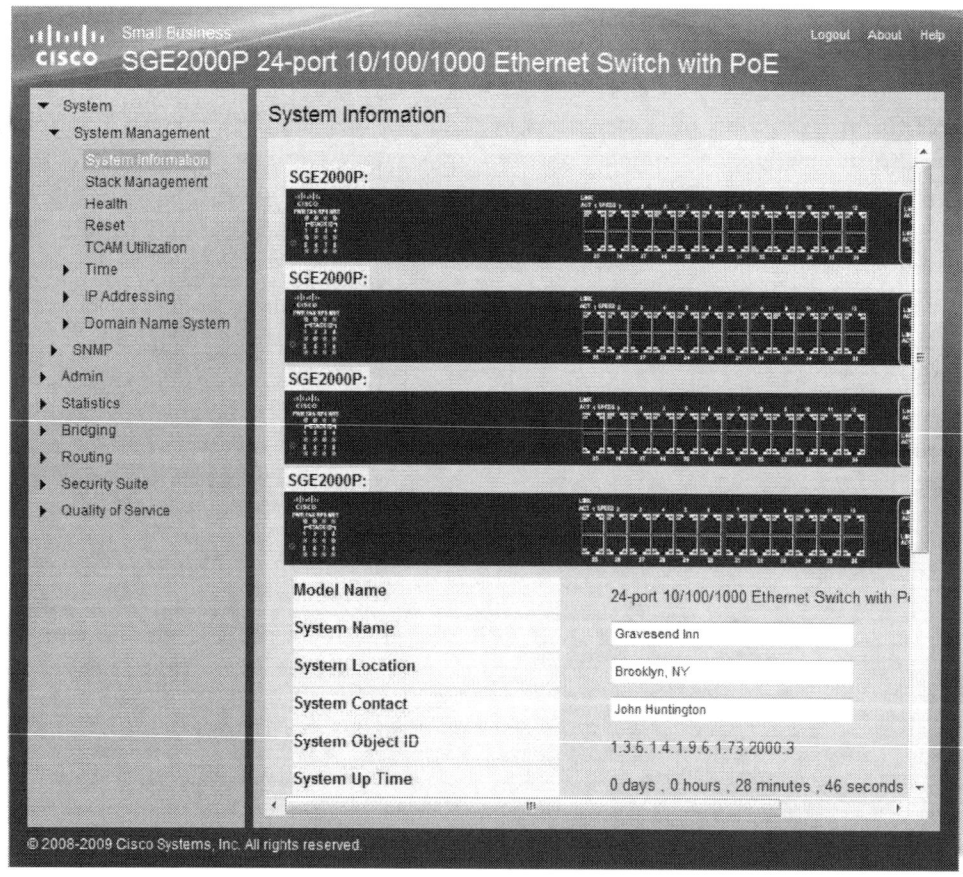

A **multilayer switch** is one that offers some functionality on Layer 3 or higher (typically simple routing, etc.); most multilayer switches also need to be managed in some way.

Routers

While switches and hubs are used to create LANs, **routers** are used to connect LANs together. Routers generally operate in OSI Layer 3 (network) and are covered further in Chapter 18.

(Note: In home networking, the device called a "router" typically does far more, and incorporates a switch, possibly a "wireless access point (see the "IEEE 802.11 "Wi-Fi"" section, on page 175) and the ability to assign network addresses automatically using DHCP (see "Dynamic Host Configuration Protocol (DHCP)," on page 185).)

Media Converter

A media converter is typically a two-port device, with one type of media on one port and another type on the other. For instance, a media converter might convert UTP cable on one port to fiber on the other.

Bridges

Sometimes, network segments need to be connected together without a router, and hardware to do this is called a **bridge**. For example, an optical bridge (see photo) essentially replaces a Cat 5e cable with a laser beam, connecting different segments of the network. Without a router, all packets are typically sent across the bridge.

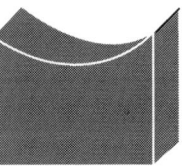

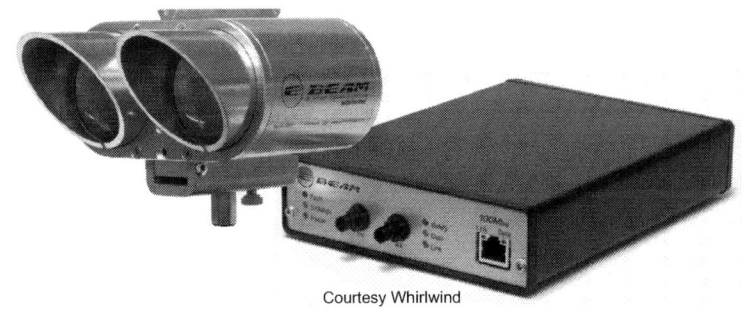

Courtesy Whirlwind

Power over Ethernet (PoE)

Since so many devices these days have an RJ-45 (8P8C) Ethernet connection, the IT industry decided that it would be great to deliver power along with the data connection, so they developed IEEE 802.3af **Power over Ethernet** (PoE). The original scheme can deliver up to about 13W of DC power at 48V by sending power in the same way that 48 VDC phantom power is delivered to a microphone; the 2009 version can deliver about 25 watts using all four pairs of cable. In the larger IT industry, probably the widest use of this technology is for Voice Over IP (VOIP) phones; a single cable run from a PoE switch can provide power and a data connection. In live entertainment, we might see PoE used for interfaces (see below), Wireless Access Points, and various kinds of control devices.

There are two types of devices that provide PoE. For example, if an Ethernet switch can provide PoE, it is called an "endspan" device. A device that sits inline and provides PoE, is referred to as a "midspan" device.

There are two primary power delivery modes, but each uses the same approach. Power Sourcing Equipment (PSE) provides 48 VDC on both conductors of one pair of the Cat 5e cable, and the return is on both conductors of another pair. DC power is injected into the center-taps on the output windings of the Ethernet data transformers on the PSE end, and is extracted on the center taps on the input side of the data transformers in the Powered Device (PD). To ensure maximum compatibility with 10BASE-T and 100BASE-T, only two pairs of the Cat 5e cable are used.

Also, to ensure that non-PoE equipment is not damaged if connected, a procedure is implemented at startup to ensure that power will only be delivered to connected

---

Entertaining Ethernet

Our industry has grown to the point that our people are now involved from the outset with some of these new networking technologies. For example, Steve Carlson, a member of the development team for the first computerized lighting console used on Broadway and one of the originators of DMX, chaired the IEEE 802.3af standards group, which developed the Power over Ethernet (PoE) standard. The fact that someone from the entertainment technology business contributed in such a significant way to the development of a major computing industry standard shows that our little industry has matured enough to get some notice from the "big boys." Carlson said that he got involved through PLASA at the request of the larger Ethernet standards effort, because "PLASA is considered the expert in the use of Ethernet in nontraditional and hostile environments."

devices configured to accept it. To implement this, the PSE checks to ensure that the connected device has a "signature resistance," indicating that the device is a PD. The PSE may optionally check (by looking for specific current returned when a classification voltage is applied) for the PD's power class. There are five classes of power, with one reserved for future expansion.

IEEE 802.11 "WI-FI"

The open IEEE **802.11** "wireless Ethernet" standard has revolutionized the portable computer market, and has found many applications in our market as well, especially during the programming of a show. IEEE 802.11, which is also called "Wi-Fi®" (after the trade association for 802.11 manufacturers) offers high-speed, easy-to-use connections to mobile computer users. Of course, that also means it can provide convenient access to us in entertainment control, as long as we use the systems carefully and prudently.

The basic operational concepts of 802.11 for typical applications are simple. Typically, a **Wireless Access Point** (WAP) is connected (through a standard Cat 5e cable) to a network, and through that network to the Internet. The WAP then acts as a "bridge" to the network, allowing portable devices to connect to it using data transmission over radio frequency (RF) electromagnetic radiation. The various 802.11 standards operate in bands allocated by the FCC for industrial, scientific, and medical (ISM) use.

The field of wireless networking is constantly evolving, but as of this writing, the most common 802.11 network types include the following:

IEEE Type	Max Speed	Frequency Band
802.11a	54 Mbit/s	5 GHz
802.11b	11 Mbit/s	2.4 GHz
802.11g	54 Mbit/s	2.4 GHz
802.11n	150 Mbit/s	2.4, 5 GHz

(Note: These frequency ranges are for the US; other countries use different ranges, so be careful if you're touring internationally).

Basic Structure

There are four basic components of an 802.11 network: stations, access points (AP or WAP), the wireless medium, and the distribution system. Stations are simply the computers you want to connect to the network. Access points are the "bridge" between the wireless and wired networks (or, in some cases, between wireless devices). APs also provide the critical functionality of converting wired MAC frames to wireless, and vice-versa, and manage radio transmission. The wireless medium is, of course, the physical medium (radio waves rather than electrons on a cable) that is used to transport the data. Finally, the distribution system, which is somewhat confusingly labeled, is the logical process by which potentially moving, connected stations are tracked as they move through and between various parts of the wireless network.

Radio transmission is, by its nature, a broadcast medium; every transmitter on a particular frequency will occupy that bandwidth while transmitting, and any receiver is capable of receiving any transmission (although, of course, it's good practice to encrypt the data so it can't be read). It is then the receiver's responsibility to sort out the transmissions and decide whether or not to act on that data. Like wired Ethernet, 802.11 uses a layered structure, and the LLC layer in 802.11 is the same as wired Ethernet. The 802.11 wireless MAC layer operates differently, but performs the same functions; the 48-bit MAC address of each station is used to identify the station to the network and for routing of frames. The PHY layer, of course, is the radio network.

The physical layer operations of 802.11 are so complex that they lie mostly outside the scope of this book, but let's go through a quick overview of each of the popular types of 802.11. The original 802.11 used frequency hopping spread spectrum (FHSS). Spread spectrum radio transmission (introduced on page 143) is a scheme whereby the transmitter rapidly changes its transmission frequency in a pseudo-random pattern. Of course, the receiver must also constantly switch channels to stay in sync with the transmitter. This approach spreads out the transmitted data signals across a wider range of RF spectrum and found early military applications, where security and monitoring are a significant concern. Spread spectrum techniques also offer significant benefits in the crowded RF spectrum, such as is found in the unlicensed bands in which 802.11 operates.

Wireless Network Layout or Configuration

Communications in an 802.11 network take place over a "basic service area", through the use of a basic service set (BSS). The BSS, which is identified by a 48-bit BSSID, is simply a group of machines on the network. This group could be an "ad hoc" or "Independent" group where stations communicate directly with one another; more common are "infrastructure" BSSs, where all communications (even between two machines in the same wireless network) take place through an access point. In addition, access points can operate in a wireless "bridge" mode, where two hard-wired connected nodes are linked wirelessly.

In a common infrastructure BSS, stations must join, or "associate," with an access point. The association request is initiated by the mobile station, and the access point can accept or deny the request based on whether the station has proper authority to access the network, and so on. Stations can be associated with only one access point at a time. If a single BSS is not enough to cover the required area, than multiple BSSs can be joined into an Extended Service Set (ESS). (Network engineers love acronyms!)

Each wireless network is given a **Service Set IDentifier** (SSID), which identifies that network through an ASCII string up to 32 characters long. This is the "network name" that users see when they turn their laptops on and select a wireless network. In simple wireless networks, the SSID is configured into a single access point; in more complex systems, the SSID is shared across multiple access points that make up an ESS.

Basic Connection Process

There are two basic "scanning" methods a machine can use to find and connect with the wireless network. One type is the "passive" scan, where the station doesn't turn on its transmitter (saving power on mobile devices), but instead listens on each 802.11 channel for "beacon" messages from an access point or station. The beacon message contains enough information for the wireless station to connect and begin communication.

In "active" scanning, the station that wants to connect sends a "probe request" message either to the broadcast SSID address (looking for a response from all SSIDs in range) or to a specific SSID. The device that receives the probe will then (if it will allow communication) respond with a probe response message, offering all the information needed to connect.

Some users will disable SSID beacons and configure their WAPs to not respond to broadcast probe request commands. This does offer a bit of security, in that casual users cannot see or access the network; however, for the wireless network to function, the WAP must still respond to a valid probe request command for its SSID and anyone monitoring the channel (albeit with more technical skill than the average user) will be able to see the SSID in the probe request/response interchange. Hiding the SSID can also cause problems with some equipment.

Security Issues

Wired networks have some built-in security, in that you can exert some control over who plugs into your network. However, as soon as you connect an open, unsecured access point to your wired network, you are opening up your network—and therefore everything connected to it—to anyone within radio range since, by definition, wireless frames are broadcast to everyone in range. So, if you are not using any sort of encryption, then almost everything you transmit on the radio link is easily viewable by anyone with an 802.11 receiver. There are many wireless security schemes out there; here's an overview of a few that are current as of this writing.

Wired Equivalent Privacy (WEP) was one of the early schemes to cryptographically encrypt 802.11 traffic. As it turns out, the encryption scheme could be cracked relatively easily, but it is certainly better to use WEP than nothing at all. The Extensible Authentication Protocol (EAP) and Wi-Fi Protected Access (WPA) are more recently developed authentication schemes that overcome many of WEP's shortcomings. In 2004, the IEEE standardized 802.11i, which is also known as WPA2. As this is a constantly evolving area, it's worth doing a bit of research before selecting a security approach for your system. Keep in mind, however, that enabling security is usually a simple matter of configuring the WAP using its setup procedure, and of course any security—even if flawed—is better than none at all.

Should You Use 802.11?

My advice for show applications: *if you can use a wire, use a wire!* 802.11 offers fantastic utility that is especially useful for the (noncritical) programming phase of any project; being able to walk around a venue with a wireless laptop or hand-held device offers amazing flexibility and ease of use. However, this utility comes with some security risks, as detailed above, and tremendous potential reliability issues, since it operates in unlicensed bands, meaning that you have absolutely no way to know (or regulate) who else is operating in that band. You may be able to keep them out of your system using security procedures, but you can't keep them from (unintentionally) interfering with your system: devices such as microwave ovens and cordless phones can cause significant interference, all of which can cause delays or even failures in data transmission. In addition, radio is, by definition, a "shared" media, meaning that it has none of the advantages of a switched, full-duplex structure that is so important to our market.

If there is no possible way to use a wired connection, then use wireless, but be sure to carefully research and implement your system, and also include in your design a way to monitor transmission and interference issues. Again, if you can use a wire (or fiber, of course), use a wire!

ETHERNET IN OUR INDUSTRY

Ethernet rightfully got off to a slow start in our industry. The nondeterministic nature of CSMA/CD, with its random back-off collision recovery time, was enough to scare off many entertainment engineers. However, over the years, the IT industry has (fortuitously for us) solved this and many of our other objections, and full-duplex, switched Ethernet is an excellent control communications solution for shows. It offers the following:

- Open standard
- Very high bandwidth
- Low cost
- Near "real-time" delivery using (high-quality) full-duplex switches
- Electrical isolation due to Ethernet's use of transformers for twisted pair connections (or fiber)
- High-quality CRC error check of every single frame transmitted across the network
- "Guaranteed" delivery through the use of TCP or other protocols

Now, lets move on to issues of network systems and look at some practical examples.

Chapter 17: Show Networks

Now that we have covered the general concepts of networking, let's look at some of the underlying components that make networks work.

TCP, UDP, AND IP

The **Transmission Control Protocol** (TCP), **User Datagram Protocol** (UDP), and **Internet Protocol** (IP) are the backbone of networks, including of course, the network of networks: the Internet. TCP, UDP, and IP are separate protocols, but are often used together in a protocol "suite" or "stack," typically TCP/IP (see figure) or UDP/IP.

These protocols were developed[1] in the late 1970s, before the OSI standard (page 157) was finalized, but TCP and UDP can be thought of as occupying OSI Layer 4 (transport), with IP fitting into OSI layer 3 (network). This means that TCP/IP or UDP/IP sit *above* networks like Ethernet, which occupies OSI Layers 2 and 1, and *below* user processes or programs such as a Web browser, which uses **HyperText Transfer Protocol** (HTTP) or **File Transfer Protocol** (FTP), or an e-mail program running Simple Mail Transfer Protocol (SMTP).

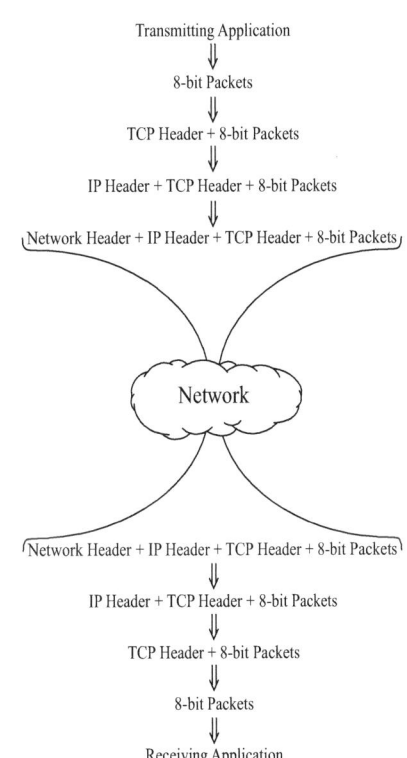

---

1. These protocols were developed by the IETF. TCP is RFC 761, UDP is RFC 768, and IP is RFC 791.

Here is a simplified table showing how the various protocols align with OSI, and the units they transmit:

Layer	Name	Function	Unit
7	Application	DNS, FTP, HTTP, IMAP, IRC, NNTP, POP3, SIP, SMTP, etc.	Data
6	Presentation	MIME encoding, data compression, data encryption, etc.	
5	Session	full-duplex, half-duplex, etc	
4	Transport	TCP, UDP, etc.	Segments
3	Network	IP	Packets/ Datagrams
2	Data link	Ethernet MAC and LLC layers	Frames
1	Physical	Ethernet PHY layer	Bits

While those in entertainment may have encountered these network protocols in conjunction with Ethernet, they were actually designed to handle traffic across a variety of networks, platforms, and systems, and accomplish this, for the most part, transparently to the user. Their work is so transparent, in fact, that many of the operational details of these protocols are beyond the scope of this book; however, the general principles and characteristics of the protocols are important to understand.

TRANSMISSION CONTROL PROTOCOL (TCP)

IP makes no guarantee to services residing on higher layers that a packet will get to a particular destination; it simply makes its "best effort." TCP, on the other hand, is a **connection-oriented**, **reliable** protocol, meaning that it can guarantee that packets will arrive intact and can be reassembled in the right order. TCP provides this reliability by creating a "virtual circuit" connection to the receiver, which must be specifically established and terminated. Using the connection (and other features of the protocol), TCP guarantees to the layers above it that a message will make it to its destination somehow, or else it will notify the upper layers that the transmission failed. TCP is used in many entertainment applications where the integrity of the transmitted data is of paramount importance, such as a motion control or pyro command.

USER DATAGRAM PROTOCOL (UDP)

UDP takes an **unreliable** approach to **connectionlessly** deliver **datagrams**. Without TCP's overhead packaging packets for delivery and dealing with the connection, UDP can be much faster and more efficient. Of course, "unreliable" UDP may, on the right network, be quite reliable, and UDP is used in many entertainment applications where speed of delivery is of the essence and the network is secure and not heavily loaded, which ensures for the most part that the packets will be sent across intact and in order. Alternatively, UDP can be used with other protocols (e.g., ACN,) handling the reliable transmission aspects of the communication process.

INTERNET PROTOCOL (IP)

To be delivered properly to its destination, a postal letter needs an address unique to that destination. Similarly, each packet of data on a network needs a destination address, and handling this information is one of the key functions of the Internet Protocol (IP). IP version 4 (IPv4) is the backbone of most networks,[2] and it provides a universal addressing scheme that can work within or between a wide variety of networks (hence the "inter" name), providing unique IDs for the connected hosts. An "IP address" is simply a 32-bit number uniquely assigned to a machine on a particular network, and 32 bits gives us 4,294,967,295 unique addresses. 32-bit binary IP addresses like `00001010 00000000 00000001 11011110` are pretty unwieldy for humans to deal with, so we usually express IP addresses as **dotted decimal** numbers,[3] with each decimal number representing 8 bits of the 32-bit address, like `10.0.1.222`. In dotted decimal format, the IP address range goes from `0.0.0.0` to `255.255.255.255`.

Network Classes

IPv4 addresses break down into two parts: a **Network IDentifier** for a particular network (remember that IP is designed to connect networks together), and a **Host IDentifier** unique for a particular device (server, workstation, etc.) on that network. IPv4 addresses are also divided into five classes, with varying numbers of bits assigned to the Network ID and the Host ID. Class A networks, for example, assign 7 bits to the Network ID and 24 bits to the host; this type of address was intended for very large organizations with many hosts. Class C assigned 21 bits to the Network ID, and only 8 bits to hosts, which allowed many small organiza-

---

2. As of this writing, IPv6 also exists (see page 233) and is gaining some acceptance, but IPv4 will likely be used for many years to come.
3. Of course when humans use the net, they usually don't enter numeric addresses, but instead enter text names such as "`www.controlgeek.net`." These text names are mapped to numeric IP addresses by the Domain Name System; see "Domain Name System (DNS)," on page 232.

tions, each with a few hundred hosts at a maximum (like most show networks), to be assigned their own Network ID.

Here's the network class breakdown:

Class	IP Address	Networks	Hosts
A	0nnnnnnn hhhhhhhh hhhhhhhh hhhhhhhh 7-bit Network ID, 24-bit Host ID	128	16,777,216
B	10nnnnnn nnnnnnnn hhhhhhhh hhhhhhhh 14-bit Network ID, 16-bit Host ID	16,384	65,536
C	110nnnnn nnnnnnnn nnnnnnnn hhhhhhhh 21-bit Network ID, 8-bit Host ID	2,097,152	256
D	1110mmmm mmmmmmmm mmmmmmmm mmmmmmmm 28-bit multicast address		
E	1111rrrr rrrrrrrr rrrrrrrr rrrrrrrr Reserved for future use		

When IP was designed in the 1970s, this scheme worked well, and would likely easily allowed just about every computer on the planet to be connected. However, as the Internet grew, this scheme proved to be inefficient,[4] since addresses were often assigned in blocks that might not match the needs of an organization—a small company with only 100 hosts could be allocated an address range that blocked out tens of thousands of IP addresses. To get around this problem (and others), things like Classless Inter-Domain Routing (CIDR) and Network Address Translation (NAT, page 232) were developed, but these are mostly outside the scope of this book.

Private/Nonroutable IP Addresses

The IP standard set aside a number of addresses within each class for **private networking**. These addresses are also called **nonroutable**, since they do not belong on the Internet and are generally blocked by routers. For shows, we often build closed (non-Internet connected) networks and, therefore, could probably just use whatever IP address we felt like. However, this is not a good practice, because if any of those machines were to be accidentally connected to the Internet, conflicts could occur, causing problems.

---

4. In fact, in January, 2011, the last full blocks of IP addresses were allocated.

Instead, it's good practice to use the following ranges of nonroutable addresses in any private show network:

Class	Start of Range	End of Range
A	10.0.0.0	10.255.255.255
B	172.16.0.0	172.31.255.255
C	192.168.0.0	192.168.255.255

Note: Not all addresses are available for use by hosts. Addresses like 192.168.255.0 are reserved for the "network address" or "network ID"; addresses like 192.168.255.255 or 255.255.255.255 are reserved for "broadcast address usage; more in Chapter 17.

Multicast IP Addresses

Class D is reserved for multicasting (page 160), where one device may need to send packets to multiple receivers simultaneously. Multicasting might be used for streaming media (like a live video) where there may be many receivers of the transmission, or by a lighting console that needs to send its data to a number of processing units that do something with that data. The range of IP addresses assigned for this purpose is 224.0.0.0 through 239.255.255.255; receivers can join the multicast using Internet Group Management Protocol (IGMP, see page 232).

Loopback/Local Host IP Address

Another reserved address is the **loopback** address, which is typically 127.0.0.1.[5] The loopback addresses allows a message to be sent from the **local host** to itself. This can be used for testing, or when an IP process needs to communicate with another IP process on the same machine.

DYNAMIC HOST CONFIGURATION PROTOCOL (DHCP)

In the vast majority of commercial and consumer networks, the IP address of the connected hosts is automatically assigned using the **Dynamic Host Configuration Protocol** (DHCP), which allows a node to obtain an IP address "lease" automatically. When a device is shut off, disconnected from a network, or the lease expires, that IP address can be released for use by others.

The DHCP process starts when the TCP/IP protocols are initiated on a machine connected to the network (in Windows®, for example, this happens very early dur-

---

5. In IPv6, the loopback address is ::1

ing machine start up). The device requesting an IP address sends a UDP packet to 255.255.255.255, the **broadcast IP address** which all nodes will receive (more on that later in "Broadcast Domain," on page 213). If a DHCP server is available on that network, it replies with an offer of an IP address (and some other information like the Default Gateway, covered later). If the host accepts the offer, the IP address is "leased" for a period of time.

Here are four exchanges captured from an actual DHCP negotiation:

```
818 241.162420 0.0.0.0              255.255.255.255    DHCP    342 DHCP Discover  - Transact
820 241.211344 192.168.1.111        255.255.255.255    DHCP    336 DHCP Offer     - Transact
823 241.228277 192.168.1.111        255.255.255.255    DHCP    336 DHCP ACK       - Transact
838 241.380147 Broadcom_8a:4d:45    Broadcast          ARP      60 who has 192.168.1.103?
```

The first line shows a host sending out a "DHCP Discover" message, looking for a DHCP server on the network. In this case, there is a DHCP server available at 192.168.1.111, and it responds with an offer of the IP address 192.168.1.103 (some details omitted for clarity), and the host accepts it. It then broadcasts a "Gratuitous ARP" command (page 200) to the whole network just to inform all the other connected hosts that this machine has now been assigned this address. The whole process takes about 200 milliseconds.

IPCONFIG/IFCONFIG COMMAND

It's easy to see what IP information has been assigned by a DHCP server to your system; from Windows, you can use a command-line utility called `ipconfig` (on a Mac it's `ifconfig` from the terminal). Typing `ipconfig` from the Windows Command Prompt (go to Start | Accessories to find it), for example, you would get something like this, enabling you to verify (or discover) the configured address (excerpt only shown for clarity):

```
C:\>ipconfig

Ethernet adapter Local Area Connection:

    Connection-specific DNS Suffix  . :
    IPv4 Address. . . . . . . . . . . : 192.168.1.101
    Subnet Mask . . . . . . . . . . . : 255.255.255.0
    Default Gateway . . . . . . . . . : 192.168.1.1
```

(Note: More on the subnet mask starting on page 194; the default gateway is covered in Chapter 18.) From the `ipconfig` command line, you can also configure and view a number of other parameters about your IP connection. Typing `ipconfig ?` will give you a complete list of options.

EXAMPLE NETWORK USING DHCP AND IPCONFIG

Let's take a look at a small show network that uses DHCP to automatically assign IP addresses. Here, we want to connect up two computers (one Mac and one PC) to a mixer, using Ethernet, IP, and a network audio transmission system. The audio networking system is sophisticated enough to be able to locate machines on the network without knowing their IP addresses in advance, so in this case we will use a small Ethernet switch that has a built in DHCP server.

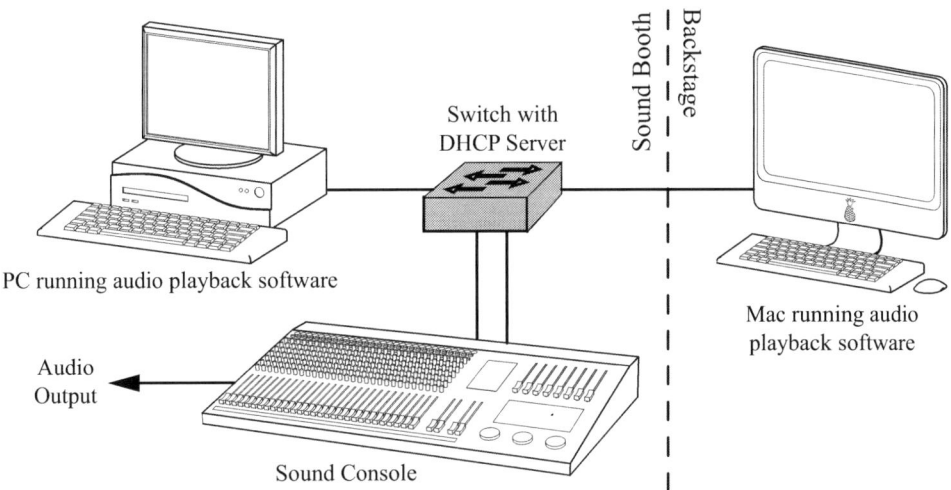

As the machines start up, the DHCP server automatically assigns the IP addresses; let's see what IP address was assigned by the DHCP server to the Mac. We go to "System Preferences," and click Network, and see the screen shown at right.

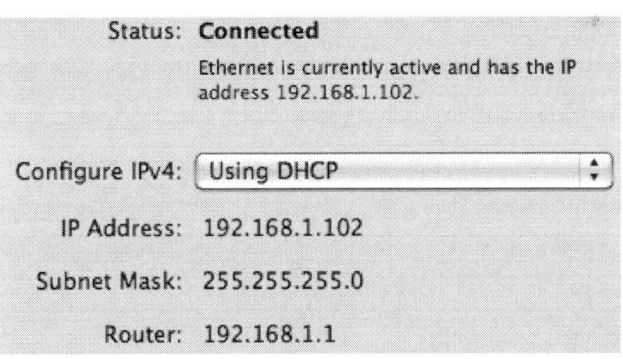

The Mac has been assigned the IP address 192.168.1.102.

We can also see the results by going to the Mac terminal and typing ifconfig (excerpt only shown for clarity):

```
ENT$ ifconfig
inet 192.168.1.102 netmask 0xffffff00 broadcast 192.168.1.255
```

Now, let's check the address assigned to the PC. We go to Control Panel, Network and Internet, Network Connections, and then right-click on the Ethernet adaptor, release on Status, and click Details and we get the screen you see here:

DHCP Enabled	Yes
IPv4 Address	192.168.1.103
IPv4 Subnet Mask	255.255.255.0
Lease Obtained	Monday, April 09, 2012 10:27:34 AM
Lease Expires	Wednesday, April 11, 2012 10:27:33 AM
IPv4 Default Gateway	192.168.1.1
IPv4 DHCP Server	192.168.1.1

This machine has been assigned `192.168.1.103`.

We can also use `ipconfig` from the PC's command prompt to see what addresses was assigned to it (Excerpt of results only shown for clarity):

```
C:\>ipconfig

Windows IP Configuration

Ethernet adapter LAN 1:

   IPv4 Address. . . . . . . . . . . : 192.168.1.103
   Subnet Mask . . . . . . . . . . . : 255.255.255.0
   Default Gateway . . . . . . . . . : 192.168.1.1
```

Because of channel limitation issues on this sound console (each console slot can only accommodate 16 audio channels and we need 32), it has two networking cards, and they also negotiate their own IP addresses.

We can view their addresses by using the audio networking system's management software:

DM1000--S1-MY16-030a08	MY16	3.3.6	192.168.1.101	100Mbps
DM1000-S2-MY16-030a94	MY16	3.3.6	192.168.1.100	100Mbps

All these addresses are managed by the DHCP server built into the switch, which also maintains its own table of IP addresses, correlated with the associated Ethernet MAC addresses:

DHCP Server IP Address:	192.168.1.1	
Client Hostname	IP Address	MAC Address
	192.168.1.100	00-1D-C1-03-0A-94
	192.168.1.101	00-1D-C1-03-0A-08
ENT-PROD-081	192.168.1.102	00-23-32-96-23-3A
ENT-PROD-203	192.168.1.103	00-10-18-8A-4D-45

Note that the DHCP server has assigned private, Class C addresses to our small, private system.

LINK-LOCAL ADDRESSES

If a DHCP server is not available, a host can determine its own unique **link-local** address, intended only for use within the local network segment (packets containing these addresses are not forwarded by routers). IPv4 addresses 169.254.1.0 through 169.254.254.255 are assigned for this purpose. These addresses can be manually assigned, but more typically the host's operating system sets them automatically using "stateless address autoconfiguration," which selects an address using a random procedure, and then checks on the network using ARP (page 200) to make sure the address is not already in use. If a DHCP server becomes available, the host should release the link-local address and use the automatically assigned address. In Windows, link-local address assignment is also called Automatic Private IP Addressing (APIPA).

STATIC/FIXED IP ADDRESSES

Rather than being configured automatically by DHCP, as is common with most home and office networks, networks with "static" or "fixed" IP addresses need to be configured manually by the network administrator. Fixed-IP addressing schemes can simplify the structure of the network and can offer some advantage for show applications. For example, broken devices are easy to replace—the replacement device can simply be set to the old, broken device's IP address and the system should be able to continue operating. As of this writing, most show networks are using fixed IP addresses; good network practice means using the non-routable, private IP address ranges.

PING COMMAND

Once you have all your network devices connected and configured, an easy way to verify that they are working properly on the network is to **ping** them. `ping` sends some special packets, called Internet Control Message Protocol (ICMP) echo requests, to the device at the IP address[6] that you want to test. It then tracks responses from that machine and gives you some timing information.

For example, let's ping the machine at `192.168.1.21`:

```
C:\>ping 192.168.1.21

Pinging 192.168.1.21 with 32 bytes of data:
Reply from 192.168.1.21: bytes=32 time<1ms TTL=64
Reply from 192.168.1.21: bytes=32 time<1ms TTL=64
Reply from 192.168.1.21: bytes=32 time<1ms TTL=64
Reply from 192.168.1.21: bytes=32 time<1ms TTL=64

Ping statistics for 192.168.1.21:
    Packets: Sent = 4, Received = 4, Lost = 0 0% loss),
Approximate round trip times in milli-seconds:
    Minimum = 0ms, Maximum = 0ms, Average = 0ms
```

"<1ms" is the time measured for the destination machine to respond and for the packets to transfer down the network. 1 ms is pretty fast, but this is not unusual for a closed, small network. If an IP address (or other configuration information) is not valid, or a machine on the network is not responding properly, you can see an error like this:

```
C:\>ping 10.0.0.26

Pinging 10.0.0.26 with 32 bytes of data:

Request timed out.
Request timed out.
Request timed out.
Request timed out.

Ping statistics for 10.0.0.26:
    Packets: Sent = 4, Received = 0, Lost = 4 100% loss),
```

---

6. It's also possible to ping a domain, such as `www.controlgeek.net`, instead of a numeric IP address; although for this to work you also need a domain name server (page 232)—and a gateway to the Internet (page 231).

There are many options for the ping command—type `ping` with no arguments to see a complete list.

EXAMPLE NETWORK USING FIXED IP ADDRESSES AND PING

Let's say we have a lighting console in the booth, and we need a DMX lighting control output backstage to run some dimmers and moving lights. We could just run a DMX line backstage, but it turns out in this venue that we already have Cat 5 lines run, so we will use the console manufacturer's lighting data distributor backstage, and connect the two systems using an inexpensive, unmanaged Ethernet switch. We will also include a laptop for testing.

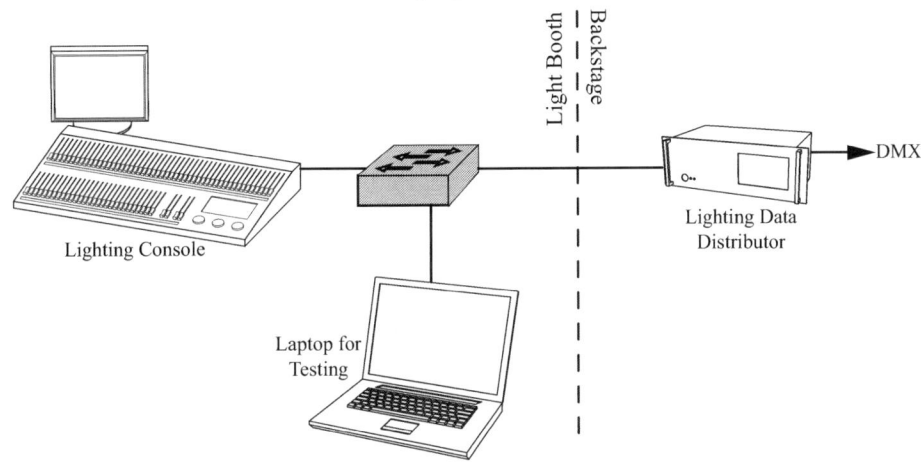

Many lighting controllers need fixed IP addresses to work[7], and these addresses have to be entered manually. Since our example network is not connected to the Internet, we could use just about any address we want, even something like `2.3.4.5`. However, this is not good practice because the `2.x.x.x` address range has been permanently assigned to other organizations, and if someone accidentally connected our closed network to the Internet, we could have some conflicts. It's much better practice instead to use one of the nonroutable, private IP address ranges (page 184). We could use any of the private address ranges, but since this is a very small network it fits most closely with a "Class C" address, so we can pick anything in the range from `192.168.0.1` to `192.168.255.254`.

---

7. As of this writing, newer protocols are becoming available (like sACN page 255) which can work with DHCP.

I picked these addresses:

Host	IP Address
Lighting Console	192.168.1.101
Lighting Data Distributor	192.168.1.102
Laptop (Testing Only)	192.168.1.111

Notice that the first three octets—192.168.1—are all set the same. Since this is a closed, private network, we could make these just about any IP addresses, but it's good network practice (in terms of leaving the ability to scale up for the future), to make these consistent by putting them all on the same subnet (we will cover this more in "Subnets and Network Masks," on page 194).

Assigning Fixed IP Addresses

We need to set the IP addresses in the lighting console and the lighting data distributor. Each of these has a touch screen, and through a special "calculator" screen we can enter the addresses: 192.168.1.101 in the console, and 192.168.1.102 in the lighting data distributor.

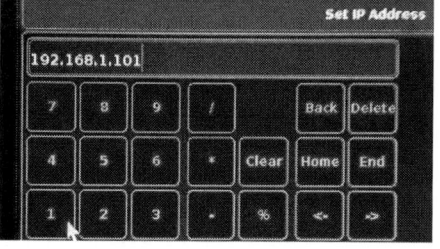

The laptop is a Windows machine, so we use the screen shown at right.. Because we are using fixed IP addresses, we won't click the "Obtain an IP address automatically"—that would enable DHCP. Instead, we click the "Use the following IP address" radio button, and enter the IP address of 192.168.1.111, and accept the default subnet mask of 255.255.255.0 (again, we'll cover this more on page 194). For this simple network, we will leave blank the "default gateway" (page 219) and the Domain Name System

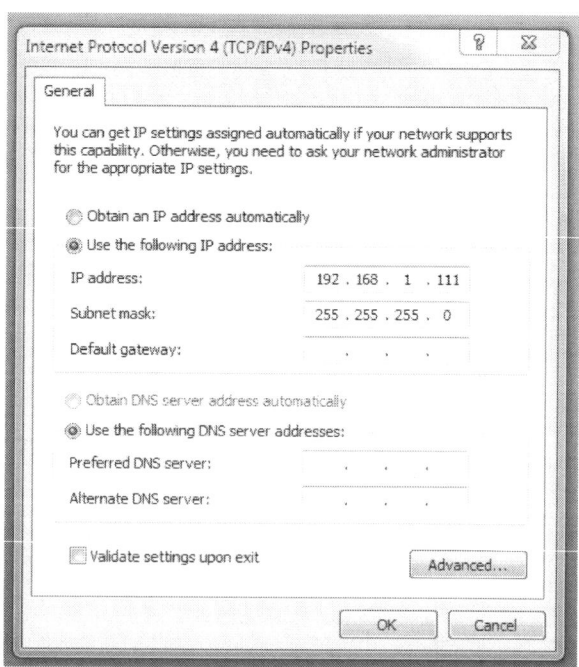

DNS (page 219). When we click "OK," and close out the window, the IP address should be set for this laptop.

Pinging to Test Connectivity

Now that we've got everything up and running, and the IP addresses configured, we can do a quick test to make sure everything's configured and connected correctly. What's the easiest way to do this? The `ping` command, of course. Unfortunately, the lighting console and the data distributor don't have the ability to ping; this is why we included the laptop in this network.

From the Command prompt, we enter the `ping` command and the IP addresses of the devices we want to test:

```
C:\>ping 192.168.1.101

Pinging 192.168.1.101 with 32 bytes of data:
Reply from 192.168.1.101: bytes=32 time=9ms TTL=64
Reply from 192.168.1.101: bytes=32 time=6ms TTL=64
Reply from 192.168.1.101: bytes=32 time=4ms TTL=64
Reply from 192.168.1.101: bytes=32 time=5ms TTL=64

Ping statistics for 192.168.1.101:
    Packets: Sent = 4, Received = 4, Lost = 0 0% loss),
Approximate round trip times in milli-seconds:
    Minimum = 4ms, Maximum = 10ms, Average = 6ms


C:\>ping 192.168.1.102

Pinging 192.168.1.102 with 32 bytes of data:
Reply from 192.168.1.102: bytes=32 time=10ms TTL=64
Reply from 192.168.1.102: bytes=32 time=4ms TTL=64
Reply from 192.168.1.102: bytes=32 time=6ms TTL=64
Reply from 192.168.1.102: bytes=32 time=5ms TTL=64

Ping statistics for 192.168.1.102:
    Packets: Sent = 4, Received = 4, Lost = 0 0% loss),
Approximate round trip times in milli-seconds:
    Minimum = 4ms, Maximum = 10ms, Average = 6ms
```

Everything looks fine—we see no "Lost" packets, and the times are all relatively short; we can now likely assume that the cable is good and connected and everything's powered up and correctly configured.

SUBNETS AND NETWORK MASKS

As networks grow, they can become difficult to manage and operate, so it typically makes sense to break them down into smaller **subnets**. Because we are increasingly using general purpose computing equipment designed to work on any size network, we too have to deal with concepts such as **subnet masks**, even when we are connecting only a handful of machines in a private, closed network. As a result, we need to wade into this widely misunderstood topic.

When a subnet is used, some bits of the host part of the IP address range (see "Network Classes," on page 183) are used to create a Network ID number, which is sort of analogous to an area code in the phone system, or a zip code in the postal address scheme—these define a subset of the larger address space. Because this number of Network ID bits is variable, the system must be told how many of the bits of the IP address actually make up the Network ID and how many are used for the Host ID. Indicating this number of bits is the simple purpose of the subnet mask.

While we may use the familiar dotted-decimal format to define the subnet mask, the computer, of course, deals with the subnet mask in binary, and because it is simply defining the number of bits used for the Network ID, the mask must start with a contiguous block of 1 bits and finish with a contiguous block of 0 bits. Nodes on the network can quickly check to see if a particular address is part of its subnet by applying a bitwise AND logical operator (page 104) to the address; the result is the network ID without the host portion of the address (hence the "mask" terminology).

This is likely getting very confusing, so here is an example of the subnet mask `255.255.255.0` applied to the IP address `192.168.1.107`:

192	168	1	107	IP address
255	255	255	0	Subnet mask
11000000	10101000	00000001	01101011	IP address in binary
11111111	11111111	11111111	00000000	Subnet mask in binary
11000000	10101000	00000001	00000000	Result of bitwise AND: the network ID

The result of the bitwise AND gives us the network ID portion of the IP address, which in this case is `192.168.1`. In this case, the whole last octet of the IP address is available for hosts, meaning that we can address up to 254 hosts using the address range `192.168.1.1` to `192.168.1.254` (`192.168.1.0` is reserved for the subnet identifier; 192.168.1.255 is the broadcast address for this subnet).

The contiguous block of 1 bits makes for some seemingly strange dotted-decimal subnet masks; another way of indicating the subnet mask is to simply append a "/" and the number of subnet mask bits to the IP address. For example, the IP address shown above, when used with the 255.255.255.0 subnet could also be shown as 192.168.1.107/24. You will likely see both forms of nomenclature.

Let's look at another example, with the subnet mask 255.255.240.0 (/20) applied to the IP address 10.43.26.222:

```
10       43       26       222     IP address
255      255      240      0       Subnet mask
00001010 00101110 00011010 11011110 IP address in binary
11111111 11111111 11110000 00000000 Subnet mask in binary
00001010 00101110 00010000 00000000 Result of bitwise AND:
                                     the network ID
```

This subnet mask leaves 12 possible bits for host IDs and the host range for this subnet would be 10.43.16.1 to 10.43.31.254. I didn't calculate this in my head; I simply went online and used a free subnet calculator (see figure).

Looking at Subnet Masks in a Different Way

As of this writing, most show networks are small enough that we don't usually have to think too much about subnets. However, many hosts will *require* you to enter a subnet mask, even if you are connecting only two machines, because the vast majority of networking equipment

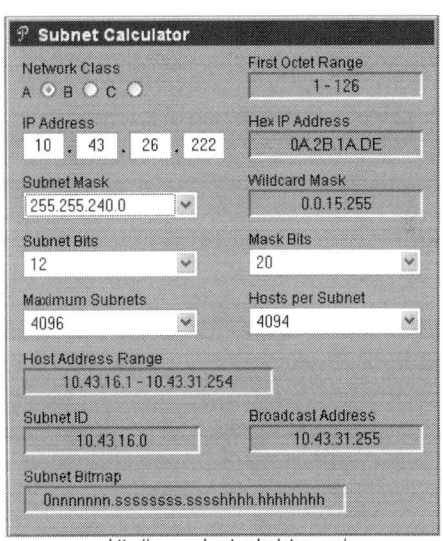

http://www.subnet-calculator.com/

used in our industry is designed (at least in part) for use on larger networks. For small, closed, simple networks, the key thing to keep in mind is that *every IP address you want to access on your network has to match exactly the range of the subnet.*

For example, let's say you have two devices on a simple network connected via a small switch. One host is set to an IP address of `192.168.1.33`; the other is set to `192.168.101.2`. If you have the subnet mask set on both hosts to `255.255.255.0`, these two devices will *not* be able to communicate because, with the subnet mask of `255.255.255.0`, the first *three* octets of all the IP addresses on the subnet have to match exactly, and we have one device set to a subnet of `192.168.101`, while the other is set for `192.168.1`. So, the first three octets do *not* match, and these systems will not be able to communicate with each other on an IP network:

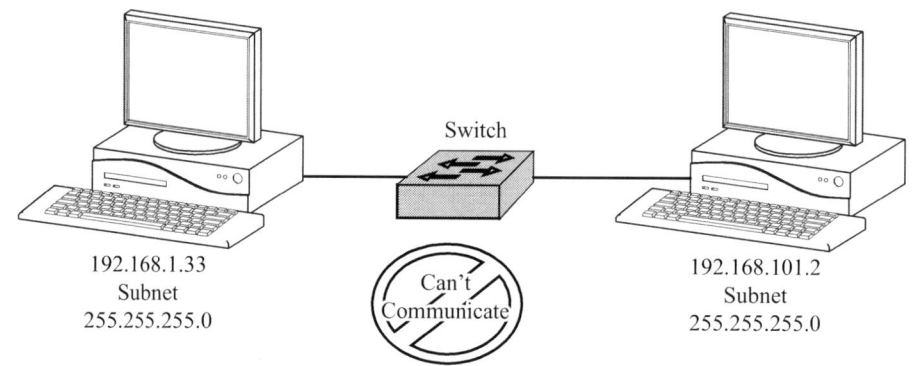

There are two easy solutions to this problem. First, you could change the subnet mask on both machines to `255.255.0.0`, meaning only the first two octets have to match:

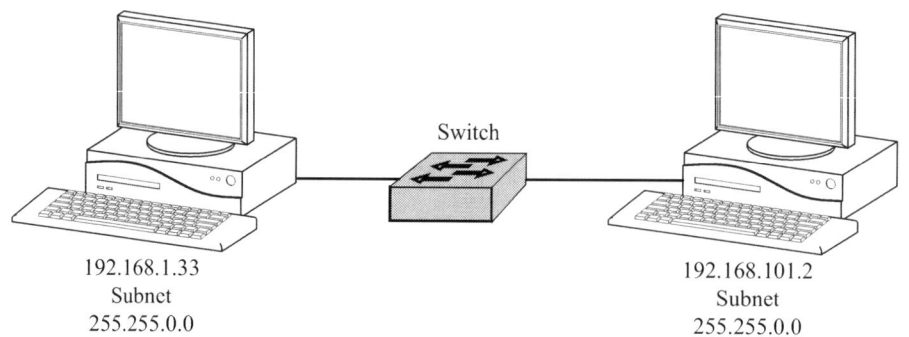

This approach means, though, that you effectively have one big `192.168.0.0` subnet, and you can't divide it further.

Alternatively, you could keep the original subnet mask of `255.255.255.0`, and change one host's IP from `192.168.101.2` to `192.168.1.2`:

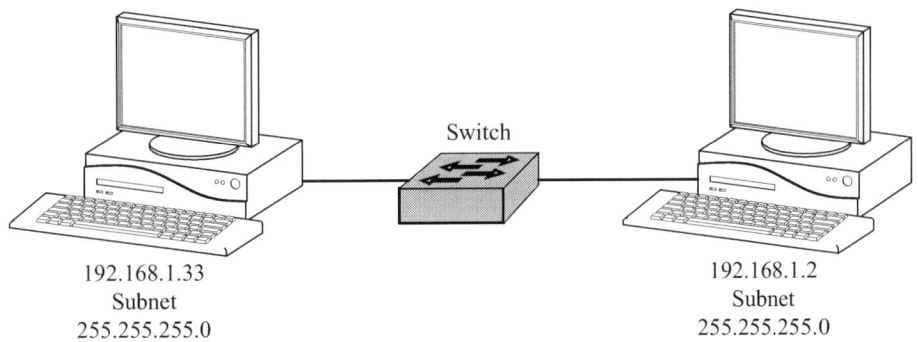

Or, you could change the other host from `192.168.1.33` to `192.168.101.33`:

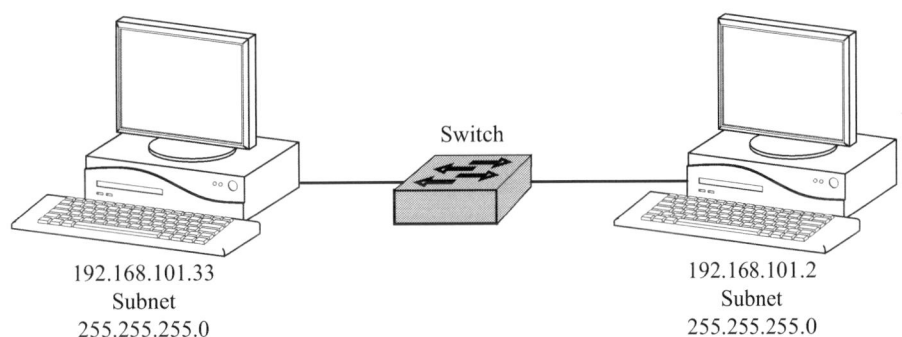

EXAMPLE NETWORK WITH TWO SUBNETS

The little lighting control network we built on page 191 is working great for our show; now it turns out we're going to be adding video, so we need to add three more hosts to our network: one video controller and two video display units, all of which run on standard PC's.

Chapter 17: Show Networks • 197

We'll update our IP address table:

Host	IP Address
Lighting Console	192.168.1.101
Lighting Data Distributor	192.168.1.102
Video Controller	192.168.2.101
Video Display 1	192.168.2.102
Video Display 2	192.168.2.103
Laptop (Testing Only)	192.168.1.111

You'll notice in the table that the new video system is on a different subnet: 192.168.2.x. Both the lighting and video systems use proprietary multicasting for their own communications, and the systems are not interoperable so there's no reason for these two systems to talk to each other. So to avoid any conflicts, and just to keep things organized, it's good practice to put them on two different subnets.

We assign the IP addresses, and now we have this:

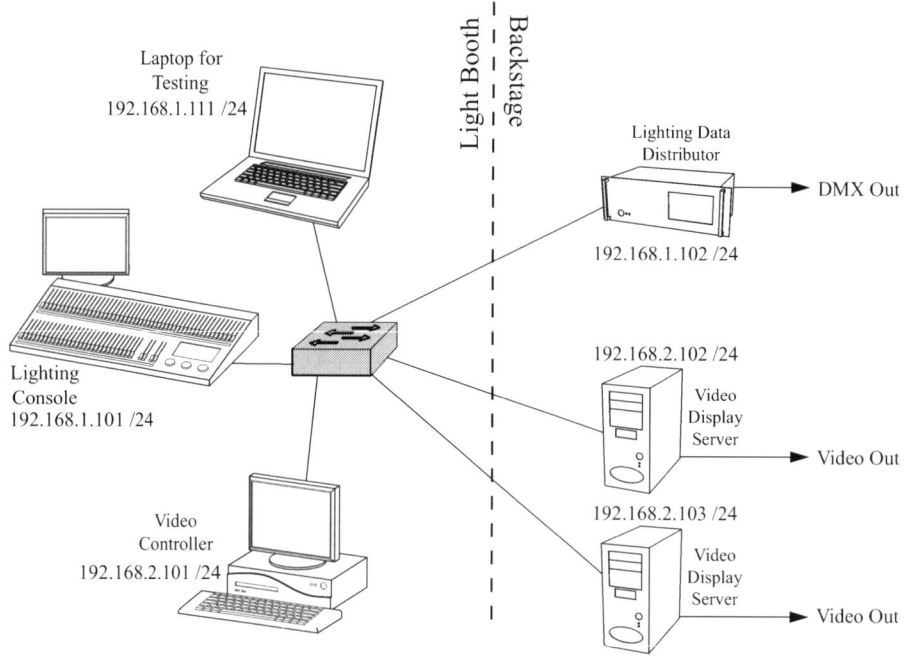

We've got everything configured so let's ping the new hosts from the laptop:

```
C:\>ping 192.168.2.101

Pinging 192.168.2.101 with 32 bytes of data:
Destination host unreachable.
Destination host unreachable.
Destination host unreachable.
Destination host unreachable.

Ping statistics for 192.168.2.101:
    Packets: Sent = 4, Received = 4, Lost = 0 0% loss),
```

Why is the destination host "unreachable"? It's plugged in and the Ethernet activity lights on all the connections are lit, so we know we have a physical connection. We try pinging the lighting console at `192.168.1.101`, and that works just fine. So that tells us our cable is good, the network is working, etc. So what's wrong?

It's one of the most common mistakes I see in network setup—our laptop is on the lighting `192.168.1.0` subnet, and we are trying to ping into the `192.168.2.0` subnet, with a subnet mask of `255.255.255.0`, which means that the first three octets of any host we are trying to reach must be exactly the same as the IP address of the machine from which you are sending the `ping` command. So we change the address of the laptop to `192.168.2.111/24`—moving it onto the video subnet—and now we can ping from the laptop to the video machines just fine:

```
C:\>ping 192.168.1.101

Pinging 192.168.1.101 with 32 bytes of data:
Reply from 192.168.1.101: bytes=32 time=8ms TTL=64
```

Physically Separated Topology

Let's address one final issue before moving on. The network as shown—with two IP subnets on a single switch—can work just fine. However, both this lighting system and the video system output a lot of multi-cast traffic and a simple, cheap switch might get overloaded handling all those packets and cause some delays or other glitches.

There's an easy solution to that—add another switch, and physically separate the networks:

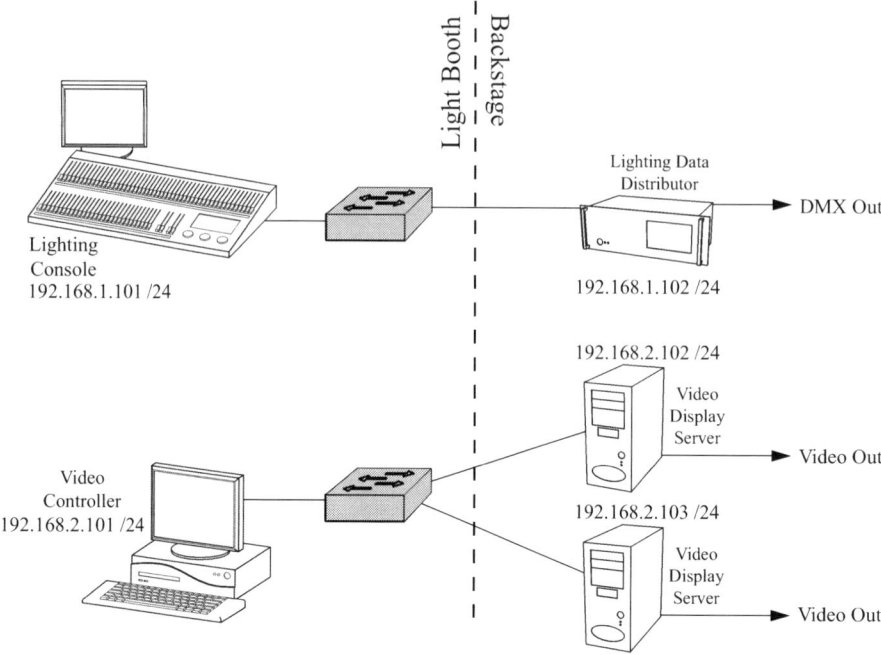

You could also buy a managed switch and implement this same system using "virtual" LANs (page 221).

ADDRESS RESOLUTION PROTOCOL (ARP)

IP addresses were designed for inter-networking purposes and, as we have seen, IP addresses can be assigned to machines either manually by the system administrator or automatically by a process such as DHCP. Lower-level network communications, however, need to communicate down at a lower hardware level, and for this purpose most machines have an additional unique physical "hardware address" such as the Ethernet MAC address (page 157). To resolve these physical addresses with IP addresses, the **Address Resolution Protocol** (ARP) was created.

To understand the basic ARP process, let's consider an example. A host on a private network needs to communicate with another host on the same private network. Unique IP addresses have been configured into both machines, but the system needs to find the *physical* address of the second machine in order to put that low-level address into the frame used for communication. Because the sending system doesn't know the physical address, the IP address is considered "unresolved," so the first machine sends an "ARP request" out onto the network. This ARP message is broadcast to all connected hosts, asking the owner of the target IP address to respond with its physical address.

Let's take a look at an actual message capture. A user at a Dell host at `192.168.1.104` pings an Apple host machine at `192.168.1.102`. Because the network was just turned on, the system doesn't yet know the Ethernet MAC address of the machine assigned the `192.168.1.102` IP address, so before it sends the ping messages, it sends an ARP request message out:

Time	Source	Destination	Protocol	Length	Info
1 0.000000	DellEsgP_88:e7:b6	Broadcast	ARP	106	Who has 192.168.1.102? Tell 192.168.1.104
2 0.000828	Apple_96:23:3a	DellEsgP_88:e7:b6	ARP	60	192.168.1.102 is at 00:23:32:96:23:3a

The first ARP request of "`Who has 192.168.1.102?`" is broadcast to the entire network segment because the sender has no idea of the Layer 2, Ethernet MAC address of the target. The Apple machine, which had been assigned the `102` address, now responds with, "`192.168.1.102 is at 00:23:32:96:23:3a`", which is the Apple's Ethernet MAC address in hex. Throughout this process, other hosts that see the ARP conversation can also note the same information "ARP cache" so they, too, will know the correlation between physical and IP addresses.

This ARP process takes less than a second, and the ping commands then go through.

Time	Source	Destination	Protocol	Length	Info
1 0.000000	DellEsgP_88:e7:b6	Broadcast	ARP	106	Who has 192.168.1.102? Tell 192.168.1.104
2 0.000828	Apple_96:23:3a	DellEsgP_88:e7:b6	ARP	60	192.168.1.102 is at 00:23:32:96:23:3a
3 0.000831	192.168.1.104	192.168.1.102	ICMP	74	Echo (ping) request id=0x0200, seq=1280/5,
4 0.000832	192.168.1.102	192.168.1.104	ICMP	74	Echo (ping) reply id=0x0200, seq=1280/5,
5 0.997754	192.168.1.104	192.168.1.102	ICMP	74	Echo (ping) request id=0x0200, seq=1536/6,
6 0.997756	192.168.1.102	192.168.1.104	ICMP	74	Echo (ping) reply id=0x0200, seq=1536/6,
7 1.997645	192.168.1.104	192.168.1.102	ICMP	74	Echo (ping) request id=0x0200, seq=1792/7,
8 1.997647	192.168.1.102	192.168.1.104	ICMP	74	Echo (ping) reply id=0x0200, seq=1792/7,
9 2.997533	192.168.1.104	192.168.1.102	ICMP	74	Echo (ping) request id=0x0200, seq=2048/8,
10 2.998470	192.168.1.102	192.168.1.104	ICMP	74	Echo (ping) reply id=0x0200, seq=2048/8,

ARP messages are also used when IP addresses are set to ensure that each host's IP address is unique on that network—having more than one host with the same IP address would cause all kinds of problems, and most operating systems will detect this and deliver to the user an error message.

For example, the simple network we laid out starting on page 187 (block diagram duplicated here) contains a Mac computer with the IP address of `192.168.1.102`.

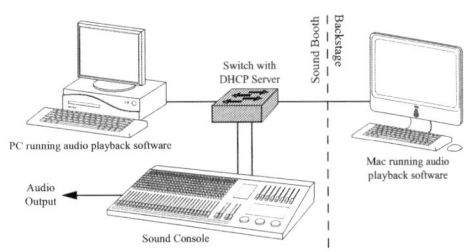

Chapter 17: Show Networks • 201

If I configure a different machine with that same IP address and connect it to the same network, the newly configured host will send a "Gratuitous ARP" for 192.168.1.102:

```
    Time         Source              Destination         Protocol  Length  Info
12  22.160020    DellEsgP_88:e7:b6   Broadcast           ARP       106     Gratuitous ARP for 192.168.1.102 (Request)
13  22.160022    Apple_96:23:3a      DellEsgP_88:e7:b6   ARP       60      Gratuitous ARP for 192.168.1.102 (Reply)
14  22.160023    Apple_96:23:3a      Broadcast           ARP       106     Gratuitous ARP for 192.168.1.102 (Request) (dupli

Frame 14: 106 bytes on wire (848 bits), 106 bytes captured (848 bits)
Ethernet II, Src: Apple_96:23:3a (00:23:32:96:23:3a), Dst: Broadcast (ff:ff:ff:ff:ff:ff)
[Duplicate IP address detected for 192.168.1.102 (00:23:32:96:23:3a) - also in use by 00:0b:db:88:e7:b6 (frame 13)]
Address Resolution Protocol (request/gratuitous ARP)
```

When the Apple computer sees the Gratuitous ARP asking about its own address, it sends back a "192.168.1.102 is at 00:23:32:96:23:3a" (message details omitted here) and then the system immediately detects a "duplicate IP address" error. The PC on which I'm trying to change the IP address indicates to me, "The static IP address that was just configured is already in use on the network. Please reconfigure a different IP address". This process ensures that every IP address on a network segment is unique.

ARP Command

The `arp` command gives you the ability to view the ARP cache, showing you which IP addresses are resolved to which physical (MAC) address. In Windows, typing `arp -a` will show you the current ARP cache.

Here's an example from a small home network:

```
C:\>arp -a

Interface: 192.168.1.101 --- 0x2
  Internet Address        Physical Address        Type
  192.168.1.1             00-18-f8-7e-98-3e       dynamic
  192.168.1.100           00-0c-ce-6d-d2-fc       dynamic
  192.168.1.102           00-09-5c-1c-cd-78       dynamic
  192.168.1.107           00-0f-35-ae-18-52       dynamic
```

It's also possible, using the ARP command, to clear the arp cache, remove specific computers, and so on. Type `arp` with no arguments to get a complete list of possibilities.

PORTS AND SOCKETS

While IP can provide a unique address for a specific host on a network, a single machine could, of course, have multiple, separate software processes running on a single host, each of them communicating simultaneously and separately with other machines on the network. For example, you might use your computer to surf the Web while in the background your e-mail program checks for e-mail, and another process simultaneously downloads a podcast. To allow this type of operation, **ports** were developed, which are simply identifying numbers assigned to particular software processes. Ports can be either formally defined or configurable; a **socket** in networking terminology is the combination of an IP address and a port number.

The port number is embedded in messaging header information to further steer a particular packet and is used in both TCP and UDP. In the example above, your machine would be using (in addition to dozens of others) port 80 for the Web browser running HTTP, while in the background your e-mail client uses port 143. Port numbers are typically in one of three ranges: "Well known" ports, which are registered with the **Internet Assigned Numbers Authority** (IANA) run in the range between 0 through 1023; "registered" ports, which are also assigned by the IANA, go from 1024 through 49151; and finally, port numbers ranging from 49152 through 65535 are called "dynamic" or private.

Here's a list of some common well known port numbers.

Port Number	Used By	Protocol
20	TCP, UDP	File Transfer Protocol (FTP)
25	TCP, UDP	Simple Mail Transport Protocol (SMTP)
37	TCP, UDP	Time Protocol
43	TCP, UDP	Whois
53	TCP, UDP	Domain Name System (DNS)
67	UDP	BootStrap Protocol (BOOTP) and also Dynamic Host Configuration Protocol (DHCP)
68	UDP	BootStrap Protocol (BOOTP) and also Dynamic Host Configuration Protocol (DHCP)
80	TCP	Hyper Text Transfer Protocol (HTTP)
110	TCP	Post Office Protocol version 3 (POP3)
123	UDP	Network Time Protocol (NTP)
143	TCP, UDP	Internet Message Access Protocol 4 (IMAP4)

Port Number	Used By	Protocol
161	TCP, UDP	Simple Network Management Protocol (SNMP)
443	TCP, UDP	HTTPS – HTTP Protocol over TLS/SSL (encrypted transmission)
546	TCP, UDP	DHCPv6 client
547	TCP, UDP	DHCPv6 server

netstat Command

`netstat` is another helpful utility that can give you a variety of information on what network connections are in place on a particular machine and what ports are in use. Typing `netstat -a` on a PC will give you a complete list, including the protocol (TCP or UDP in this case) and the port number. There are many options for the netstat command; type `netstat ?` for a complete list.

TESTING NETWORKS

The traffic of the network is pretty much invisible without the use of some sort of diagnostic tool. There are many useful tools and techniques on the market; here are a couple that I use.

Wireshark

One of the most widely used network troubleshooting tools is free, multi-platform, open source **Wireshark**[8] Wireshark can capture any packets on a network segment and then present them in a time-stamped list, sorted out by protocol. You've already seen Wireshark in action, since I created the packet captures in this book using the analyzer. It's pretty easy to get Wireshark up and running, and you can learn a lot just by looking through the traffic, even if you don't know what all the protocols do.

Back in the old days using hubs, we were able to view all the network traffic simply by running Wireshark on any machine connected to the network—the hub forwarded all traffic to every connected interface. A switch, however, only forwards packets to an interface when it thinks that interface should get them. So, if you can run Wireshark on the machine that is receiving the traffic you are concerned with, then that will work just fine.

---

8. http://www.wireshark.org/

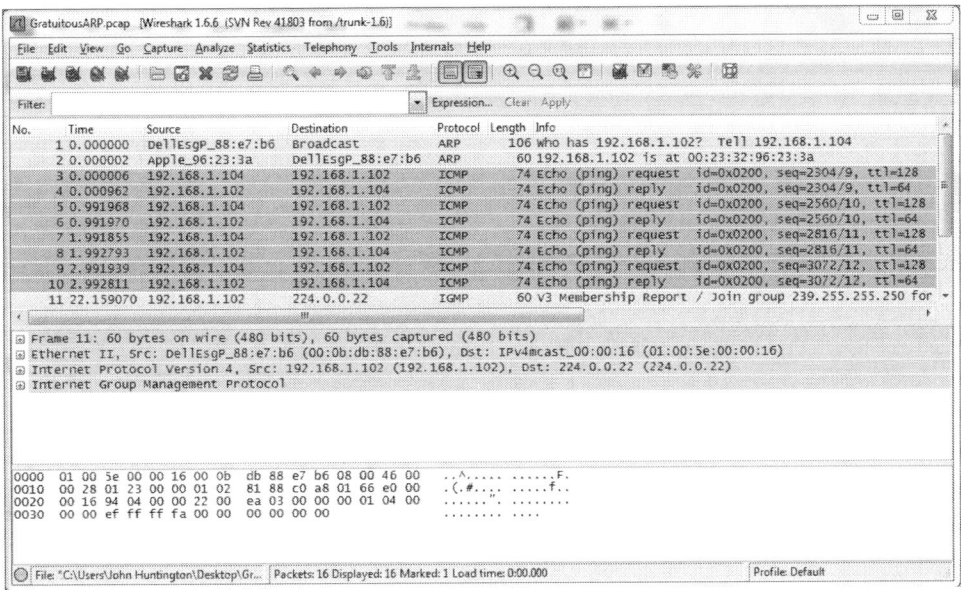

But if that's not possible, another technique would be to connect your target machine and your analysis machine (I generally use my laptop) using a hub, and then connect the hub to the network switch. Another technique is to use **Port mirroring**, which copies all traffic from one interface to another, where it can be monitored. For example, if you're having problems on the video control computer connected to interface #3, you could mirror that traffic to interface #8 and use Wireshark to do troubleshooting on a separate machine connected there.

Essential Net Tools

Essential Net Tools[9] is a commercial product for Windows that I find to be very useful, especially for testing communications with a new target device I've never controlled before.

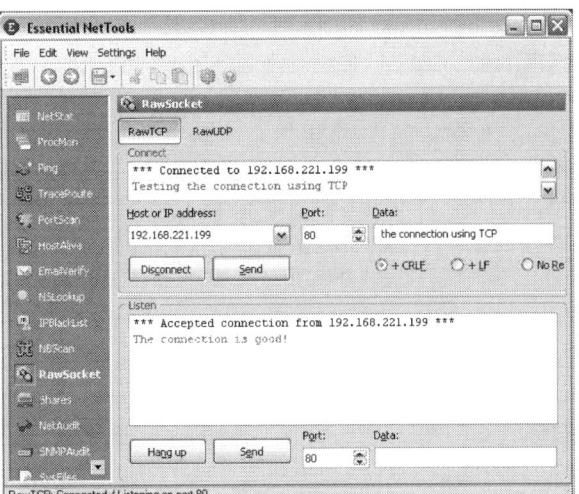

Tera Term

Tera Term,[10] another free tool, can also do testing of network connections; you can see it in action in "Serial Connection Example," on page 152.

PuTTY

Another popular program is PuTTY, which is freely available[11] and can either run the telnet protocol on port 23 or send raw ASCII to any other port. This is particularly useful for initial configuration of network controlled devices, troubleshooting, and diagnostics, since you can emulate any system by typing the protocol you wish to emulate. An example using PuTTY follows in the next section.

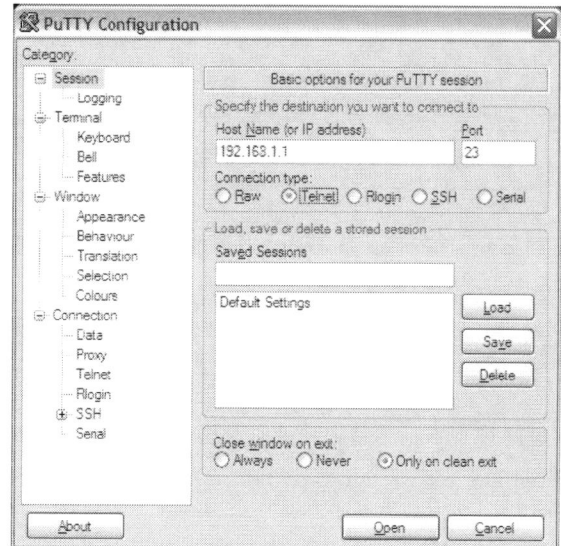

TelNet Command

Telnet (TELetype NETwork) uses TCP port 23 as a way to establish virtual client-server (page 82) "terminal" sessions over a network, enabling a user to type and receive characters just as if he or she were sitting at the remote computer. The telnet protocol can operate in three modes: half-duplex (rarely used), one character at a time, or one line at a time. To allow interoperability over a wide variety of systems, terminal types, or "emulations", are used since in the old days, each terminal manufacturer would map out their own way of dealing with characters and keys. Telnet is both a protocol and a program, so people often get confused between the two, although telnet programs will typically default to the telnet protocol port 23 and use the telnet protocol. However, you can also communicate using ports other than port 23, and this makes some telnet programs useful as a general, low-level diagnostic tool. So, someone saying "I telnet'd to that piece of gear over the network" might actually be saying that they used the telnet *program*

---

9. http://www.tamos.com/products/nettools/

10. http://ttssh2.sourceforge.jp/index.html.en

11. http://www.chiark.greenend.org.uk/~sgtatham/putty/

to send and receive raw ASCII to and from a remote device, with the characters packaged up into TCP/IP messages by the telnet program and using some port other than 23.

Windows and other OSs can be enabled to run a version of telnet (as of Windows 7, you have to install it separately); type `telnet` from the command line to start it. However, (in Windows anyway) this telnet program is limited, so most people prefer to use another program, such as PuTTY or Tera Term. Below is a communications exchange with a computer-based sound server device,[12] done through PuTTY. This kind of exchange can be done manually, for testing and configuring, or as part of a cue or other automated exchange.

```
OK                      *********************************
OK                      ** Welcome to Soundman-Server! **
OK                      *********************************
config get interfaces
Interfaces 2
Interface 0 "ASIO 2.0 - Maya 7.1"
Interface 1 "MOTU FireWire Audio"

OK
config set interface 1 inputs 18 outputs 18 playbacks 200;
OK
config get interface
InterfaceInfo 1 "MOTU FireWire Audio" Inputs 18  Outputs 18
DefaultSampleRate 48000

OK
set chan p1 track file "T:\Fountain\Left Main.wav"
OK
set chan p2 track file "T:\Fountain\Right Main.wav"
OK
set chan p3 track file "T:\Fountain\Left Surround.wav"
OK
; now set up the matrix for playback
OK
set chan p1-p3 px1.1 px2.2 px3.3 px 4.4 px5.5 px6.10 px7.12 gain 1
OK
set chan out1-7 gaindb -3 chan o10 o12 gaindb 0
OK
; setup is complete, can now play whenever we want to
OK
play p1-p3
```

---

12. Thanks to Loren Wilton for providing this exchange with Soundman-Server, an audio sound server distributed by Richmond Sound Design.

```
OK
get chan p1-3 status
Channel Playback 1 Status Playing, Time 00:01:12.045
Channel Playback 2 Status Playing, Time 00:01:12.045
Channel Playback 3 Status Playing, Time 00:01:12.045
OK
```

Traceroute Command

Traceroute (`tracert` on Windows) is a very useful troubleshooting command, especially if you are having issues connecting to a distant host and you are trying to track down a source of delays. Since tracing the route of our small, closed network isn't very interesting, let's take a look at a `tracert` command from my computer in Brooklyn to Portland, Oregon's city Web page at `portlandonline.com`:

```
C:\>tracert portlandonline.com

Tracing route to portlandonline.com [74.120.152.100]
over a maximum of 30 hops:

1   1 ms    <1 ms   <1 ms   192.168.1.1
2   6 ms    13 ms   7 ms    gig-2-11-nycmnys-rtr1.nyc.rr.com
                            [25.30.151.117]
3   16 ms   11 ms   11 ms   184-152-112-73.nyc.rr.com [185.151.113.73]
4   8 ms    10 ms   23 ms   bun6-nyquny91-rtr002.nyc.rr.com
                            [24.29.148.254]
5   14 ms   15 ms   15 ms   ae-3-0.cr0.nyc20.tbone.rr.com [66.109.6.76]
6   9 ms    9 ms    9 ms    107.14.17.169
7   9 ms    8 ms    11 ms   xe-9-0-0.edge2.Newark1.Level3.net [4.59.20.29]
8   8 ms    12 ms   8 ms    ae-31-51.ebr1.Newark1.Level3.net [4.69.156.30]
9   9 ms    9 ms    10 ms   ae-2-2.ebr1.NewYork1.Level3.net [4.69.132.97]
10  9 ms    10 ms   9 ms    4.69.141.18
11  10 ms   8 ms    9 ms    ae-1-100.ebr2.NewYork2.Level3.net
                            [4.69.135.254]
12  45 ms   43 ms   49 ms   ae-2-2.ebr1.Chicago1.Level3.net [4.69.132.65]
13  39 ms   41 ms   48 ms   ae-6-6.ebr1.Chicago2.Level3.net [4.69.140.190]
14  67 ms   73 ms   74 ms   ae-3-3.ebr2.Denver1.Level3.net [4.69.132.61]
15  91 ms   90 ms   91 ms   ae-2-2.ebr2.Seattle1.Level3.net [4.69.132.53]
16  92 ms   225 ms  219 ms  ae-21-52.car1.Seattle1.Level3.net
                            [4.69.147.163]
```

```
1794  ms   95 ms   95 ms  EASYSTREET.car1.Seattle1.Level3.net
                          [4.53.146.70]
1895  ms  112 ms  96 ms   209.162.220.74 Portland, OR)
19 102 ms  96 ms  97 ms   69.30.59.54 Beaverton, OR)
20 104 ms 103 ms 104 ms   2.152.ptldnet.portlandoregon.gov
                          [74.120.152.2]
21 104 ms 102 ms 107 ms   www.portlandonline.com [74.120.152.100]
Trace complete.
```

From the `tracert` results, you can see the general path between the two machines. The first thing you see is that the `portlandonline.com` URL resolves to the actual IP address of `74.120.152.100`. Packets then travel to my network's default gateway of `192.168.1.1` bounce around within my cable provider's network, then go from NYC to Newark, back to NYC, then to Chicago, Denver, Seattle, Portland, Beaverton, and then finally to the Portland Web server.

I then pinged the `74.120.152.100` address directly:

```
C:\>ping 74.120.152.100

Pinging 74.120.152.100 with 32 bytes of data:
Reply from 74.120.152.100: bytes=32 time=104ms TTL=233
Reply from 74.120.152.100: bytes=32 time=104ms TTL=233
Reply from 74.120.152.100: bytes=32 time=104ms TTL=233
Reply from 74.120.152.100: bytes=32 time=102ms TTL=233

Ping statistics for 74.120.152.100:
    Packets: Sent = 4, Received = 4, Lost = 0 0% loss),
Approximate round trip times in milli-seconds:
    Minimum = 102ms, Maximum = 104ms, Average = 103ms
```

It never ceases to amaze me, even after doing this for so many years, that the average time for a packet to be routed and passed across the country and back is only 104 milliseconds!

MOVING ON

The basics we have covered is this chapter will allow you to build a large variety of networks useful for all kinds of different applications on shows. But there are additional things to cover for larger networks, so the next chapter outlines the basics of more advanced show networks.

Chapter 18: Advanced Show Network Topics

Before moving on to more sophisticated network systems, techniques, and practices, we need to expand on some of the basic concepts we covered in Chapter 16, "Networking Basics" and Chapter 17, "Show Networks". To do so, let's take a look at traffic forwarding on a simple network[1] that connects together seven hosts using a couple switches, an old hub, and a router.

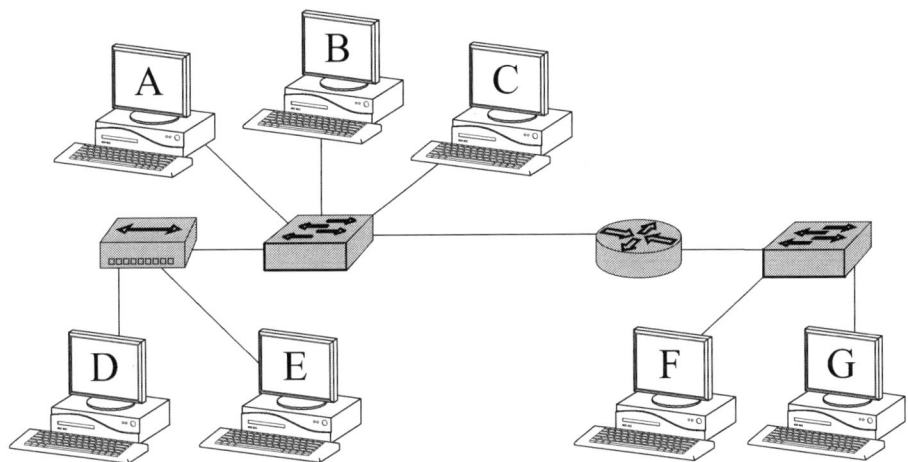

A couple notes: the hosts here are depicted as generic computers, but they could be lighting or sound consoles, machinery control systems, computers running video server software, or just about anything else with an Ethernet jack and an IP address. In addition, hubs are generally obsolete for show networks, and routers are pretty rare in our field (as of this writing, anyway). But it's important to have at least a basic understanding of each of these devices and the impacts they have on network design, so I'm including them here.

---

1. See Chapter 16 if you need a review of the symbols in the diagram.

Let's say that host A wants to unicast a message to host C. It sends the message out its Ethernet port, and the switch forwards the packet to host C:

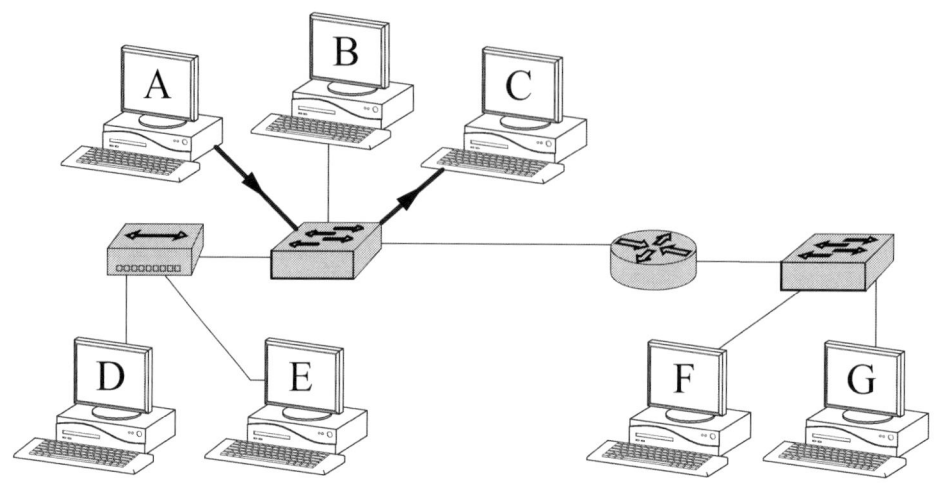

Next, let's say that host A wants to unicast a packet to host E:

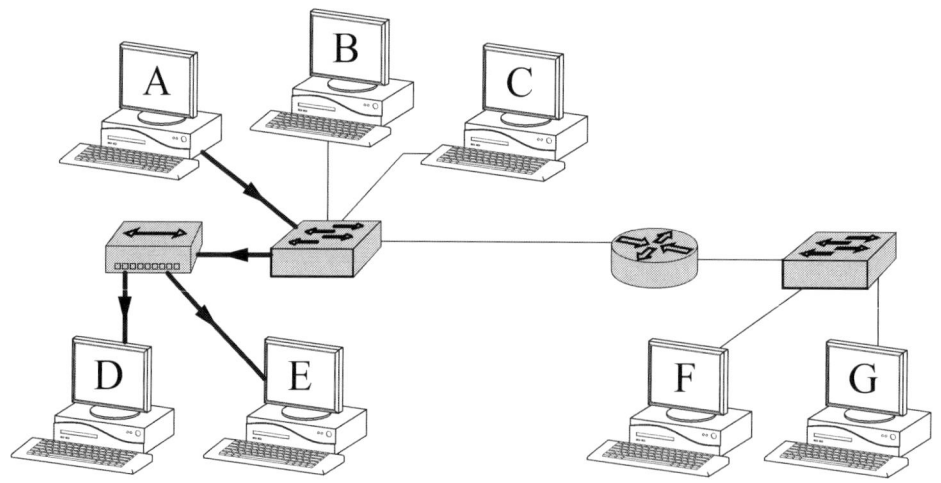

This message was unicast only to E, so why is a copy also forwarded to host D? Hosts D and E are connected to a hub, which is connected to the switch. The hub is a dumb device—all it can do is send out anything that comes in to every other connected host.

Now let's say that host A wants to send a packet to host G.

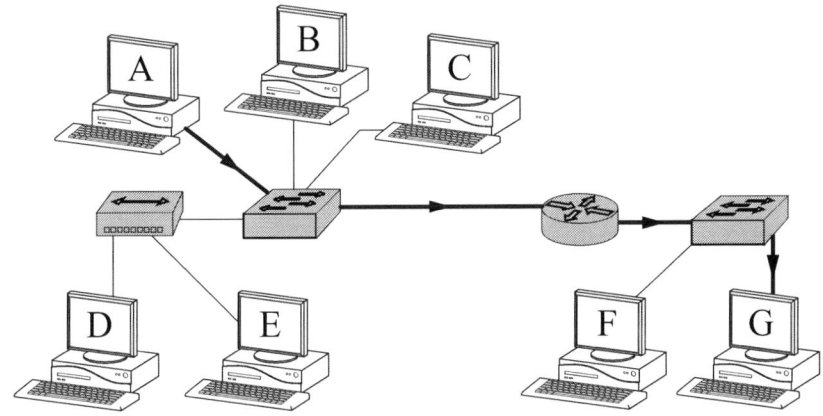

Host A sends its packet to the switch, which then forwards it to the router; the router understands the address of host G, and knows that it's on another network, so it sends it out in that direction; the switch connected to host G then makes the final transmission.

BROADCAST DOMAIN

In my simplified example above, I left out the Layer 3 IP addresses, and assumed that all the hosts had already communicated at least once with each other, populating the system's ARP tables (page 200). So let's add IP addresses, subnet masks (page 194), and look at the network as if it had just been turned on:

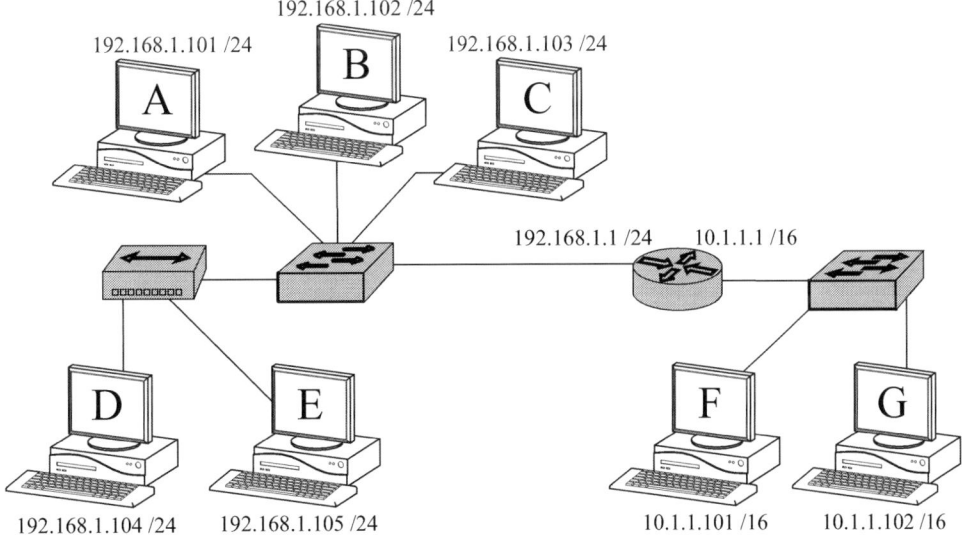

Chapter 18: Advanced Show Network Topics • 213

We can see here that on the left side of the block diagram, we have one network with five machines connected, assigned non-routeable Class C addresses (`192.168.1.x`), and a subnet mask of `/24`. This means, in simple terms, that the first three octets of the address of each host on this network must match exactly in order for those machines to communicate. All five machines are correctly assigned to this subnet.

On the right hand side of the diagram is a router, which connects to a separate, larger network (we can't see all of it here). This network is using Class A addresses in the range of `10.1.1.x`/16. Notice that the router itself has two addresses assigned to it, and remember that routers are designed to connect networks together. Here the router is connecting the `192.168.1.0/24` network with the `10.1.0.0/16` network (more in "Layer 3 Routing," on page 220).

Now let's go back to where I said that host A, with an IP address set to `192.168.1.101`, wanted to send a message to host C at `192.168.1.103`. We just turned the system on, and no traffic has been passed, so it doesn't know the physical address of host C.

What does host A do? It sends an ARP command, to find the Layer 2, physical address of the device with the Layer 3 IP address of `192.168.1.103`:

Source	Destination	Protocol	Length	Info
DellEsgP_88:e7:b6	Broadcast	ARP	106	Who has 192.168.1.103? Tell 192.168.1.101

Where does it send it? It sends ARP messages out to the broadcast address, meaning the message will reach this network's "broadcast domain":

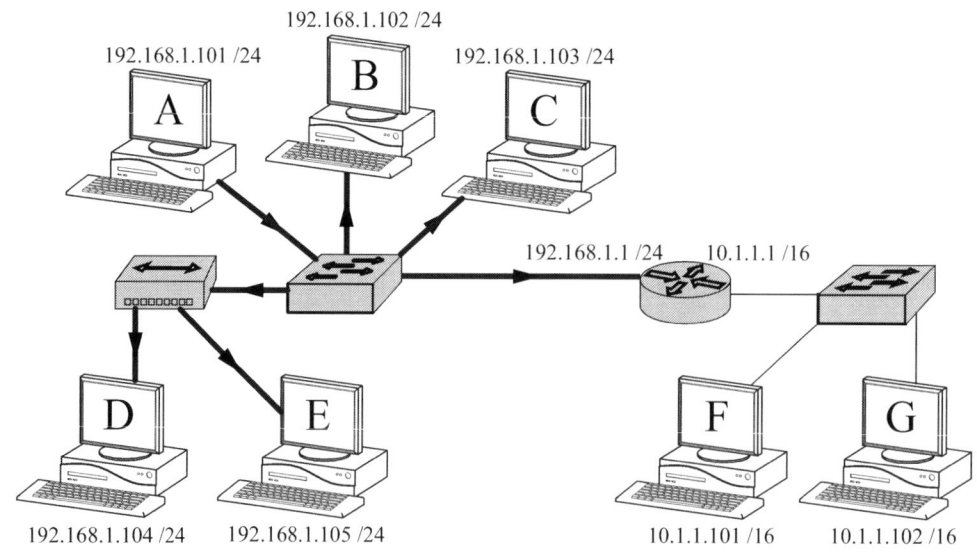

What is a **broadcast domain**? It's simply a part of a network (or all of a smaller network) where all the connected nodes in the domain can reach each other using broadcast messages and Layer 2, Data Link Layer communications.

Take a look at the diagram again. Why don't the hosts on the `10.1.1.0` network see the broadcast? Because routers do not forward broadcast messages, and for good reason. Can you imagine if every machine on the Internet got all the broadcast transmissions from every other machine in the world? Segmenting the network is important for management of traffic, security, and maximizing bandwidth, but also requires Layer 3 techniques to forward traffic between the networks.

Back to our example. The ARP message comes back from host C, which received the broadcast, to host A, carrying its Layer 2 MAC address.

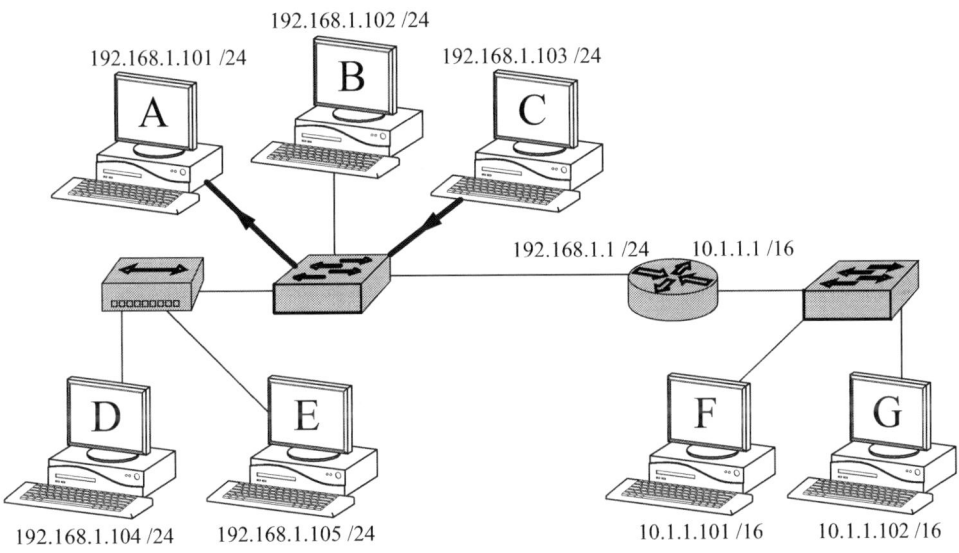

Now host A has the physical MAC address of host C, and it can start its communications process. Keep in mind, while we've been talking here about Layer 3 IP addresses, the switches don't know or care about IP addresses—they've been working with Layer 2 MAC addresses the entire time.

NETWORK TOPOLOGY AND BROADCAST STORMS

In the mid 1990s when Ethernet started gaining acceptance in our industry, we built LANs using hubs, which operate at OSI Layer 1 and (as we discussed on page 170) simply broadcast every incoming frame to every other device on the network. To expand the LAN, we would just connect a second hub to the first, and so on, until we hit some limitations.[2] This simple scheme was effective and easy to use, but wasted bandwidth and had problems with data collisions that degraded network performance (page 163). Fortunately, we are no longer likely to encounter hubs in our market because full-duplex switches (page 170) operating at OSI Layer 2 became so powerful and so cost effective, ranging from inexpensive units that configure themselves automatically for simple applications to complex, powerful units that can incorporate routing and other functionality.

With unmanaged switches and straightforward, closed show networks, as long as you keep within the 100 meter distance limitation for individual segments, you can generally still expand a network by simply adding additional switches.[3] However, with unmanaged switches there is one important thing to keep in mind: you have to avoid loops or redundant pathways between switches. Why?

A switch only has three options for an incoming frame: it can **forward** it on to another interface, based on its Layer 2 destination address; it can discard it if there's a problem; or it can **flood** any frame sent to the broadcast address, or to an address it doesn't already know and have in its forwarding tables. Flooding typically sends a copy of the frame out to all interfaces except the one on which the frame arrived; this prevents frames from bouncing back and forth if two switches are connected. However, duplicate pathways between switches can end up overriding this safeguard. How does this happen? Let's look at an example.

A new, eager technician puts together a network made up of two unmanaged switches, A and B, which connect a video controller (host A in the diagram) to a

---

2. The limits included the "5-4-3" rule, which managed collision domains and network diameters, and said you could have a maximum of five segments tied together with up to four repeaters, with no more than three segments containing active senders. Fortunately, in our market you are not likely to even encounter a hub (we use switches), so we don't have to worry about this much anymore. I will also skim over related issues with the "collision domain," since these days we use full-duplex switches almost exclusively.

3. In larger "enterprise" networks, switches are usually not connected directly to other switches of the same type, but instead connect in a hierarchy with hosts connected to "access" switches, which connect up the line through "distribution" switches to high performance "core" switches, which typically connect to a router. In a typical show network (as of this writing, anyway), the use of this kind of hierarchy is rare.

sound FX playback system (host B). The technician got the system working with a single connection cable between the switches, but then figured that running a second cable between the switches would give redundancy and double the bandwidth, which would be especially useful when transferring large media files between the two systems. So he connects a second cable from switch A switch B, and tells the video controller to send a (unicast) "Next cue Go" message to the sound FX system via switch A. Switch A now has two valid pathways to send this frame to the sound FX machine, so it sends copies of the frame out both (indicated by different line styles in the diagram):

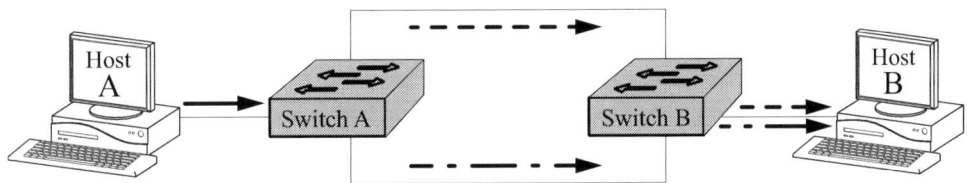

Switch B now gets both messages and forwards them to host B. Receiving two "Next cue Go" messages, the sound FX system executes the next two cues when you only wanted to execute one.

As bad as this situation is, though, it likely would never happen, because with the two direct connections between the unmanaged switches you have a loop, and something far worse would have already incapacitated the network: a broadcast storm.

Broadcast Storm

Take another look at the diagram above. Let's say the system has just been fired up, and the video controller (host A) needs to reach the sound FX system (host B) to send the control message, but it only has host B's Layer 3 IP address. To find the Layer 2 MAC address needed by the network, the operating system of host A sends out a broadcast ARP message (as we discussed on page 213). Switch A then floods this broadcast frame to every other interface, and because there are two connections between the switches, switch B now receives two identical, incoming broadcast ARP request messages. Switch B does the only thing it knows to do: it floods both broadcast frames to all its interfaces. One of those interfaces is connected to host B, which will receive both ARP requests and probably just respond twice, which likely wouldn't be a problem.

Chapter 18: Advanced Show Network Topics • 217

But what about the connections heading back to switch A? The switch, as we discussed, is smart enough not to return frames back to the incoming interface, but because we got two frames on *different* interfaces, switch B sends the frame that arrived on the top connection in the diagram to the bottom connection, and vice versa:

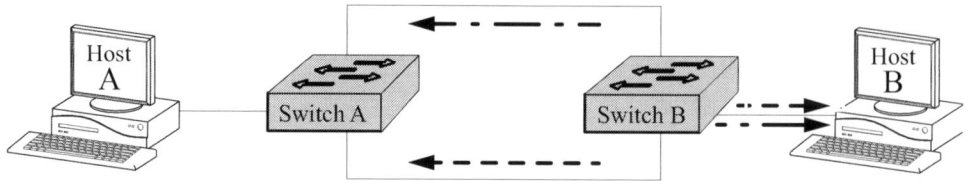

Switch A now sees two broadcast frames coming in on two interfaces that need to be flooded, so it takes the frame coming in on the top connection and as part of its flood sends it to the bottom connection in the diagram, closing the loop back to Switch B, and vice versa.

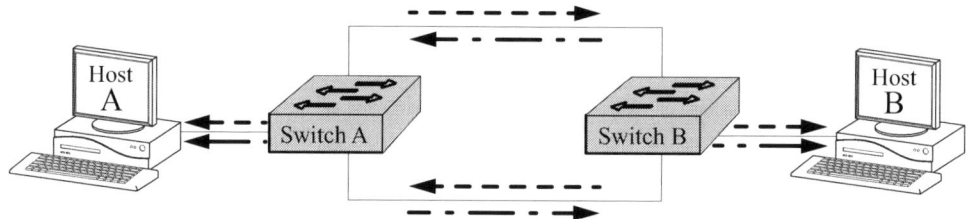

This is called a **broadcast storm**, and the frames could end up looping around the system indefinitely, causing all kinds of problems. To avoid broadcast storms in simple systems, loops and redundant pathways must be avoided. But in our show-must-go-on industry, we often want redundancy, so how could we implement it in a network?

Spanning Tree Protocol (STP)

To shut down redundant links before storms could happen, the **Spanning Tree Protocol** (STP) was developed; more advanced (and most managed) switches typically come with STP turned on by default. The problem with STP—developed as a precautionary measure to avoid traffic storms—is that it can take some time to "converge" after the network topology is changed, either intentionally or through a failure. The Rapid Spanning Tree Protocol (RSTP) speeds up this process, and can "converge" a network within a few seconds, meaning that we can actually use it to our advantage to manage redundant connections.

DEFAULT GATEWAY

Let's go back and take a look at our example network from above. I showed a packet going from host A to host G, which we now know is on a separate network with completely different IP addresses:

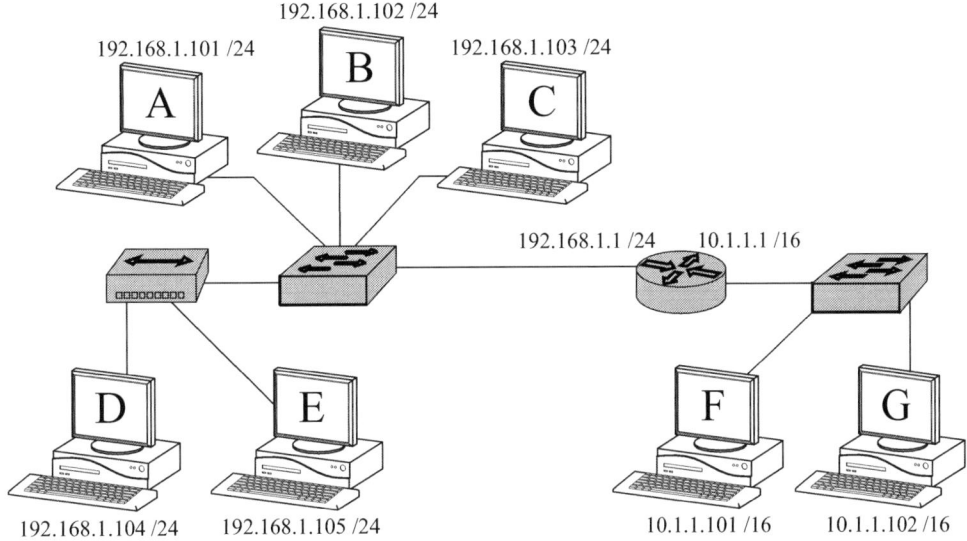

How do packets get from one network to another? Through the use default gateway and Layer 3 routing.

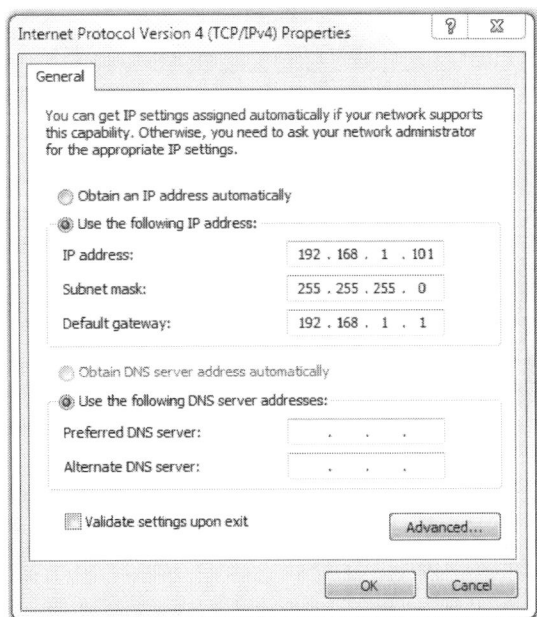

A **default gateway** (or "router address" on a Macintosh) is simply the configured IP address to which a host will forward packets for which it can't otherwise find a route. In this case, the gateway out of our `192.168.1.0` network is `192.168.1.1`, which is an interface on the router. So, in order to tell host A where to send these packets, we configure the default gateway to this address (see screen capture).

Now, when host A attempts to send a message to host G at

Chapter 18: Advanced Show Network Topics • 219

10.1.1.102, the sending machine will immediately realize that this address is not in its subnet, and it will instead send the message to its default Gateway of 192.168.1.1. What happens next? The packet is routed forward towards the target.

LAYER 3 ROUTING

Routing can be complex; entire books are written on the subject. But it's important to understand it at least in basic terms. Simply put, a router (page 173) connects networks together, and in this example network, the router has at least two physical Ethernet interfaces (there could be more not shown in our diagram), with one interface physically connected to each network, and each assigned an appropriate IP address/subnet mask in the software configuration of the router.

The router keeps a **routing table** telling it where to find (in this case, by physical interface) each network. When the packet comes in on the 192.168.1.1 interface destined for 10.1.1.102, the router looks in its routing table and then it knows exactly where to send it:

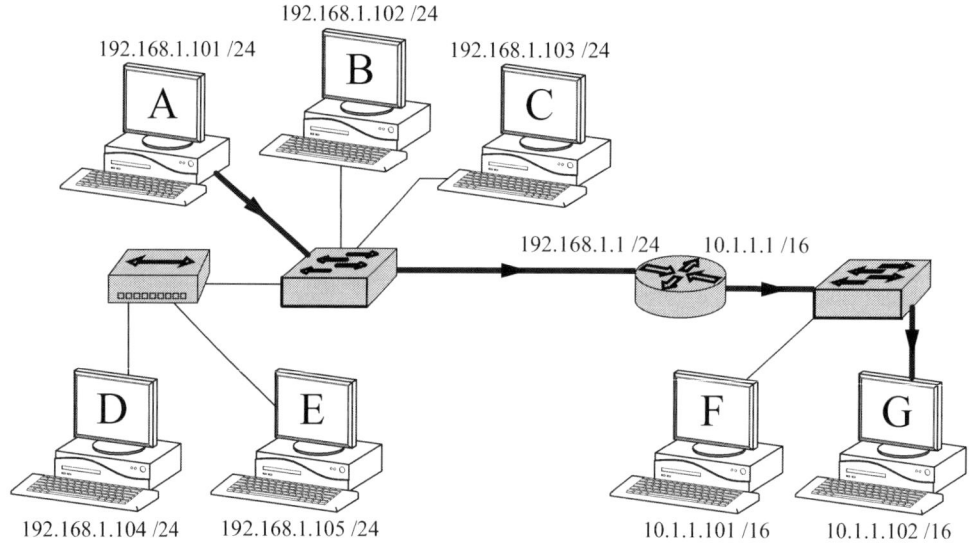

In this case, the message from A to G was sent using TCP, so host G now has to acknowledge the TCP transaction back to 192.168.1.101. But it knows that this target IP address is outside its network, so where does it send it? To its own default gateway for the 10.1.0.0 network: 10.1.1.1.

When it receives the packet, the router looks in its routing table, and forwards the packet back to the appropriate interface:

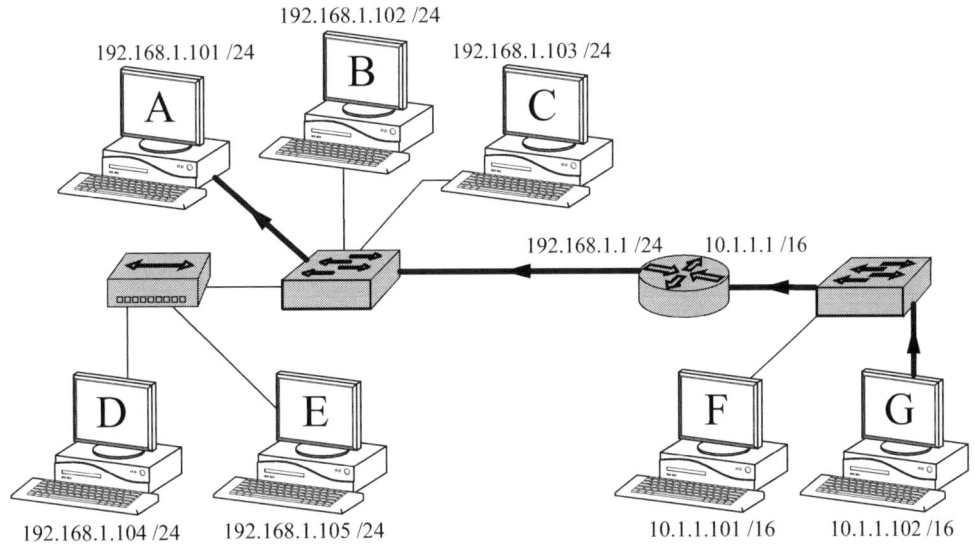

Keep in mind that most switches don't understand the IP addresses, operate at a low level, and work very fast. Routers, with all the processing they have to do, are inevitably a bit slower (although still very fast!); this is something to keep in mind when designing your network. You can literally study routers for years and still have more to learn; see my website for more resources, and the "Example Network with VLANs and Managed Switches," on page 226, for an example of routing used on a real show.

VIRTUAL LANS (VLAN)

It's often good network design to separate different types of traffic. For example, let's say on a show (as in the example on page 199), we have lighting and video controllers at front of house; backstage we need two lighting data distributor units for lighting, and two video display servers. While the video system uses a modern and efficient multicast data distribution approach, the lighting system is using an older system that simply broadcasts all its control data.

We'd like to keep the traffic segregated, so the broadcast lighting protocol traffic won't interfere with the video control protocol, and vice versa. We can simply get two sets of switches[4] and physically separate the networks:

---

4. We could build these networks each with a single switch, but I've used two here to leave room for future expansion at front of house.

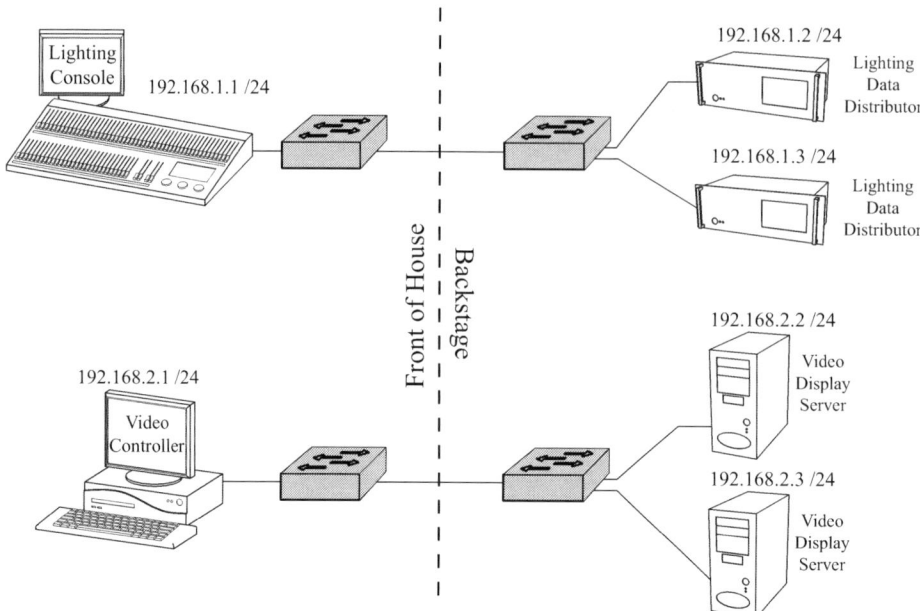

We now have two completely separate broadcast domains, meaning the video system will never see the lighting traffic:

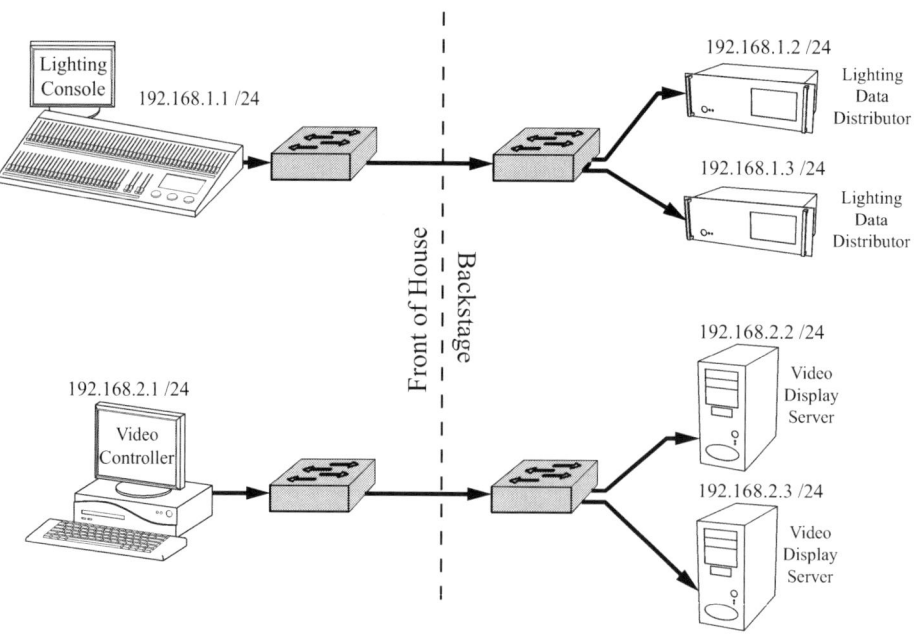

With this design, we need to purchase four switches, and manage two sets of cables, etc. But we are very good at cable management in our industry, so this isn't hard to do, and this approach gives us some redundancy (always a good thing). With the ever-dropping prices of network hardware, the cost for implementing this approach for a small network is likely pretty reasonable, and this physically separated approach may be the best solution for a small, straightforward system.

However, what if we need to scale this up to span a larger system, a large facility, or even a theme park, where it might be expensive and a maintenance burden to manage multiple, physically separated networks? Let's look at another possible approach, putting all the connected devices onto a single physical network:

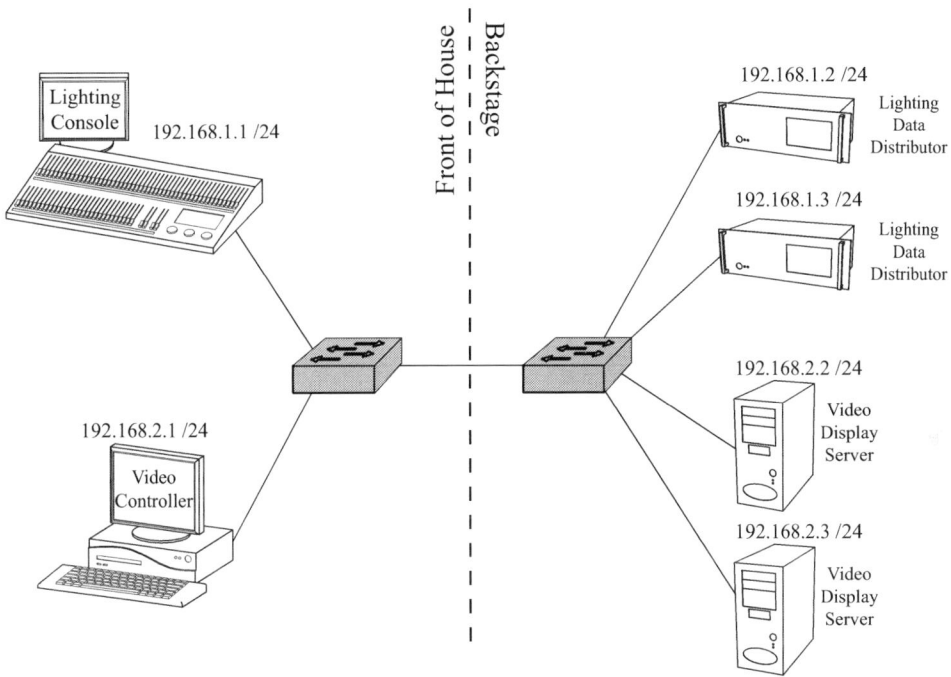

Take a look at the IP addresses. Because we used good networking practices in building our existing network, lighting is on one subnet, video is on another. Lighting won't be able to communicate with the video machines, and vice versa. But what about the broadcast domain?

Remember, standard switches operate on Layer 2 and do not even understand Layer 3 IP addresses. As we mentioned, the video system here uses a modern multicast data approach and the lighting console is broadcasting its data. That means that each of the switches will forward every frame of lighting control data (and there is a lot of it) to all connected interfaces.

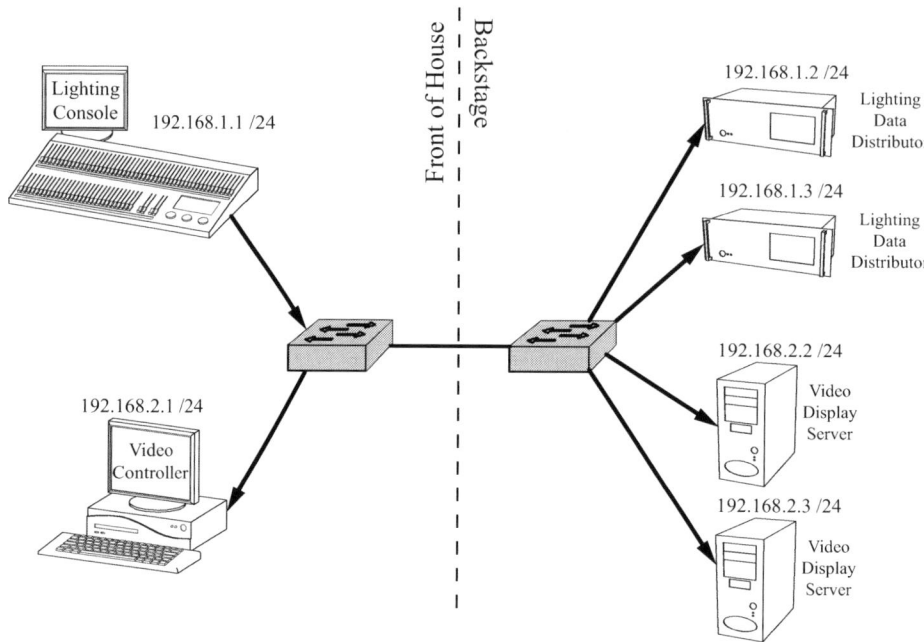

The receivers in all the video devices will then have to process each broadcast lighting data frame, wasting time, even though the data was never intended for those units. This configuration could be a problem, especially for the video system, which is using its proprietary, multicast protocol on the network to keep the video display servers tightly synchronized. What if, instead, we could still use a shared physical infrastructure—saving hardware costs and simplifying system management—but separate the networks virtually?

We could implement this using **Virtual Local Area Networks** (VLANs), which use the hardware of the switch itself to segregate all the connected devices in each VLAN into their own *virtually* separated broadcast domain. With the cost-effective switching horsepower we have available today, this approach allows a single physical network to simultaneously serve a number of purposes, while unifying cable management and configuration:

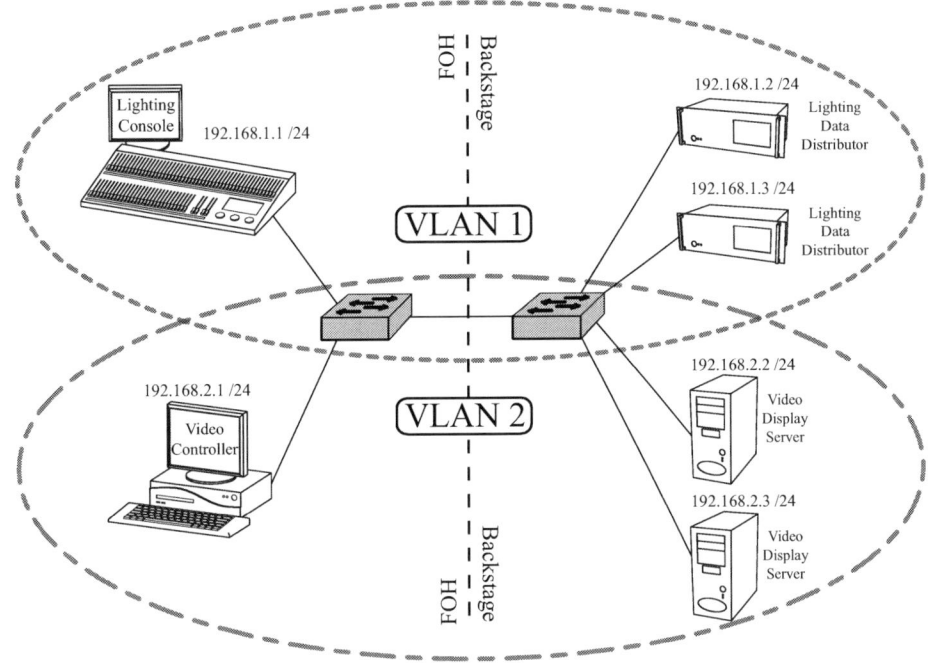

With this VLAN system, we have two virtually separate broadcast domains:

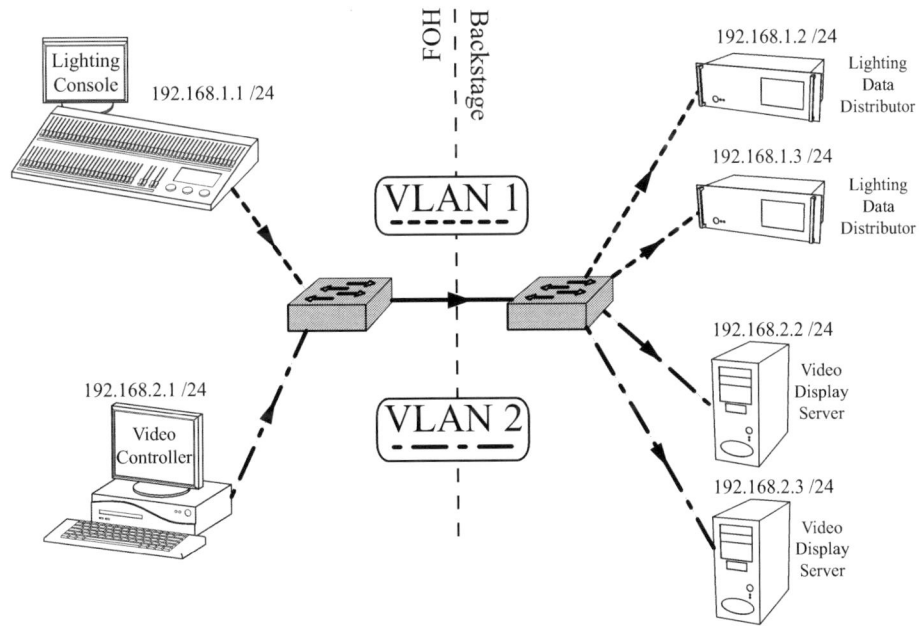

Chapter 18: Advanced Show Network Topics • 225

Of course, we need to make sure that the connection line between the two switches—called a VLAN trunk—has enough bandwidth to handle the combined traffic, but with gigabit switches and bandwidth-efficient control data, this often is not a problem.

The VLAN traffic is sorted out in most networks by "tagging" the packets with a VLAN number, using the IEEE 802.1Q standard; un-tagged packets run on the switch system's "native" VLAN. So the trunk line shown above would carry tagged traffic from both VLANs; with that information the receiving switch would be able to sort out the traffic appropriately. Typically, physical interfaces on switches come from the factory configured for the default VLAN, and then the network administrator sets each specific Ethernet interface to a particular VLAN using management software of the managed switch.[5]

One other note on VLANs: Switch hardware works at Layer 2 to implement VLANs, but it's good networking practice to put all the hosts in a particular VLAN into their own Layer 3 IP subnet (as shown in the example above).

Let's take a look at VLANs used on a real show.

EXAMPLE NETWORK WITH VLANS AND MANAGED SWITCHES

To bring all this together, let's look at a real show network that uses VLANs extensively: our annual *Gravesend Inn* interactive haunted hotel attraction at New York City College of Technology (Citytech, where I teach). The *Gravesend Inn* started in 1999 and was built around a hardware-based show control system, with all the sensors connected using contact closures and RS-232 based input/output (I/O). Over the years, as we added more and more technology to the attraction, the network grew into a mess of isolated, physically separated, unmanaged Ethernet switches: one for lighting, one for sound, one for video playback, one for show control, etc. (similar to what you see on "Physically Separated Topology," on page 199). I eventually moved on to a system of fully managed, layer 3, 24-port Gigabit PoE Switches (we need PoE to power the video surveillance cameras and other items), and integrated the entire network into one physical network infrastructure, using a combination of cables we ran permanently, and a few temporary runs as well.

---

5. It's good security practice to use the default VLAN only for management and assign everything else to a different VLAN.

We're using Cisco Small Business switches, and their proprietary "stacking" system, which handles redundant lines automatically, and takes four switches spread around the venue and effectively makes them into one 88 interface[6] switch. This approach made things a lot more straightforward and eliminated a lot of extra Cat 5 cable length, since I simply allocated the closest available physical switch port to any device that needed a connection, rather than having to home run back to the appropriate physically separated switch.

Here's a screen capture showing the status and interface allocation of part of "Switch D," which is in the middle of the attraction

:

#	Interface	Description	Port Type	Port Status	Port Speed	Duplex Mode
1	4/g1	TCP-I/O B (CSA)	1000M-copper	Up	1000M	Full
2	4/g2	Relay Box B (CSA)	1000M-copper	Up	1000M	Full
3	4/g3	Relay Box C (CSA)	1000M-copper	Up	1000M	Full
4	4/g4	DCH Watchout	1000M-copper	Up	1000M	Full
5	4/g5	CON Watchout 2A	1000M-copper	Up	1000M	Full
6	4/g6	CON Watchout 2B	1000M-copper	Up	1000M	Full
7	4/g7	CON Watchout 2C	1000M-copper	Up	1000M	Full
8	4/g8	Camera: Coat Check	1000M-copper	Up	1000M	Full
9	4/g9	Camera: Dining Chamber	1000M-copper	Up	1000M	Full
10	4/g10	Camera: Tipsy's Parlour	1000M-copper	Up	1000M	Full
11	4/g11	Camera: Washroom	1000M-ComboC	Up	1000M	Full
12	4/g13	Camera: Suite 13	1000M-copper	Up	1000M	Full
13	4/g14	Camera: Cranny	1000M-copper	Up	1000M	Full

This one switch handles streaming video, video playback, and show control components; it distributes out all the traffic to the other switches through the "stacking" lines.

---

6. These are 24 interface switches, but the stacking feature takes up two interfaces on each switch.

To keep all the signals separated from each other, I segregated the traffic into seven separate VLANs:

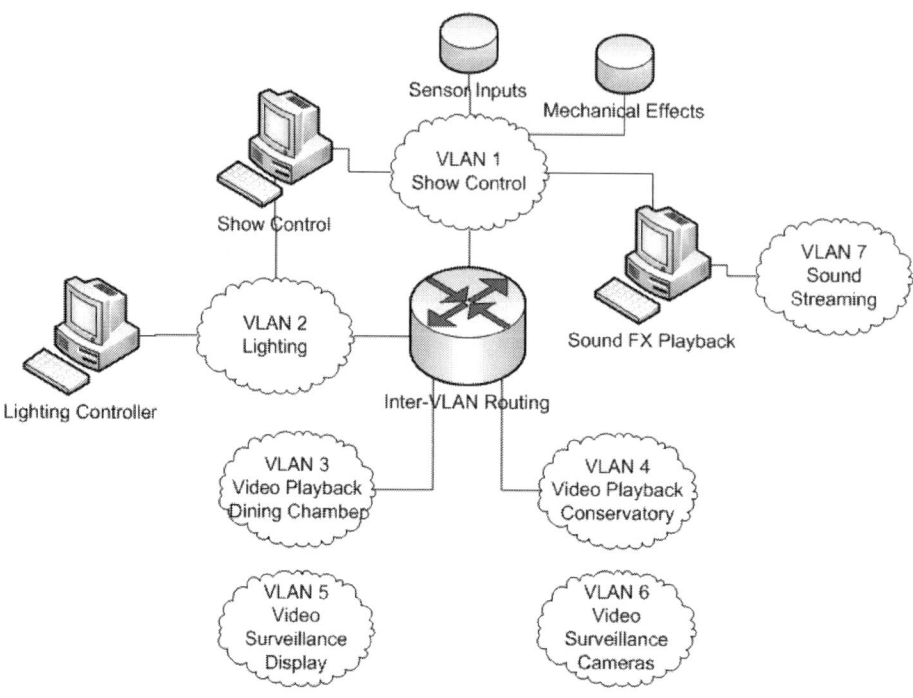

While the entire network shares a common physical infrastructure, each of the VLAN "clouds" on this diagram is, effectively, a separate virtual network. The seven VLANs are configured into the switch system using its Web-based management interface, and physical Ethernet interfaces are assigned to a particular VLAN though a simple web-based operation. Here, gigabit Ethernet interface 6 on switch 1 of the system is being assigned to VLAN 6.

VLAN/Discipline Overview
To explain how this network worked, let's go through each VLAN.

The lighting VLAN serviced the lighting console, a backup console, a network processing unit that distributed DMX, and two wireless access points for programming (one in our theatre and one in the basement). All the Ethernet switch interfaces used by these devices are assigned to VLAN 2, and work on a subnet of `192.168.2.0`, with a subnet mask of `255.255.255.0` (`/24`).

Sound effects for the show come from a PC running a sound effects playback system, which distributes dozens of channels of digital audio over a network audio interface, connected through VLAN 7 and assigned IP addresses in the `192.168.7.0/24` subnet. Audio comes through the VLAN from the PC to one mixer in the theatre for the main system, and another in the basement (each mixer then sends out analog audio to the various amps or self-powered speakers throughout the attraction). With this VLAN heavily loaded with dozens of channels of streaming, high sample rate digital audio, I wanted to keep the control signals firing the actual cues separate. So how did I fire sound cues from show control?

Take a look at the system diagram again and you will see *two* lines leading out from the sound effects system; one of those goes to the audio streaming network (VLAN 7); the other goes to the show control VLAN (1). To make this work, we had to equip the sound effects machine with two physical Ethernet interfaces; one resides on the sound streaming subnet and has an IP address in the `192.168.7.0/24` network; the other resides on the show control VLAN in the `192.168.1.0/24` main show control subnet. The sound effects PC's operating system internally sends its packets to the proper interface based on each packet's IP address; audio samples flow out on the interface connected to the sound subnet; while control messages come and go on the main show control network interface.

Video playback for the attraction is done using a distributed video server system, which uses a proprietary protocol to synchronize multiple video displays on the network. To keep these two systems separated (for ease of programming and control) I used two VLANs. VLAN 3 and subnet `192.168.3.0/24` was for a single "dining chamber" video server; VLAN 4 and subnet `192.168.4.0/24` was for a three-screen bay window effect.

Next up is the video surveillance system which, with 16 IP Power over Ethernet (PoE, see page 174) cameras, generates an enormous amount of streaming traffic from the cameras to a commercial, security-grade Digital Video Recorder (DVR) server. The cameras, of course, were spread all over the attraction and simply connected via whichever of the four switches was physically closest. In addition to the camera streaming traffic, the DVR server also streams video to two "search client" PC's, (used by operators to do instant replay and recall). To keep the search clients' traffic separate from the streams from the 16 cameras, the solution was—again—two VLANs: VLAN 5 and subnet `192.168.5.0/24` for the display and search systems; VLAN 6 and subnet `192.168.6.0/24` for the cameras and their streams and the DVR server. The server comes with two physical Ethernet interfaces, so two cables connect that one physical computer to the same switch; each of the Ethernet interfaces is, of course, assigned to the appropriate VLAN with an IP address on that subnet.

Layer 3 Routing

I've skimmed over one important VLAN: VLAN 1, which contains the computer running the show control system, several input boxes that read in sensor status from all over the attraction, and Programmable Logic Controller (PLC) output boxes that control the animated effects throughout the attraction. In addition to these sensors and actuators, the show control system also needs to communicate with other systems—lighting, sound and video—which all reside on different VLANs. How can they communicate? We use inter-VLAN routing (remember, a router connects separate networks together).

To see why, let's say we try to send a "play" command from the show control system at `192.168.1.111` to the video display server machine at `192.168.4.112`. These systems are on different subnets (`192.168.1.0/24` and `192.168.4.0/24`) and therefore will not be able to connect, even if we replace the entire switch system with a bunch of hubs or put everything on the same VLAN. And even if we backed out the subnets to something like `192.168.1.0/16` and `192.168.4.0/16`, it still wouldn't work because each VLAN is a separate broadcast domain, and the broadcast ARP commands that would be issued as part of the communications process would not be able to reach across the VLANs.

The Cisco Small Business switches we use have some Layer 3 capabilities, meaning that they can understand not only the raw Layer 2 Ethernet addresses of each frame, but also the associated IP addresses of the packets. So, the "stacked" system as a whole offers a feature that (when enabled) not only a switching infrastructure, but also a router that can pass traffic back and forth between the VLANs.

This is pretty easy to use—you just enable the feature (through the "console" port), and when the switch system sees IP traffic on a particular VLAN, it automatically creates a routing table entry to forward packets to the correct destination on each VLAN. Note "Local" is indicated as the route type, telling us that the switch system as a whole doesn't have to forward the packet anywhere beyond its local physical switch infrastructure.

IP Static Routing

	Dest. IP Address	Prefix Length	Next Hop	Route Type
☐	192.168.1.0	/24		Local
☐	192.168.2.0	/24		Local
☐	192.168.3.0	/24		Local
☐	192.168.4.0	/24		Local
☐	192.168.5.0	/24		Local
☐	192.168.6.0	/24		Local
☐	192.168.7.0	/24		Local

Default Gateways

When you assign an IP address on these switches to a VLAN for administrative purposes, that also becomes the default gateway for that VLAN. So we use the appropriate default gateway when configuring each host. For example, the video playback systems on VLAN3, which have addresses in the `192.168.3.0` network, are all configured with a default gateway of `192.168.3.1`. All the video streaming systems in VLAN 5/`192.168.5.0` subnet were configured with a default gateway of `192.168.5.1`, and so on. In this way, the host can properly send packets to the correct gateway, and then the router can correctly forward packets between the VLANs.

IP Address	Mask	Interface
192.168.1.1	255.255.255.0	VLAN 1
192.168.2.1	255.255.255.0	VLAN 2
192.168.3.1	255.255.255.0	VLAN 3
192.168.4.1	255.255.255.0	VLAN 4
192.168.5.1	255.255.255.0	VLAN 5
192.168.6.1	255.255.255.0	VLAN 6
192.168.7.1	255.255.255.0	VLAN 7

Unfortunately, not every device on our network could be configured with a default gateway, and this is the reason you may have noticed a second connection in the diagram from the show control computer to VLAN 2, the lighting VLAN. Packets from show control to lighting would be send to the show control VLAN default gateway (`192.168.1.1`), and the router could then forward these packets to the lighting VLAN. However, the lighting console, without a default gateway, wouldn't know what to do with packets returning to show control, which is on a different subnet. The solution to this? A second Ethernet card in the show control PC (just as in the sound FX machine) and two IP addresses; one on VLAN 1, show control, and the other on VLAN 2, lighting.

In the end, this integrated network approach worked really well for us, but it would certainly be just as effective to build the system using multiple unmanaged switches, one for each network. That would mean, however, that you'd have to put multiple Ethernet adaptors into the show control machine and manage all those issues. I'll stick with the VLAN'd system—it's just more fun!

OTHER NETWORK SYSTEM PROTOCOLS

As of this writing, most of the networks we use in entertainment control are small, unmanaged, and not connected to larger networks or to the Internet. But since our networks get more sophisticated all the time, it's important to understand some of the issues involved in—and techniques and systems used in—larger networks, so I'll offer a brief overview here.

Internet Group Management Protocol (IGMP) and IGMP Snooping

Some traffic on a show network may use multicasting to efficiently transmit its data out from a single control console to multiple control devices. But a switch, operating only on Layer 2, only forwards traffic out to devices that it knows need to see that traffic. One solution would be to simply flood (broadcast) all the multicast packets to every connected device; this, however, negates any of the traffic management benefits of multicasting. The multicast subscription process, however, uses a protocol called **Internet Group Management Protocol** (IGMP) operates at Layer 3, and switches with **IGMP Snooping** can "snoop" into the Layer 3 IGMP traffic to determine whether or not a particular connected interface on the switch should receive multicast traffic.

Domain Name System (DNS)

In show networks, we (as of this writing, anyway) rarely use the **Domain Name System** (DNS), but it is a critical part of the Internet, so it's worth a brief mention here. When you type a Universal Resource Locator (URL), such as `http://www.controlgeek.net/` into your Web browser, it submits the name to the DNS, which then returns the IP address of the text-based name. Packets can then be routed to the correct IP address.

Network Address Translation (NAT)

A technique commonly used to maximize the available IP addresses is **Network Address Translation** (NAT). NAT is also mostly outside the scope of this book, but here's a brief introduction. Using NAT, a router (like the one you likely have at home) connecting a LAN (like your home network) to the Internet changes the source and/or destination addresses of packets that travel through it, so that a single public IP address on the Internet can be shared by many hosts on the private network serviced by the router. For example, let's say we have a small private network with 10 hosts connected by a router/switch, which assigns *private*, Class C addresses. If one of the 10 machines needs to communicate with another machine on the private network (e.g., to transfer a video file from your desktop to the laptop), the machines simply talk directly to each other through the switch. However, if a user wants to browse a favorite Web page, he or she now needs to connect to the Internet, and now the user's machine needs a *public* IP address in order for its data to travel successfully on the public Internet. However, because there are so many computers out in the world now, IP addresses have become increasingly precious—if every machine were to require a globally unique IP address, we would have run out of addresses long ago.

So, using NAT, the router changes the source address of the packets from the sending machine to match the public (WAN) IP address of the router's gateway. When packets return for the machine running the Web browser, the router also changes the destination of returning packets back to match the private address of the destination host. In this way, many hosts can share a single, precious public IP address.

Virtual Private Network (VPN)

In these days where security is a primary concern, a **Virtual Private Network** (VPN) offers users (typically at large organizations) a way to use the public, unsecure Internet to access their own secure networks. Typically, traffic on a VPN connection is encrypted, and connections to the main system are often through a security firewall. Details of VPNs are also outside our scope here.

Quality of Service (QoS)

Quality of Service (QoS) is very confusingly named. It basically defines who wins and who loses when a network runs out of bandwidth by defining levels of performance and prioritization in a network, and can be assigned to a particular user or a type of traffic. It's especially useful in things like Voice Over IP (VOIP), which is the backbone of many office and home telephone systems. VOIP packets, for example, typically need a higher priority of service than an e-mail. Not all networks support it, and the details of how QoS works are outside the scope of this book, but it is something that you might see in a sophisticated show network, since we are always concerned with latency.

IPV6

The 4,294,967,295 unique addresses offered by 32 bit Internet Protocol Version 4 (IPv4) seemed like at lot in the 1960s, when the Internet was being developed, since the network at that time consisted of only a few university and government computers. However, with the proliferation of so many connected devices (even your smart phone needs an IP address to browse the Web or send and receive e-mail) the last blocks of unique IPv4 addresses were handed out in January, 2011, and the Internet is only able to continue on through the use of techniques like Network Address Translation (NAT, page 232).

IP version six (IPv6)[7] uses 128 bit addresses, which gives us 340,282,366,920,938,463,463,374,607,431,768,211,456 unique numbers. That's 340 undecillion, 282 decillion, 366 nonillion, 920 octillion, 938 septillion, 463 sextillion, 463 quintillion, 374 quadrillion, 607 trillion, 431 billion, 768 million, 211 thousand, 456 unique addresses.[8] That is something like 4 billion addresses for every living human.

While IPv4 is going to be around for a long time to come, as of this writing, IPv6 has gained some acceptance, so it's good to at least have a basic understand of this updated protocol.

A New Address Format and Shorthand

With so many address bits, the "dotted decimal" scheme (page 183) that we are so familiar with really would become unwieldy, needing 16 decimal numbers. Instead, the designers of IPv6 use 32 hex digits, broken down into eight "quartets" separated by colons; this approach is called **colon-hexadecimal**. For example, here's an IPv6 address used by Google:

```
2001:4860:8002:0000:0000:0000:0000:0069
```

To make the number a bit more manageable, leading zeros can be discarded, as in the last quartet, where the `0069` can change to just `69`. In addition, any quartet of contiguous zeros can be replaced by a single `0,` so this address could also be represented as:

```
2001:4860:8002:0:0:0:0:69
```

To make it even shorter, when contiguous *quartets* of zeros take place in the address, that whole block can be replaced by double colon `::` nomenclature, which just means "some number of `0000` quartets":

```
2001:4860:8002::69
```

---

7. IPv5 was the "Internet Stream Protocol" that offered a different service, not the expanded address space.
8. Thanks to some contributor to Wikipedia for figuring this out.

But what if you had an address like this, with multiple noncontiguous quartets of continuous zeros?

```
2002:0000:0000:7003:0000:0000:0000:0070
```

The :: can be used only once in any IPv6 address; using two would be ambiguous), so this address could be shown as either:

```
2002::7003:0:0:0:70
```

or

```
2002:0:0:7003::70
```

We can ping using IPv6:

```
C:\>ping -6 www.v6.facebook.com
Pinging www.v6.facebook.com [2620:0:1cfe:face:b00c::3] with 32
bytes of data:
Reply from 2620:0:1cfe:face:b00c::3: time=80ms
Reply from 2620:0:1cfe:face:b00c::3: time=87ms
Reply from 2620:0:1cfe:face:b00c::3: time=80ms

Ping statistics for 2620:0:1cfe:face:b00c::3:
    Packets: Sent = 4, Received = 3, Lost = 1 25% loss),
Approximate round trip times in milli-seconds:
    Minimum = 80ms, Maximum = 87ms, Average = 82ms
```

(Note the clever `face:b00c` hex address.)

Prefix/Subnet

In the same way that IPv4 addresses have a subnet mask, IPv6 addresses also break down using a variable "prefix" length. For example, if you see a IPv6 address with a /64 prefix length, that would mean that the first 64 bits of the IPv6 address are the network prefix, with the remaining 64 out of the 128 available bits to be assigned to hosts.

The Internet Corporation for Assigned Names and Numbers (ICANN) assigns a "registry prefix" to one of several regional authorities, who then assign addresses to an Internet Service Provider (ISP); a company or organization that wants IPv6 addresses then can get a prefix group from their ISP. A large organization, who could need a massive number of host addresses, can then—just as in IPv4—"borrow" some of the host bits to form subnets. The last part of the host address is called the "interface ID," and is a unique ID in the subnet, typically created by splitting the Ethernet MAC address and inserting a special hex string in the middle: `FFFE`. This process now gives each IPv6 address a regional location, an ISP, an organization, and an ID for the host, and that all makes a globally unique address like this:

Registry Prefix	ISP Prefix	Customer Prefix	Subnet Prefix	Interface (Host) ID

With this prefix structure, IPv6 replaces the several ranges of classful, IPv4 non-routeable or private addresses with **Unique Local Addresses** (ULA) and a prefix of `fc00::/7`. In addition, prefix `ff00::/8` has been designated for multicasting, and IPv6 also reserves `fe80::/10` for link-local addresses (page 189). These unique addresses (try `ipconfig` or `ifconfig` [page 186] on your machine to see some) are created locally, and can be used for initial exchanges on startup and for other local purposes, since routers do not forward packets from these addresses.

Three Types of Transmission

Like any network, IPv6 networks have to accomplish various things for different purposes, and they have three general types of messages: Global Unicast, Multicast, and Anycast.

Global Unicast

There are so many addresses available in IPv6 that it's possible to have completely unique "global unicast" addresses that allow direct communication from one host to another across the Internet without any Network Address Translation (NAT, page 232). For this purpose, IPv6 reserves addresses whose first quartet starts with `001` (indicated in IPv6 addressing shorthand as `2000::/3`) for global unicasting. (Put another way, the only options for this quartet would be `0010` and `0011`, or `2` or `3` in hex, which is why all the IPv6 addresses I've listed so far start with 2).

Multicast

Because you can now have direct, NAT-free global unicasting, broadcasting doesn't make much sense anymore; it's been replaced with multicasting.

Anycast

IPv6 has a new type of transmission that allows multiple servers to use the same unicast address; this would be especially useful in a large Web hosting infrastructure where traffic loads need to be balanced across many servers.

IPv6 Network Systems

As in IPv4, IPv6 systems can still use static IP addresses for critical infrastructure like routers or default gateways, and hosts can still use a Dynamic Host Configuration Protocol (DHCP) server for automatic address assignment (however, because the broadcast address has been eliminated, DHCP in IPv6 uses a special multicast address of `FF02::1:2`).

IPv4's Address Resolution Protocol (ARP) uses broadcast transmission, and has been replaced in IPv6 with the **Neighbor Discovery Protocol** (NDP), which can allow a device to discover the local network's IPv6 prefix through some special multicast messages. NDP can also be used to find the address of a Domain Name Server (DNS) connection for translating text-based names into IP addresses, and a default gateway to know where to send packets going outside the network.

IPv6 in Entertainment Control?

IPv6 offers some tremendous advantages to the structure of the Internet and, as of this writing, its adoption is beginning. However, with its gigantic installed base, IPv4 will be with us for some time to come. And for many of the small, closed networks we are building in our industry (as of this writing, anyway), IPv6 solves many problems that we don't have (very few if any of our systems use NAT, for example; most systems at this point don't even have a router). Time will tell...

Part 4: Standards and Protocols Used in Entertainment

We have now covered the basics of entertainment control systems, and also datacom and networking systems for transmitting show data. Now, we will move on to cover the details of some specific protocols and standards used in entertainment control systems.

Chapter 19: **Digital MultipleX (DMX512-A)**

DMX512-A started humbly, but is now the most successful and widely used entertainment lighting control standard in the world. DMX is a simple but effective standard, released in 1986 by the United States Institute for Theatre Technology (USITT) for control of dimmers; "DMX" stands for Digital MultipleXing; "512" is the number of control "slots" available for transmission. DMX is now used to control all kinds of gear, some not even in existence in 1986: moving lights, LED fixtures, and even video servers. In 1998, USITT transferred maintenance of DMX to the Control Protocol Working Group (CPWG) of the Entertainment Services and Technology Association (ESTA). In 2004, the CPWG released a major (and backward compatible) update to the standard, and updated the standard again in 2008, so the full title as of this writing is: "E1.11–2008 USITT DMX512-A, Asynchronous Serial Digital Data Transmission Standard for Controlling Lighting Equipment and Accessories." In 2011, ESTA merged with the Professional Lighting and Sound Association (PLASA), so the standard is now managed by the PLASA CPWG.

DMX'S REPETITIVE DATA APPROACH

DMX can send anything from 1 to 512 data slots using a repetitive control approach (page 73). To explain this, let's take a look at four updates (columns moving in time left to right) of DMX data, with the changes in **bold**. In our application, data is received by a small six-channel dimming system, patched to listen to DMX data slots 1 through 6:

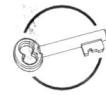

Slot	Update 1	Update 2	Update 3	Update 4
1	00	00	00	00
2	5F	5F	5F	5F
3	39	**3A**	**3B**	3B
4	7F	**7E**	**7C**	**7A**
5	FF	FF	FF	FF
6	26	**27**	27	**26**

In this example, dimmer 1 is turned off, but DMX can only send data repetitively, so a 00_{16} is sent over and over and over, every update. Dimmer 2 is set, unchanging, at a level of $5F_{16}$, so that data is sent repeatedly as well. Starting with update 2, dimmer 3 fades up from 39_{16} to $3B_{16}$; dimmer 4 fades down, so at each update, those levels change slightly. Dimmer 5 is at maximum the whole time; dimmer 6 fades up and then fades down. This simple approach has a major advantage in that if an update is lost along the way (caused by a loose connector or wire, etc.) it will be quickly overwritten, and with incandescent filaments (for which the standard was designed), losing a few updates might not even be visible to the audience. This approach does waste bandwidth, but since DMX was designed to be sent over a dedicated serial interface (page 247), bandwidth was not much of a concern.

ADDRESSING

DMX broadcasts its data to all receiving devices connected to a particular data link, and each device on that link must be configured in some way to read data starting at a user-configured **address**. For example, let's say you have a system with two 6-channel dimmer packs and one 12-channel dimmer pack, all connected to the same DMX link. You configure the first six dimmer racks to listen to DMX slots 1–6, the second to slots 7–2, and the 12-channel unit to listen to slots 13–24. Single-channel color scrollers and moving lights are addressed in a similar manner (see figure).

A Bit of DMX History

DMX was originally developed within the USITT at the prodding of a number of individuals in the entertainment lighting industry, particularly Steve Terry, then of the rental house Production Arts Lighting, and USITT Engineering Commissioner Mitch Hefter, then of Strand Lighting. At that time, several competing multiplexing protocols were in the marketplace. Colortran had its proprietary digital multiplexing protocol, known as CMX, which had been developed by Fred Foster (now of ETC fame). Lighting Methods Incorporated (LMI) had developed the Micro 2 protocol to connect the then-dominant Kliegl Performer control console to LMI dimmer racks. Kliegl had its own K96 protocol; Strand had CD 80 (an analog protocol, later to become AMX192); Electro Controls and Avab also had systems. In a session at the 1986 USITT conference in Oakland, representatives from all the major lighting companies of the time discussed the possibility of creating an open, digital, multiplexed lighting-control standard. Remarkably, by the end of the session, the basics of the standard had been worked out, and the compromise was to use the CMX data format running at a Micro 2–like data rate.[1]

The address is set in the target devices using a scheme determined by the manufacturer. Some have the user set the actual starting DMX slot number as the fixture address, as described above; other manufacturers think it's easier to have the user set a "rack," "device," or "unit" number. With this "rack number" approach, the dimmer packs above would be set so that the first is rack 1, the second, rack 2, and the third, rack 3. This approach makes things easier in small systems, but can get confusing for larger systems.

Each manufacturer decides how the DMX address will be configured in their products. Many use buttons and an electronic display; some older units use "thumbwheel" switches, which mechanically display the setting, or "DIP"[1] switches, which force the user to set the address in binary.

UNIVERSES

Installations needing more than 512 slots use multiple DMX **universes**,[2] with second, third, or higher universes being referred to as DMX slots 513–1024, 1025–1536, and so on. This can cause quite a bit of confusion, because a unit with the address we are calling 513 is actually listening and addressed to *slot 1 in the second universe*.

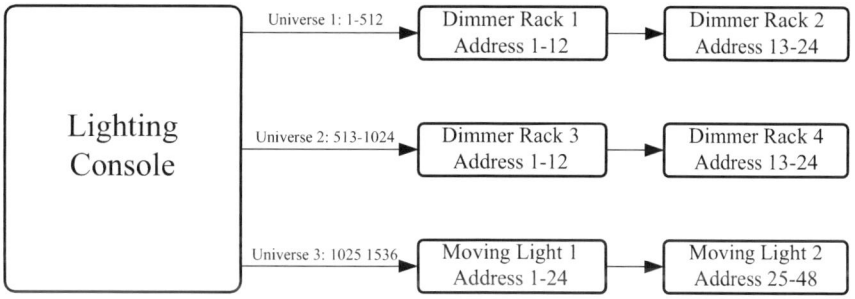

CONTROLLING EQUIPMENT OTHER THAN DIMMERS

DMX, designed primarily for dimmer level control, is not really optimal for control of more sophisticated devices like moving lights, since an eight-bit control slot can only represent 256 possible levels. This means, for example, that for a

---

1. DIP is an electronics term standing for Dual Inline Package.
2. Mitch Hefter, who chaired the PLASA task group E1.11-2004 (USITT DMX512-A) reports, "During the revision process, it was suggested that the term universe should be replaced by the term galaxy, since there is only one universe, but multiple galaxies. I pointed out that we are in the business of entertainment and make-believe, and that there are multiple universes in science fiction. Therefore, there was no pressing reason to change the term universe."

moving light, a single DMX slot couldn't achieve control of even 1° resolution in a 360° axis. But DMX is a good example of the power of standardization, since virtually every moving light now has a DMX jack, because every manufacturer wants their gear to be controlled by popular lighting control consoles, and they all have DMX interfaces.

There is no standard for mapping DMX slots for non-dimmer-level data. Many moving-light manufacturers, for example, use two DMX slots for control of high-precision functions such as pan and tilt; going from one 8-bit slot to two gives 16 bits of resolution, and that increases the number of possible steps from 256 to 65,536. But while one manufacturer may assign its first two octets to pan, another might use those same octets for tilt. So, there are nearly as many DMX implementations as there are brands and models of equipment. To give you an idea of how a manufacturer might use DMX for its fixture, let's look at one commonly available fixture—the Phillips Vari*Lite VL3500Q Spot, which is commonly used as of this writing.

DMX Slot	Parameter	Range
1	Intensity	0-255
2	Hi Byte Pan	0-65535
3	Lo Byte Pan	0-65535
4	Hi Byte Tilt	0-65535
5	Lo Byte Tilt	0-65535
6	Edge	0-255
7	Zoom	0 (small) - 255 (big)
8	CTO Mixer	0 (open) - 255 (diffused)
9	Blue Mixer	0 (open) - 255 (full saturation)
10	Amber Mixer	0 (open) - 255 (full saturation)
11	Magenta Mixer	0 (open) - 255 (full saturation)
12	Color Wheel	0-216 / 217-255 (spins)
13	Gobo Wheel 1	0-216 / 217-255 (spins)
14	Hi Byte Gobo 1 Index/Rot	0-65535
15	Lo Byte Gobo 1 Index/Rot	
16	Gobo Wheel 2	0-216 / 217-255 (spins)
17	Hi Byte Gobo 2 Index/Rot	0-65535
18	Lo Byte Gobo 2 Index/Rot	
19	Gobo Wheel 3	0-216 / 217-255 (spins)
20	Hi Byte Gobo 3 Index/Rot	0-65535

DMX Slot	Parameter	Range
21	Lo Byte Gobo 3 Index/Rot	
22	Beam Iris	0 (small) - 255 (open)
23	Strobe	0 (open) - 255 (max)
24	Focus Time	0-255
25	Color Time	0-255
26	Beam Time	0-255
27	Gobo Time	0-255
28	Control	0-255

The first data slot in this fixture is used for intensity, and the next two are used for "coarse" and "fine" pan adjustment; tilt is controlled by slots four and five in a similar way. Those parameters map out fairly directly, but after that, things get much more complicated, as functions such as color and gobo are dealt with, and these vary wildly amongst various fixtures and manufacturers. In the end, this single light eats up 28 DMX slots, which means that only 18 of these fixtures will eat up an entire DMX universe. Thankfully, most consoles keep all this configuration information in "fixture libraries" or "personality files" so you don't have to keep track of it.

PHYSICAL CONNECTIONS

DMX was developed before layered, network control approaches became commonplace in our market, and so the standard defines not only what the bits mean, but also the timing, electrical interface, connector, etc. DMX is based on the differential, multidrop EIA-485-A standard (page 150) and a single, daisy-chained link is designed to accommodate 32 connected physical devices (although you probably don't want a daisy chain this long—see the "Splitters, Isolation, and Signal Grounding" section, on page 249). Portable cable for DMX is standardized in "E1.27-1-2006 (R2011) Entertainment Technology Standard for Portable Control Cables for Use with ANSI E1.11 (DMX512-A) and USITT DMX512/1900 Products." This standard states that cable must use twisted pairs, and be designated by its manufacturer to be compliant with EIA-485 (microphone cable is NOT compliant with 485, and definitely not at the data rates used in DMX). Individual pairs must be shielded or the cable can have an overall shield. A maximum cable length is not specified, since it can be determined by many factors, and the standard specifies that the cable must be rugged enough to take the abuse that is part of portable use. Permanent cable is standardized in ANSI E1.27-2 – 2009, Entertainment Technology - Recommended Practice for Permanently Installed Control Cables for Use with ANSI E1.11 (DMX512-A) and USITT DMX512/1990, and Cat 5 or higher cables are allowed in addition to cable rated for 485.

Connectors

The DMX standard specifies that connectors, if used, shall be the five-pin XLR type, with receiving devices using male XLRs, and transmitters using female XLRs. Three of the five pins are used for a primary DMX link; the other pair of pins was designed for a second, optional data link, which might be used, for example, to send dimmer rack over-temperature or other information back to a console.

Here are the wire functions and pin numbers for the five-pin XLR:

5-Pin XLR Pin Number	Function	Use
1	Data link common	Common reference
2	Data 1–	Primary data link
3	Data 1+	
4	Data 2–	Optional secondary data link
5	Data 2+	

Three-Pin Connectors

Since two pins in the five-pin XLR connector defined by DMX were designated for an *optional* secondary data link, some manufacturers (especially moving-light manufacturers) ignored the rigid DMX specification of five-pin connectors and, instead, used three-pin XLR connectors to carry only the primary data link and the shield. This created a nightmare for rental houses and end users, who have to mix and match such non-standard equipment with DMX equipment that follows the standard and uses five-pin connectors. Either a separate inventory of three-pin XLR cable must be maintained, or five-pin to three-pin adaptors must be used with every such moving light or other device. Many reputable moving-light manufacturers have now abandoned this practice, or are at least including both three-pin and five-pin XLR connectors on their equipment.

Alternate Connectors

Under certain special circumstances, such as if the product is too small to physically fit a five-pin XLR, the new standard allows alternative connectors (however, it still does *not* allow the use of any XLR other than five-pin, so three pins are still out). RJ-45 (8P8C) connectors and related punchdown blocks are allowed for permanent installs, where the connection is accessible *only* to qualified personnel (to keep from plugging into an Ethernet network, for example, which could cause damage). The RJ-45 (8P8C) pin out for standard "Category" cable is:

Pin	Category Wire Color	Function	Use
1	White/orange	Data 1+	Primary data link
2	Orange	Data 1–	
3	White/green	Data 2+	Optional secondary data link
4	Green	Data 2–	
5	Blue	No connection	Not used
6	White/blue	No connection	Not used
7	White/brown	Data link common for data 1	Common reference
8	Brown	Data link common for data 2	Common reference

DATA TRANSMISSION

DMX transmits eight-bit data words asynchronously, in a serial format, at a rate of 250 kbit/s, with no parity and two stop bits. After a DMX "start code," the data for slot 1 is sent first, then data octets are sent sequentially for any number of slots, up to the maximum of 512.

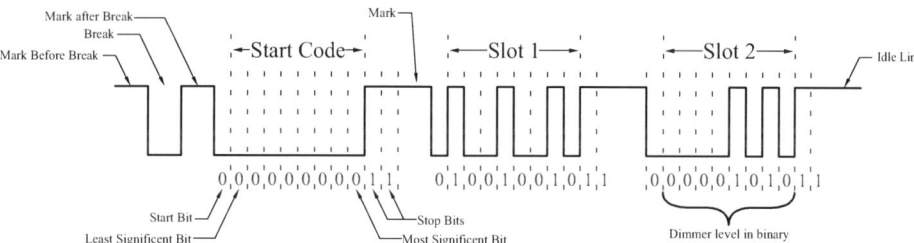

In its idle state, the DMX line is held at a "Mark" level with the data line in a high condition. When a DMX packet is to start, a "Reset Sequence" is sent, which consists of a "Break," a "Mark After Break," and then the "Start Code" (slot 0). The Break simply drops the line to a low level for a short period of time. The line is then returned by the transmitter to the Mark (high) state for the Mark After Break (MAB) signal, which must be at least 88 microseconds, and is the general sync pulse for the start of the data portion of the DMX packet, warning the receiver that the next transition from high to low will be the start of data. Next, the Start Code is sent. Standard dimmer level data uses a "Null" Start Code of eight 0s; Alternate Start Codes are allowed; more on page 251.

After the start code, the actual eight-bit slot data words are sent, each preceded by a start bit and followed by two stop bits. Any number of data slots, from one to 512, can be sent. The receiver is responsible for determining which slot or slots to listen to based on the user-configured address, and the data, with eight bits, has a range of 0–255. The DMX standard does not specify how to map the 256 possible levels onto a typical lighting scale of 0 to 100 percent.

Here is a notated oscilloscope screen capture of an actual DMX link:

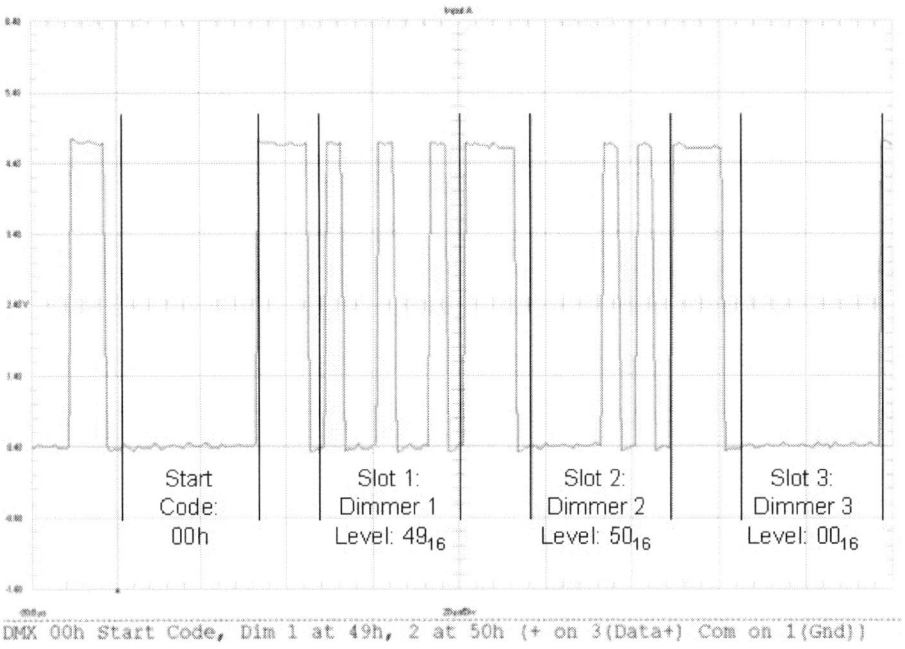

Refresh Rate

The number of times in a second that a controller (typically a console) updates a slot's data is called the "refresh rate," which is measured in updates per second. Because any number of slots, from 1–512, can be sent, the refresh rate varies with the number of slots; sending data for fewer slots allows a faster refresh rate. Keep in mind that the refresh rate here pertains only to the update rate of the DMX signal itself, not the rate at which the console updates the DMX data or how fast the receiving device can act on it. This is an important distinction: for rock-and-roll style "chases," a console and data update rate less than 20 Hz is generally considered unacceptable. The maximum update refresh rate for 513 slots (512 data slots plus the start code) under the standard is 44 updates per second (nearly one update per AC cycle), which is generally considered adequate for most applications.

DMX DISTRIBUTION AND TERMINATION

Since DMX runs at a fairly high data rate, precautions must be taken to ensure that the data is transmitted and received properly. For electrical reasons, cables run to receivers on a DMX link should only be "daisy chained" sequentially through each receiver, and never "two-ferred."

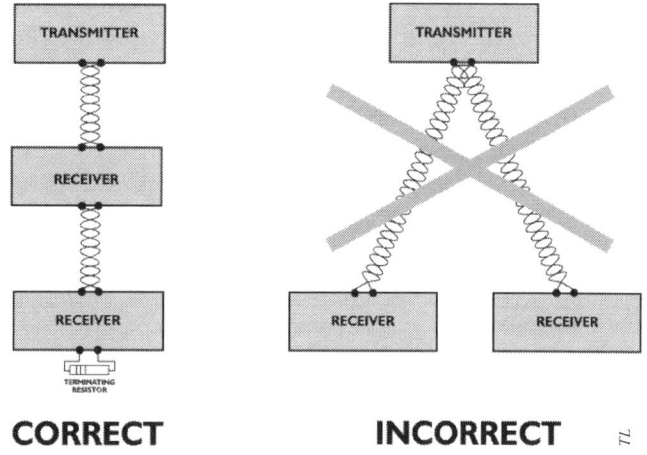

CORRECT **INCORRECT**

Also, the last physical device in any DMX link, whether it connects one device or 20, must be "terminated," typically with a 120Ω resistor; this termination is often provided as a switch that can place a resistor across the line. Running the DMX signal properly ensures good data integrity, as unterminated lines tend to "reflect" the DMX signal back down the line, possibly corrupting the data. "Stubs" off the data transmission line, as would be found in a two-fer, tend to act like antennas, and adversely affect the impedance of the transmission line and introduce electrical noise.

Splitters, Isolation, and Signal Grounding

While up to 32 devices can be connected to a single DMX link, this is not advisable for a number of reasons, but primarily because a failure in one device (even a simple power failure) can block data from being transmitted down the line. To get around this problem, DMX "splitters" (also known as repeaters or distribution amplifiers) are used, so that if the connection between the splitter and a device is damaged, the other connections will still continue to function.

DMX connects very high-power devices (such as dimmer racks) to delicate, low-power, electronic devices (such as consoles). If a high-voltage failure occurs in a dimmer rack, it is quite pos-

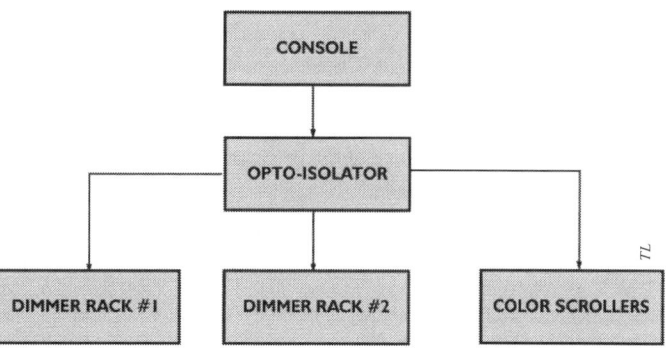

Chapter 19: Digital MultipleX (DMX512-A) • **249**

sible for the voltage to travel back up the DMX data link and blow up the console; it is also possible for a failure in one dimmer rack to damage the control electronics of another. For this reason, DMX links should, if at all possible, be electrically isolated (see "Isolation," on page 97), either optically (most common) or galvanically (i.e., with Ethernet). With an isolated system, there is no electrical connection between the DMX input and the output, and high-voltage or other electrical failures are isolated from delicate control electronics. Optical isolation is often combined with splitter functionality into units called "opto splitters."

DMX PATCHING/MERGING/PROCESSING/TEST EQUIPMENT

One of the primary advantages of standardization is that the market expands dramatically for all products using a successful standard, and it then becomes economically feasible for new types of devices to be created. DMX patching, merging, routing, and processing equipment (see photo) can do whatever you need done in a system. It can reroute the DMX data to multiple points throughout a facility, remove or add slots to a data stream, merge the outputs of multiple consoles, and so on.

Courtesy Pathway Connectivity, Inc.

USB adaptors for personal computers are available, and used in computer-only lighting consoles and other special applications. DMX test sets (see photo) greatly simplify DMX troubleshooting, allowing a user to measure and analyze the DMX signal and quickly find and troubleshoot problems.

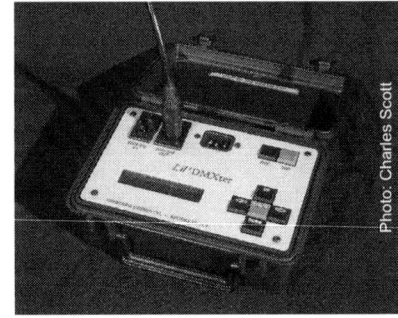

Wireless DMX Transmission

As the cost of radio data equipment got cheaper, a number of manufacturers started offering affordable wireless DMX data transmission equipment. Some use proprietary radio wireless systems, while others base their products on 802.11 (page 175).

Courtesy City Theatrical

These systems are ideal for special applications where it's impossible to run a cable, such as a moving platform with some wireless dimmers on a turntable. Otherwise, for reliability reasons, my advice remains: If you can run a wire, use a wire.

ALTERNATE START CODES

The original DMX standards assumed that data was going to be sent with a null start code, primarily to "dimmer class" data. Over time, manufacturers implemented additional functionality on top of the DMX standard, and DMX512-A created a formal procedure to register an Alternate START Code (ASC) with PLASA. Several ASCs were also defined in the standard; for example, Remote Device Management (RDM, Chapter 20) uses an ASC of CC_{16}. An ASC of CF_{16} means that slots in that DMX packet will be a System Information Packet (SIP). The SIP includes an error check of the previous null start code (dimmer) data packet, and contains information such as the manufacturer ID of the device, the universe number, and so on. Another standardized ASC is 91_{16}, which allows the use of a manufacturer ID, a two-word value assigned by PLASA to a specific manufacturer or organization. Through the use of this ID, a manufacturer can send proprietary data over a DMX link. The list of Alternate Start Codes is maintained by PLASA and is online.[3]

Simple devices, such as dimmers, should simply discard any non-null start code information they don't understand; otherwise, the device could interpret other kinds of data as being dimmer-level data and cause the levels to fluctuate wildly.

ENHANCED FUNCTION

DMX512-A specifies enhanced functions that will not damage legacy equipment through four enhanced function topologies, numbered 1–4.

- *Enhanced Function 1* allows half-duplex EIA-485 signals on the primary data link and other data to return on the same (primary) data link, provided that this is done using a registered alternate start code. No functionality is specified for the second data link. RDM uses Enhanced Function 1.
- *Enhanced Function 2* allows unidirectional DMX data on the primary link and 485 unidirectional data flowing the opposite direction on the secondary link.
- *Enhanced Function 3* is unidirectional DMX data on the primary link and half-duplex 485-compliant signals on the secondary link.

---

3. http://tsp.plasa.org/tsp/working_groups/CP/DMXAlternateCodes.php

- *Enhanced Function 4* is half-duplex 485-compliant signals on both pairs, and the return signal is on the primary data link using a registered alternate start code.

DMX OVER A NETWORK

DMX was designed as a complete, unified standard, specifying everything from the connector to the meaning of each bit in each octet. Since it took off in the market in the 1980s, DMX implementations have gotten more and more complex, while at the same time networks have become ubiquitous. So, it's now often desirable to separate DMX's data commands and protocol from its physical infrastructure, and use a layered approach to send it over an Ethernet network using IP. The benefits of doing so range from reducing the number of cables runs to the truss to saving money on an interface, since nearly every computer can now talk Ethernet, but very few speak DMX's RS-485.

As we discussed in Chapter 2, there are a number of proprietary solutions for sending DMX data using Ethernet. But these systems are generally not interoperable, so there's not much to cover here. But there are two free to use ways to send DMX over networks in widespread use: Art-Net, and Streaming ACN (sACN).

Art-Net™

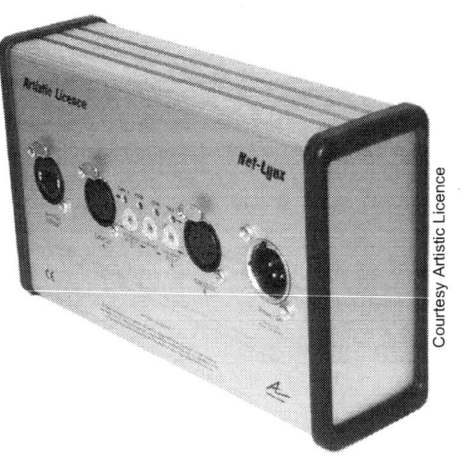

Art-Net is an easy-to-use way of transporting DMX and RDM data (see Chapter 20) over closed, private Ethernet networks using UDP and IP. Art-Net was developed and is owned by Wayne Howell of Artistic Licence, who first published the protocol in 2002 and made it free for all to use without royalty. As of this writing, the current version is Art-Net 3, which is designed to be backwards compatible with previous versions and can transport up to 32,767 DMX universes worth of data over a 100BASE-T, closed Ethernet network.

Device Types

An Art-Net device that translates between Art-Net over Ethernet and DMX over EIA-485 is called a "node." Any node can send DMX data onto the network, but a "controller" is a connected device that primarily generates control data, such as a lighting console.

Addressing

Every device on an IP network needs a unique address, and the default method for use by Art-Net devices is to start with a `2.x.x.x/8`, IP address, and then fill the last three octets of the IP address with data from the node's globally-unique Ethernet MAC address. To find other Art-Net devices on a network, the "ArtPoll" message is sent to the "directed broadcast address" of `2.255.255.255`. If a Server or Node receives this message, it should reply with an "ArtPollReply," which contains the responder's IP address and other information about the protocol type.

This method works fine on small, closed networks, but the protocol document says explicitly, "It is important to ensure that Art-Net data is not routed onto the Internet." Why? The `2.?.?.?/8` Class A addresses used in this process are publicly assigned to other users[4]. Because Art-Net addresses are based on the (effectively random, in this case) MAC address, an Art-Net network could actually create IP address conflicts with hosts around the world.

A better practice for larger networks, which can be configured into Art-Net nodes compatible with later versions of the protocol, is to use the nonrouteable, private IP address blocks. Some Art-Net nodes can be configured to use the Art-Net address assignment process with the appropriate Class A nonrouteable `10.0.0.0` address block. Alternatively, some devices allow you to assign your own private addresses, and more sophisticated Art-Net devices can use a DHCP server.

Transmission

With the network configured, Art-Net uses a proprietary packet structure sent over UDP port 6454 to transmit DMX data. In basic operation, an "ArtDmx" packet is sent whenever a DMX frame is received, so the rate at which ArtDmx packets are sent over the network should correspond to the incoming DMX data refresh rate.

The system can communicate in two modes: peer-to-peer and controller-to-peer. In peer-to-peer mode, up to 40 universes can be transmitted, and the Art-Net server sends out ArtDmx packets to the `2.255.255.255` directed broadcast address, and received by all Art-Net nodes on the network. Again, this works fine

---

4. In this case, the `2.0.0.0` network address block is assigned to to the RIPE Network Coordination Center, which is a Regional Internet Registry (RIR) assigning IP addresses for clients throughout Europe, the Middle East, and parts of Central Asia.

on small, private, unmanaged networks, but a more sophisticated switching infrastructure may see this repetitive, broadcast data as a problem or even a threat.

Art-Net 2 added a server-to-peer mode, where unicast transmission is used to transmit to the address of the node. Unicast is a better practice from a network point of view, but unicasting means that a controller would have to establish multiple unicast streams out to the various receiving nodes, meaning you would need one stream for each connected device. When more than four or five unicast streams is required, Mr. Howell suggests using broadcast transmission instead.

Art-Net can differentiate between 32,768 different DMX universes, and each universe is assigned a "Port-Address" (not to be confused with a computer port or IP address). The Port-Address is made up a "Net," which is one of 128 groups of 16 consecutive "Sub-Nets" (not to be confused with an IP subnet), made up of 16 "universes":

Bit 15	Bits 14-8	Bits 7-4	Bits 3-0
0	Art-Net Net	Art-Net Sub-Net	DMX Universe

From the chart, you can see that we have 15 bits available, which gives us the 0-32767 unique Port-Address IDs, which gives us the maximum number of universes that Art-Net can theoretically carry. Here's the maximum number of universes specified in the protocol documentation, based on different Ethernet types:

Transmission Type	10BaseT	100BaseT	1000BaseT
Broadcast (Peer to Peer)	40	40	40
Unicast (Controller to Peer)	40	400	4000+

Art-Net In The Entertainment Control Market

We've just scratched the surface of Art-Net here; the system can also merge DMX data, manage devices, and transmit RDM messages. For more information, see `http://www.artisticlicence.com/`.

Art-Net has been implemented by many manufacturers, has a large installed base, and has been very successful in the market. But it was designed at a time when users rarely built anything other than unmanaged, small, closed networks, and care must be taken by users and network administrators to make sure that it works well with various network systems; Streaming ACN (sACN) was designed to address many of these networking issues.

E1.31: Streaming ACN (sACN)

ACN (Chapter 21) was developed as a full replacement for DMX, defining a completely new, vastly improved, and more flexible control scheme that works in any sort of layered network environment. However, to accomplish this requires a very different approach to control, which needs to be adopted by both manufacturers and users.

To address the need for DMX to run on more "lightweight" processing infrastructure, address DMX's enormous installed base, and the widespread familiarity of lighting technicians with the concepts of "universes" and "slots," the developers of ACN created **Streaming ACN** (sACN), which is more accurately known as "ANSI E1.31 – 2009, Entertainment Technology – Lightweight streaming protocol for transport of DMX512 using ACN." The sACN standard allows multiple controllers and receivers to transmit up to 63,999 universes of DMX slots over an IP network, and was designed from the ground up to follow good networking practices and work efficiently, while using some structures from "full" ACN to ensure full compatibility with the more powerful parts of that standard.

A lighting console or similar originator of a DMX stream is called a "source" in sACN, and each source can be given a user definable name. There's no limitation in the standard on the number of DMX sources in a network, and therefore it's possible that a receiver may receive multiple packets carrying data for the same DMX universe. To address this issue, the protocol designers included a "priority" field, which ranges from 0-200, with a source of 200 being highest priority (100 is the priority set when this feature is not used). Alternatively, the receiver can merge data using highest takes precedence (page 16). If there is a conflict, or the receiver gets too many sources and it doesn't know what to do, it can generate a "sources exceed" error, which should be presented to the user in some way.

Transmission

Rather than wastefully broadcasting repetitive data over and over, DMX style, sACN only transmits dimmer control slot data when something changes; if the data is not changing (e.g., all the lights are off or we are sitting in a static, unchanging lighting look), the DMX source simply transmits on the network a "keep alive" packet, containing the unchanging slot data about once per second. In this way, the amount of data sent on the network is dramatically reduced, but the problem of a console getting unplugged or the network going down can also be quickly detected, since receivers time out and indicate a fault if they haven't received at least a keep alive message every 2.5 seconds.

Protocol Structure

The sACN protocol uses a straightforward packet "wrapper" approach, using ACN's Session Data Transport protocol (page 272) and its designated port of 5568. A 16-bit field in the packet defines the universe number, from 1-63999 (0 and 64000-65535 are reserved for future use), and packet transmission on the network is tied to the maximum refresh rate specified in DMX for a packet with the same number of slots. Part of the root layer is an ACN "Component IDentifier" (page 272), which is a Universally Unique IDentifier (UUID), ensuring that no two sACN devices can have the same physical address. Messages connect with any transport mechanism via ACN's Root Layer Protocol (page 273). The primary network, of course, is UDP/IP over Ethernet.

Sequence numbers are provided in the packet structure to deal with out-of-order delivery; simple receivers that don't understand the sequence number can just ignore this field. More sophisticated receivers can do a bit of math to determine if packets are arriving in order, and if they are not, then those out of sequence packets can be buffered and processed in order, or discarded if the sequence interruption is long enough.

This may all seem like a lot of wasteful overhead to add to the simple DMX structure, but keep in mind that even common 100 Mbit/s Ethernet is *400 times* faster than DMX's 250,000 bit per second data rate, so there is plenty of bandwidth available.

IP Addressing

IP addresses for sACN are assigned as indicated in ACN's "EPI-29 (Revised Rules for Allocation of Internet Protocol Version 4 Addresses to ACN Hosts)." which allows three possible schemes to automatically assign IP addresses to sACN nodes. The preferred method is to use a DHCP server on the network (page 185), but if that's not possible, the EPI allows the use of link-local addresses, which can be set without user input (page 189). A third approach is "Detecting Network Attachment in IPv4" (DHC-DNA), which was defined within the IETF (RFC 4436), but doesn't, as of this writing, seem to be widely used.

Multicast Data for Further Efficiency

Another way that sACN optimizes network utilization and ensures compliance with good networking practices is through the use of multicasting (page 185); it's also possible to unicast data if such addresses are configured into the system somehow. The multicast address `239.255.x.x` is used, with the last two octets of the IP address set to the designed universe number. For example, to transmit universe 1, the `239.255.0.1` multicast address would be used. Transmitters and receivers use IGMP (page 232) to manage the multicast streams.

RDM over sACN?

As of this writing, the task group is working on, "BSR E1.33 Extensions to E1.31 for Transport of ANSI E1.20 (RDM)," a proposal to transport RDM (Chapter 20) packets over sACN.

Universe Synchronization

As of this writing, work is underway on "BSR E1.30-12, EPI 34, E1.31 Universe Synchronization." This work is to address some timing issues in sACN when multiple controllers are used.

sACN in the Entertainment Control Market

As of this writing, sACN can be found in many larger lighting console systems and is growing in popularity.

NETWORK OVER DMX?

As of this writing, work is underway on "BSR E1.45 - 201x, Transport of IEEE 802 data frames over ANSI E1.11 (DMX512-A)." The project, launched in 2012, is set to define a simple method to send Ethernet (IEEE 802) data frames over DMX physical links using an Alternate START Code (ASC). The development of this standard is related to another fascinating effort also underway as of this writing: IEEE 802.15.7 "Visible Light Communication." This is an effort to use LED lighting fixtures to transmit Ethernet data—LEDs can be blinked on and off fast enough to modulate data and transmit it over the lighting fixtures in an office environment, thereby making wireless transmission with none of the pitfalls of radio. This obviously has potentially widespread applications in office environments, and could see applications in our industry as well. Because many kinds of fixtures already take DMX, the IEEE task group is looking at DMX as a way to carry these Ethernet frames to the luminaires. Time will tell...

DMX IN THE ENTERTAINMENT CONTROL MARKET

The open DMX512-A standard has knocked out all other past competitors (analog and proprietary approaches) to completely dominate the field of lighting control. It has also found use outside the entertainment lighting market (smoke machines, video servers, video screens, architectural lighting control), and few manufacturers would consider making a lighting product that can't generate or be controlled by DMX.

However, DMX has been stretched way beyond what it was ever designed to do, creating a situation on big shows where many universes are needed to run all the fixtures. This creates a coordination and paperwork nightmare for the production staff, but the technology and its limitations are well-known, and a lot of stunningly sophisticated work is now done.

So, while DMX will certainly be around for many years to come, it's my hope that we can eventually transition to a more extensible, sophisticated, flexible protocol like ACN (see Chapter 21).

Chapter 20: Remote Device Management (RDM)

DMX512-A (Chapter 19) dominates the world of lighting control because it is simple to use and implement, and it works well. However, as the lighting market has matured and expanded and new devices have been invented, the limitations of DMX have become apparent, especially in large or complex systems, or those made up of sophisticated devices. To help address these complexities, "ANSI E1.20-2010, Entertainment Technology – RDM, Remote Device Management over DMX512 Networks" was released by ESTA (now PLASA) in 2006, with a revision in 2010. Remote Device Management (RDM) is an extension to—not a replacement for—DMX, and the development effort was headed by Scott Blair, based on a talkback protocol he developed at High End Systems around 2000.

RDM has many features that are desirable in lighting control systems. It standardizes an open, bi-directional communication approach, bringing extended functionality to a DMX infrastructure. It also allows a control console to automatically "discover" connected devices, and then configure, monitor, and manage them from the console—all without getting out a ladder or a personnel lift in order to push buttons on that moving light up on the truss, for example. RDM can also deliver error and status information to the operator, so a problem might even be able to be corrected before it becomes visible on stage.

BASIC STRUCTURE

RDM is tied to DMX's EIA-485 physical layer, which allows half-duplex, bi-directional communication[1] between any connected devices on the same physical link. RDM uses the same pair of wires as DMX512-A, under DMX's "Enhanced Function Type 1" (page 251), and this means that older opto splitters and other devices will probably not work with RDM, since they may not be able to allow data to pass in both directions. However, EF-1 uses only one pair of wires, so it will work on older permanent installations where only one pair was pulled.

RDM packets use the standard DMX slot structure and an assigned alternate start code of CC_{16}, interleaving its packets with standard null start code (dimmer level)

---

1. It's fortuitous that the developers of DMX selected 485 instead of 422, which allows only one transmitter.

DMX packets. With this approach, compliant DMX-only equipment (that properly evaluates the start code[2]) should work fine on an RDM-enabled link, even if that DMX equipment doesn't have RDM functionality.

RDM systems follow a general process in which the "controller" (e.g., a lighting console) first goes through "discovery," where it determines what "responder" devices are connected, by searching and taking note of each RDM device's unique ID (UID), which is similar to an Ethernet MAC hardware address. Once the discovery process is complete, the controller can get the status of and manage a wide variety of connected device parameters, such as the responder's DMX address, lamp hours, pan/tilt orientation, and so on. These remote-management features alone have the potential to save a ton of time on a show, and RDM can also act as a platform for the development of sophisticated commands.

RDM manages access to its devices through a master/slave, polled method, where only the controller (lighting console) can initiate communication. The controller can communicate directly with any individual device or to all connected devices through the use of a broadcast address. Only one controller can be active on a DMX link at any given time, and the controller's polling process controls the response timing to avoid data collisions except during discovery, where data collisions are expected (more on page 263). "Inline" devices (such as opto splitters, or processors) can operate in either "transparent" or "proxy" modes. A transparent inline device simply passes data on, but does not initiate the discovery process. Up to four transparent inline devices can be included on a single RDM pathway, and the inline devices can, of course, also be responders, allowing for monitoring and configuration of their parameters.

RDM MESSAGE STRUCTURE

RDM messages break down as follows:

Start Code (CC_{16})
Sub-Start Code (01_{16})
Message Length
Destination UID (48-Bit)

---

2. Some older or even new, poorly designed DMX equipment may not actually check the start code. This kind of equipment would, therefore, treat the RDM packets as lighting control data, and this would cause wildly erratic behavior.

Source UID (48-Bit)
Transaction Number
Port ID/Response Type
Message Count
Sub-Device (16-bit)
Message Data Block (Variable Size)
Checksum (16-bit)

Start Codes
The DMX alternate start code used by RDM is CC_{16}. The sub-start code, also set in the standard, is 01_{16}.

Message Length
The message length is the number of slots used by the RDM packet, including the start code, but not including the checksum. Since all DMX data is sent in eight-bit slots, the maximum message length that can be specified is 11111111_2.

Destination and Source UIDs
The Unique ID (UID) is 48 bits, with the 16 most significant bits indicating a PLASA-assigned manufacturer's ID, and the remaining 32 a unique device ID assigned by the manufacturer (like a serial number). With this structure (just like an Ethernet MAC address), unless a manufacturer makes a mistake and assigns the same UID to two devices, device ID conflicts cannot occur. In addition, a broadcast UID is included to reach all devices; also, a structure is included to enable a manufacturer to reach only its devices on a system.

Transaction Number
The transaction number is a number generated by the controller, and is incremented with every RDM packet that the controller transmits, rolling back over to zero when the maximum number is reached. When a responder replies, it includes this transaction number, and in this way a controller can associate a past command with a current response.

Port ID/Response Type
The function of the port ID/response type changes with the type of message. If the message is sent from a controller, the port ID is set to the controller's port number,

so that this number, in conjunction with the controller's UID, can determine the exact physical port on the controller that is sending the message.

Responders, instead, use this slot to indicate the type of response. Some responses include the following:

- RESPONSE_TYPE_ACK
- RESPONSE_TYPE_ACK_OVERFLOW
- RESPONSE_TYPE_ACK_TIMER
- RESPONSE_TYPE_NACK_REASON

This range of types enables responders to work with a wide variety of data exchanges, and also indicate to the controller if there is a problem (NACK, or negative acknowledgement).

Message Count

Controllers should always set the message count field to 00_{16}, but responders use the message count slot to indicate to the controller that additional RDM messages are pending collection by the controller. The controller can then fetch these queued messages from the responder using the GET QUEUED_MESSAGE command.

Sub Device

The sub-device field is used in devices, such as dimmer racks, that have multiple components. In this way, a specific dimmer could be addressed in the rack (e.g., to set it for nondim operation).

Message Data Block

The message data block is the actual message itself, and is made up of the command class (CC), a 16-bit Parameter ID (PID), the Parameter Data Length (PDL), and the parameter data (PD).

The command class indicates what the actual message does. The defined types include:

- GET_COMMAND, which gets information from a responder
- GET_COMMAND_RESPONSE, which the responder sends in reply to a GET_COMMAND message
- SET_COMMAND, which will set a value in a responder

- SET_COMMAND_RESPONSE, which is the message back from the responder in response to a SET_COMMAND message
- DISCOVERY_COMMAND and DISCOVERY_COMMAND_RESPONSE, which are used in the discovery process described below

The Parameter ID is a 16-bit number that indicates which parameter in the responder should be addressed by any of the above commands. To ensure maximum interoperability, some parameters are set by PLASA in the standard; others are left to manufacturers to assign. The Parameter Data Length indicates how many slots are used by the parameter referenced in the command class. The parameter data is, of course, the actual data.

Checksum
The message is ended with an additive checksum of the data contained in the message's slots.

THE DISCOVERY PROCESS

In a traditional DMX control scheme, the user has to manually decide on and configure the address for each connected device, and make sure that each address does not conflict with any other device looking at the same slots. One of the key advantages of RDM is its ability to "discover" all connected devices and then allow the configuration to be done remotely. RDM does this by using a process to scan the entire address space, and then note the UID of each device that responds.

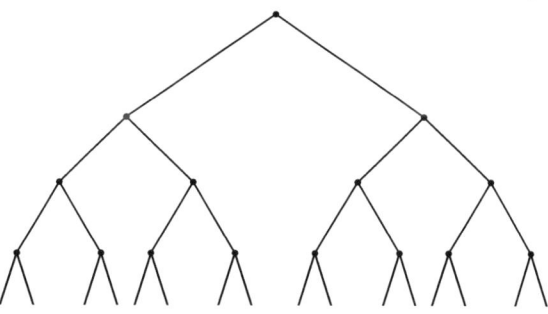

To accomplish this, RDM uses a "binary search tree." The mechanics of binary searching are outside the scope of the book, but, basically, this is a branched approach to searching a potentially large known range. RDM has a 48-bit address range, which means 281,474,976,710,655 possible addresses. Searching for each of those addresses on startup would take an enormous amount of time. Instead, the system divides the range into branches and checks each branch. If there is no response from a particular branch, it moves on to the next one; if there is a response, it continues down the branch to determine what RDM devices are connected.

To discover the connected devices, a controller first "unmutes" every device in the system, by sending out a DISC_UN_MUTE message to the broadcast UID (the address to which every connected unit should listen). Next, one branch picked by the controller is sent a DISC_UNIQUE_BRANCH message, and any and every connected RDM unit whose UID is within that branch should respond. It's likely that a branch would contain more than one device (or many devices), and if multiple units respond simultaneously, this would create a number of data collisions. However, RDM takes this into account, and if the controller can read a return message with a valid checksum, it assumes that message survived the data collisions intact—meaning there should be a valid connected unit at that UID. The controller will then send a DISC_MUTE message only to that specific device at that specific UID, and the device will then MUTE (disabling any further responses) and send an acknowledgment back the controller to verify the process.

The controller then sends another DISC_UNIQUE_ BRANCH message to the same branch, to see if any other (still unmuted) responders are available on that particular branch, and this process continues until no valid responses are received. At that point, the controller assumes all the devices on that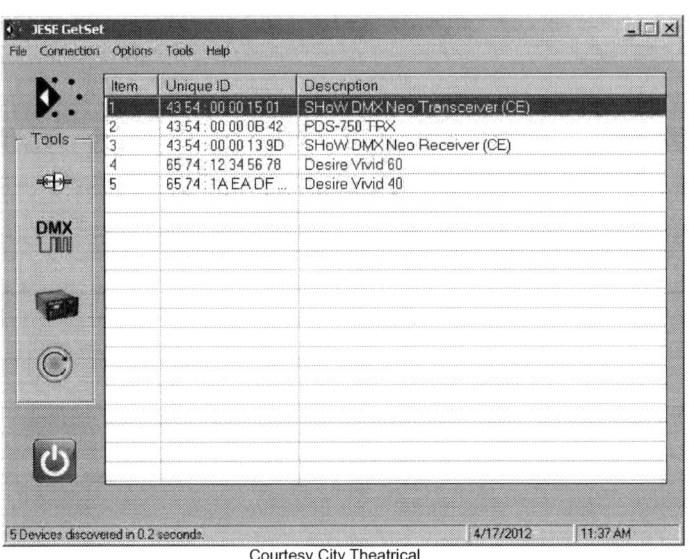

Courtesy City Theatrical

branch have been reached, moves onto the next branch, and repeats the process until there are no further responses. In this way, whole branches of address space where there may not be any addresses can be quickly checked and ruled out, enabling the system to move relatively quickly through the massive amounts of address space.

Once the device is initially configured, sending another DISC_UNIQUE_ BRANCH message to the system will cause any new devices to respond, and this is something a controller should do periodically to catch any newly connected devices. In addition, existing devices can be checked periodically to make sure they are still connected and operational.

RDM MESSAGES

With the system discovered and configured, a controller can send RDM messages to configure, set, or monitor a wide variety of unit parameters. Remember, of course, that the actual lighting control data would still be sent using DMX packets. The standardized RDM commands include:

Message	*Name*
Communication Status	COMMS_STATUS
Get Queued Message	QUEUED_MESSAGE
Get Status Messages	STATUS_MESSAGES
Get Status ID Description	STATUS_ID_DESCRIPTION
Clear Status ID	CLEAR_STATUS_ID
Get/Set Sub-Device Status Reporting Threshold	SUB_DEVICE_STATUS_REPORT_THRESHOLD
Get Supported Parameters	SUPPORTED_PARAMETERS
Get Parameter Description	PARAMETER_DESCRIPTION
Get Device Info	DEVICE_INFO
Get Product Detail ID List	PRODUCT_DETAIL_ID_LIST
Get Device Model Description	DEVICE_MODEL_DESCRIPTION
Get Manufacturer Label	MANUFACTURER_LABEL
Get/Set Device Label	DEVICE_LABEL
Get/Set Factory Defaults	FACTORY_DEFAULTS
Get Language Capabilities	LANGUAGE_CAPABILITIES
Get/Set Language	LANGUAGE
Get Software Version Label	SOFTWARE_VERSION_LABEL
Get Boot Software Version ID	BOOT_SOFTWARE_VERSION_ID
Get Boot Software Version Label	BOOT_SOFTWARE_VERSION_LABEL
Get/Set DMX512 Personality	DMX_PERSONALITY
Get DMX512 Personality Description	DMX_PERSONALITY_DESCRIPTION
Get/Set DMX512 Starting Address	DMX_START_ADDRESS
Get Slot Info	SLOT_INFO
Get Slot Description	SLOT_DESCRIPTION
Get Default Slot Value	DEFAULT_SLOT_VALUE
Get Sensor Definition	SENSOR_DEFINITION
Get/Set Sensor	SENSOR_VALUE
Record Sensors	RECORD_SENSORS
Get/Set Device Hours	DEVICE_HOURS
Get/Set Lamp Hours	LAMP_HOURS
Get/Set Lamp Strikes	LAMP_STRIKES
Get/Set Lamp State	LAMP_STATE
Get/Set Lamp On Mode	LAMP_ON_MODE
Get/Set Device Power Cycles	DEVICE_POWER_CYCLES
Get/Set Display Invert	DISPLAY_INVERT

Message	Name
Get/Set Display Level	DISPLAY_LEVEL
Get/Set Pan Invert	PAN_INVERT
Get/Set Tilt Invert	TILT_INVERT
Get/Set Pan/Tilt Swap	PAN_TILT_SWAP
Get/Set Device Real-Time Clock	REAL_TIME_CLOCK
Get/Set Identify Device	IDENTIFY_DEVICE
Reset Device	RESET_DEVICE
Get/Set Power State	POWER_STATE
Get/Set Perform Self Test	PERFORM_SELFTEST
Get Self Test Description	SELF_TEST_DESCRIPTION
Capture Preset	CAPTURE_PRESET
Get/Set Preset Playback	PRESET_PLAYBACK

In 2012, "ANSI E1.37-1 – 2012, Additional Message Sets for ANSI E1.20 (RDM) – Part 1, Dimmer Message Sets" was approved; this added the following messages:

Message	Name
General Parameter Messages:	
Get/Set Identify Mode	IDENTIFY_MODE
Get/Set DMX512 Block Address	DMX_BLOCK_ADDRESS
Get/Set DMX512 Fail Mode	DMX_FAIL_MODE
Get/Set Startup Mode	DMX_STARTUP_MODE
Get/Set Power-On Self Test	POWER_ON_SELF_TEST
Get/Set Lock State	LOCK_STATE
Get Lock State Description	LOCK_STATE_DESCRIPTION
Get/Set Lock PIN	LOCK_PIN
Get/Set Burn-In	BURN_IN
Get Dimmer Info	DIMMER_INFO
Get/Set Minimum Level	MINIMUM_LEVEL
Get/Set Maximum Level	MAXIMUM_LEVEL
Get/Set Curve	CURVE
Get Curve Description	CURVE_DESCRIPTION
Get/Set Output Response Time	OUTPUT_RESPONSE_TIME
Get Response Time Description	OUTPUT_RESPONSE_TIME_DESCRIPTION
Get/Set Modulation Frequency	MODULATION_FREQUENCY
Get Modulation Frequency Description	MODULATION_FREQUENCY_DESCRIPTION
Get Preset Info	PRESET_INFO
Get/Set Preset Status	PRESET_STATUS
Get/Set Preset Merge Mode	PRESET_MERGEMODE

Room was left in the standard for PLASA to add public messages and, additionally, manufacturers were given room to add messages of their own. Conflicts between devices from different manufacturers using the same messages are avoided because the responders associate the message with the UID, which has a manufacturer ID component. Here is a screen shot showing the actual parameters captured from a device over an RDM system:

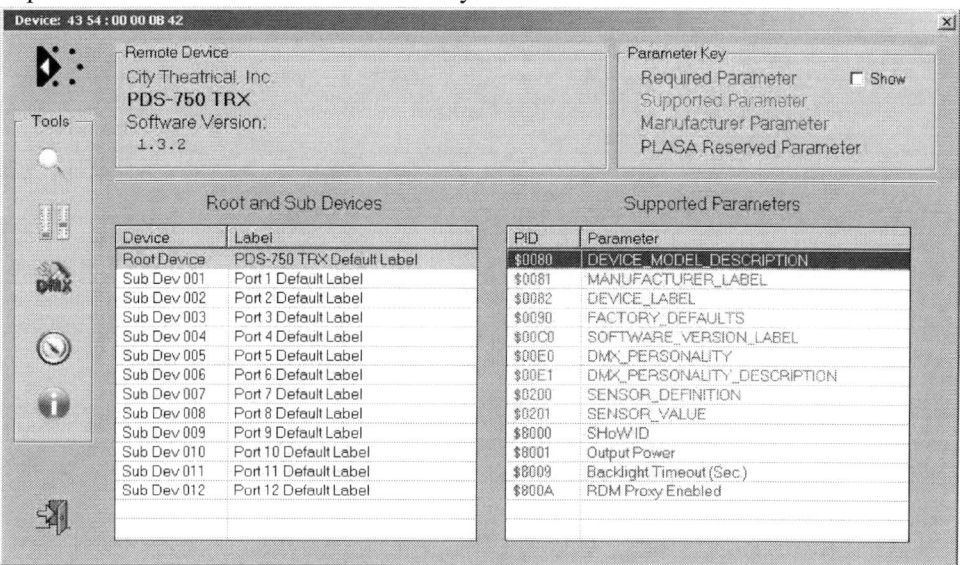

This structure allows things like this to be implemented:

RDM AND NETWORKS

While RDM was initially designed for transmission on DMX's EIA-485 physical transmission layer, in this age of networks it makes sense to carry it over Ethernet. Art-net (page 252), can transport RDM messages, and as of this writing, work has been underway for several years within PLASA on BSR E1.33 (RDMnet), Message Transport and Device Management of ANSI E1.20 (RDM) over IP Networks. Also under development is "BSR E1.37-2 Additional Message Sets for ANSI E1.20 (RDM) –Part 2, IPv4 & DNS Configuration Messages." This standard proposes new messages to allow RDM to configure a device's IP address, DNS, and DHCP settings remotely. Check with the PLASA Control Protocol Working Group CPWG), or `www.rdmprotocol.org`, for more information on the project status.

RDM IN THE ENTERTAINMENT CONTROL MARKET

As of this writing, RDM has been implemented in products from a number of manufacturers, and "plugfests" are conducted regularly to ensure that all the devices can talk to each other.

Chapter 21: Architecture for Control Networks (ACN)

"E1.17 – 2010, Entertainment Technology—Architecture for Control Networks," otherwise known as "ACN,"[1] is the most sophisticated open, consensus control standard yet developed by the live entertainment industry. ESTA (now PLASA) started work on ACN in 1997, and the standard was finally released in 2006; a revision was issued in 2010. The work took much longer than anyone expected, mostly because the task group had to invent several key protocols; it was an unprecedented collaborative effort, with direct competitors working together.

A BIT OF BLUE SKY THINKING

DMX works just fine, and RDM will let you do a lot. So why do we need this complex ACN protocol? ACN could replace DMX, giving you an open, network-based DMX replacement. But that's just the start. Let's do a bit of blue-sky thinking for a moment, and imagine:

For the Designer:
- Moving lights that precisely and automatically track moving scenery.
- A performer's actions determining which of a range of preselected colors is displayed on stage, while the lighting positions and intensity are controlled from the console.
- A moving light "hot backup" on the truss that can automatically take over for any other moving light in case of a failure.
- A touring show's console that could be monitored "live" over the Internet.
- A video clip or a lighting fade that can track the tempo of music.
- A sound image that automatically tracks a moving light (or vice versa).

---

1. Originally "Advanced Control Network," but the group wisely changed it to something a bit more future-proof.

For the Production Electrician

- Every device in a system configured using only names that make sense to you (no more universes or slots!).
- A huge lighting system that could be "discovered" in a few seconds.
- A list of selectable video clips accessible from your console, which is automatically updated whenever the projection designer updates the media, using the same network.
- The ability for the factory to diagnose a problem in your system over the Internet.
- A moving light that informs you of a stuck color wheel or a blown lamp before you bring the light up on stage.
- Multiple consoles from multiple manufacturers independently controlling the same dimmer rack over the same network infrastructure.
- 100% full tracking console backup with automatic switch-over within a second using just a network connection between the two.

ACN, with its network-centric design and incredible power and sophistication, could open the door to all of these ideas, and far, far more.

BACKGROUND AND MANDATE

The market that spurred the development of ACN was one we have seen time and time again: multiple, non-interoperable control approaches from a number of manufacturers vying in the marketplace, with users crying out for the systems to be able to talk to each other. In this case, the non-interoperable systems were Ethernet-based lighting data transmission systems from several lighting manufacturers, none of which could communicate with the others. Interoperability, as we have seen with DMX (Chapter 19) and MIDI (Chapter 22), generally induces markets to grow, so the key players within the ACN effort worked together for the betterment of all.

Interoperability, in fact, was the number one design goal for ACN, and potential applications for ACN extend beyond the lighting industry. While the standard development was certainly driven by lighting manufacturers, the task group (originally headed by Steve Carlson and later by Dan Antonuk) wisely made every effort to ensure that ACN can be applied as broadly as possible.

OVERVIEW

Basically, ACN enables a number of components to be connected on a control network. ACN is platform- and network-independent, but of course, IP running over

Ethernet is the most obvious choice at this point. Over the network, "controllers" can discover, configure, monitor, and control "devices", by "getting" and "setting" properties. Each component (or type of component) can be controlled in a way optimized for that device.

A few examples are:

- A moving light may present its pan and tilt axis properties in degrees with precision to three decimal places, while simultaneously relating those properties to XYZ target coordinates, which are presented in meters relative to a zero point onstage. The same luminaire might also present a list of the names of installed gobos, while allocating a simple on/off setting for its shutter and a percentage with precision to one decimal place for its dimmer.
- A video server might present a table filled with names of available clip and disk directories, while its current clip time code is presented as a continuously updating time value.
- The server's alpha channel might be presented as an integer value, while a rotation parameter could be allocated in degrees with precision to two decimal places and its current operating temperature is presented in degrees C.

ACN allows control for each parameter of each unit to be customized and, therefore, optimized; no longer does anyone have to shoehorn sophisticated control approaches into the unidirectional exchange of a number of eight-bit slots (as in DMX or RDM) or seven-bit integers (as in MIDI).

ACN'S ACRONYM SOUP

ACN is relatively straightforward conceptually, but (as is often the case) the ease of use it offers means that incredibly complex, low-level work is done by the system. Since this book is targeted at end users, we will only scratch the surface here, and offer a general, functional overview of ACN. If you want to know more about how the building blocks work, read the standard.

Network engineers may seem to be in love with acronyms, but in fact these abbreviations become a necessary evil with something as complex as ACN. In ACN, the main acronyms (you might as well start learning these now) are CID, DDL, DMP, SDT, RLP, and UDP/IP. Let's take a quick look at each; later we will cover some further in depth.

Component IDentifiers (CID)
Every unit connected to an ACN network needs a unique address, known in ACN as a Component IDentifier (CID). Given past experience with limited address space (i.e., IPv4), the task group wisely used a 128-bit number for this CID, allowing for an enormous possible range. Devices will typically have a single CID, but it's also possible for one physical device to use more than one—for example, if you were running multiple control applications on the same computer, each application might use its own CID.

Device Description Language (DDL)
The Device Description Language (DDL) defines the interface between device functionality and the control abstraction of ACN. A DDL is a file, written in eXtensible Markup Language (XML), which describes a device's properties in a tree-like structure for use by controllers over the ACN network. Properties listed in a DDL file are associated with "behaviors," which provide additional information about the property, such as what the property actually does, a name for the property to be used by the controller, and so on. The DDL structure also gives different controllers the ability to pick and choose which level of control they want. A simple controller may just get the address of the properties and some other, simple information, while a sophisticated controller might use the entire DDL file.

Device Management Protocol (DMP)
While the DDL contains a list of the properties associated with a particular device, the Device Management Protocol (DMP) is the part of ACN where the "getting" and "setting" of the properties takes place. Other functions include the handling of parameters that might be continuously (or irregularly) updated, through the use of "events" and subscriptions. The DMP only understands how to work with the properties; it doesn't understand what they mean.

Session Data Transport (SDT) Protocol
The Session Data Transport (SDT) Protocol solved some of the key design challenges of ACN, offering communications far more sophisticated than those available in a simple protocol like DMX. Let's look at two key differences between DMX and ACN, and how SDT addresses these issues. First, a simple protocol like DMX repeatedly transmits data, whether levels are changing or not. A continuously updated approach like this wastes bandwidth, but has the advantage of overwriting any data that gets corrupted along the way or isn't delivered, and the approach works well for a variety of straightforward applications. However, with a protocol suite as complex as ACN and the huge amount of data ACN networks could carry, a DMX-like repetitive data approach could swamp the network with unneeded, repetitive data. Instead, what is needed in show systems is "reliable"

transmission, meaning that each control packet is guaranteed to reach its destination (or the system will be alerted if a failure takes place).

Second, in DMX, because its data is transmitted over a simple, point-to-point bus connection, all the control data is broadcast to every connected device, and it's the responsibility of each receiver of the data to decide whether or not to act on that data. If three devices need to be controlled by the same data, they either all listen to the same data slots on the DMX line or (if at other times they need to be controlled independently) multiple copies of the data are sent. In a sophisticated network like ACN, this kind of transmission could flood the network with redundant, unnecessary data, so "multicasting" data is preferable, giving multiple devices a way to simultaneously receive the same stream of data from a single sender, maximizing network efficiency.

Standard approaches based on TCP/IP could solve some of these problems, since the suite offers reliable transmission, and a wide variety of multicasting approaches are available. However, TCP was generally designed to manage the connection of devices in pairs, not in groups, and TCP is optimized to handle "streams" of data, rather than short, burst-like messages. Of course, it's quite possible with TCP to simply set up dozens or hundreds of connections and still get the all-important reliable delivery, but this could make for an extremely slow network or one that needs very careful management and design, since TCP was designed more with reliability than performance in mind.

The task group looked far and wide for a protocol that could offer both reliability and multicasting, but couldn't find anything that would work within the demands of ACN. So, the task group invented from scratch the Session Data Transport (SDT) protocol, which offers reliable multicasting and guarantees to layers above (i.e., DMP) that packets will be delivered to multiple receivers and that if packets arrive out of order, the correct packet sequence can be reconstructed. SDT does this through the use of "sessions" (hence the name), through which data can flow to and from a device bidirectionally. SDT also integrates some other key, low-level functions of ACN, such as packing data into Protocol Data Units (PDU). However, most SDT protocol details relate to low-level network management and therefore are outside the scope of this book.

Root Layer Protocol (RLP)
The Root Layer Protocol (RLP) is the interface between ACN protocols (SDT, DMP, etc.) and the lower-layer network transport protocols, such as UDP/IP. RLP has been designed separately to ensure maximum network independence. ACN includes a series of Profiles for Interoperability (EPI) and one of the EPIs

addresses ACN over UDP: "EPI 17. ACN Root Layer protocol Operation on UDP." If a designer wanted to send ACN over a different network, a new RLP could be created.

UDP/IP and Ethernet

UDP stands for User Datagram Protocol; IP, of course, is the Internet Protocol, and this pair of protocols (actually part of the TCP/IP suite) is the obvious choice for an ACN transport mechanism. UDP is TCP's simpler, less complex cousin. It offers "unreliable," connectionless delivery, but this works in ACN because the reliable, ordered, multicast data delivery is handled by the Session Data Transport (SDT) protocol. UDP has another advantage for control purposes in that it is generally faster than TCP, since it sends datagrams right away. TCP needs additional time to manage connections and it also might delay an individual packet to pack it properly onto a stream. IEEE 802.3, commonly known as Ethernet, is ACN's main physical network.

PROTOCOL STRUCTURE

The ACN protocols work together in a layered approach; here is a simplified structure:

Component
(console or target device)
Component IDentifier (CID)
Device Description Language (DDL)
Device Management Protocol (DMP)
Session Data Transport (SDT)
User Datagram Protocol (UDP)
Internet Protocol (IP)
Network (e.g., Ethernet)

IDENTIFIERS AND ADDRESSES

ACN uses a number of Universally Unique IDentifiers (UUIDs), and the large number of different types of ID numbers that are spread throughout the ACN standard can be quite confusing. However, the basic idea is simple: ACN protocols create or use a series of unique identifiers that cannot occur anywhere else on the network, and the protocols can create these UUIDs without any sort of global coordination or consultation with any sort of registration authority. ACN uses several techniques for generating these UUIDs, but describing those techniques is outside the scope of this book.[2]

As detailed above, each connected ACN device has a Component IDentifier (CID), which is one UUID. Other IDs include the IP address, the component names, and the Device Class IDentifier.

IP Addressing

Each host on an IP network, of course, needs a unique IP address. Many show networks, as of this writing, require manual configuration of IP addresses, but since ACN is designed to be as self-configuring as possible, EPI13 "Allocation of Internet Protocol Version 4 Addresses to ACN Hosts" mandates automatic configuration of IP addresses through DHCP (page 185) and several related protocols.

Fixed Component Type Name (FCTN)
User Assignable Component Name (UACN)

Two text strings are included in ACN for human identification of a particular device: the Fixed Component Type Name (FCTN) and the User Assignable Component Name (UACN). The FCTN would typically be set by the manufacturer and would include the manufacturer name and a general model type (i.e., "Zircon Designs Woo Flash"). The UACN is assigned by the user and would likely include some informational text useful to them (i.e., "Unit #666, Truss # 13").

Device Class IDentifier (DCID)

The confusingly named Device Class IDentifier (DCID) should really be called "Device Name" or something similar, since the DCID is used by a manufacturer to indicate a particular equipment model or type of device to the ACN network. For instance, 20 moving lights of the same brand and type should present 20 identical DCIDs to a network; of course, the CIDs and other IDs will be unique. Consoles and other controllers can use this information for configuration, associating certain types of control libraries with the DCID; only one DCID is allowed per CID.

---

2. See RFC 4122 for more information.

In Summary

Let's add a column to our simplified protocol summary table in order to show how identifiers and addresses lay out (including some other addresses handled by the DDL and other protocols):

Protocol or Unit	Address/ID Type
Component (console or target device)	
Component IDentifier (CID)	Component IDentifier (CID)
Device Description Language (DDL)	Device Class IDentifier (DCID)
Device Management Protocol (DMP)	Protocol ID (set by PLASA for each Protocol)
Session Data Transport (SDT)	Session Number
User Datagram Protocol (UDP)	Port Number
Internet Protocol (IP)	IP Address
Network (e.g., Ethernet)	MAC Address

Of course, in UDP/IP over Ethernet, the network itself deals only with IP addresses and MAC addresses; ACN's CIDs, DCIDs, and so on are all just part of the data payload carried over the Ethernet MAC frame.

DISCOVERY

In complex DMX-based systems, electricians have to lay out and configure every device in the system, figure out what DMX "universe" each device is on, resolve any conflicts, patch and assign all those units, and then track it all with some tedious paperwork. ACN, on the other hand, was designed to be "plug and play," and it can automatically discover connected devices and present them to the user for configuration. ACN discovery is an important feature, but it is very complex, so we will only include a simplified overview here.

On IP networks, ACN Discovery uses the Service Location Protocol (SLP, RFC 2608), which is an open protocol developed under the IETF to locate "services" on a network. OK, get ready for more acronyms—SLP has three types of objects that ACN uses: Service Agents (SA), User Agents (UA), and Directory Agents (DA). In ACN, Service Agents are controlled devices (a moving light, a dimmer rack, a video server, a sound matrix) that offer their "service" to the network; User Agents are controllers that look for the Service Agents. Optional (but very useful) Directory Agents are servers that store information about known service agents and how they can be reached. The services themselves are indicated using the Uniform Resource Identifier (URI), and might take the form of something like `service:acn.esta`.

When an ACN network of connected devices is discovered, the controller (User Agent) first looks to see if there is a Directory Agent (DA) available. If it finds one, it will unicast all further service requests to the DA. If not, it multicasts a request for the service type `acn.esta`. Any connected device that can offer an ACN "service" then responds with its unique Component IDentifier (CID). If the controller (UA) recognized that unique CID as one it has dealt with before, it can simply recall that configuration and control information from that previous session and use that. If not, it issues an attribute request to that device, which responds with a DMP message called `csl-esta.dmp`. This response contains information about the device such as the DCID, the two human-readable text strings, and other information.

If the controller recognizes the device as a type it understands, it can move on to another device; otherwise, it could request the full DDL (Device Description Language) file to get all the available control parameters, and then present this information to the user for further action. When the controller has worked through all the networked devices in this manner, all this information could be presented to the user, who would then take the discovered devices and patch them to appropriate control channels or other control methods.

I'm sure your head is hurting now after reading that section, but keep in mind that the whole point of ACN is to make the user's life *easier*. If the discovery process works correctly, the user should simply see something like this (an actual ACN discovery result):

Network Devices

Device Type	Name/Component	Status	Connected	Address
MultiConsole	EOS Console Console _1 : BYE BYE BIRDIE NYC 10th Tech_	Online	Connected	10.101.102.110
CEM3	Rack: CEM3.Rack (1)	Online	Connected	10.101.101.101
RdmDmxGateway	Net3 4-Port Gateway Ray's Second Gateway	Online	Connected	10.101.50.101
RdmDmxGateway	Net3 4-Port Gateway Ray's ACN Gateway	Online		10.101.50.103
AcnRfrClient	ETC Console	Online	Connected	10.101.102.110
OtherAcnDevice	EOS Console Console	Online		10.101.102.110
OtherAcnDevice	EOS Console RVI	Online	Connected	10.101.102.112

ACN discovery result provided by Dan Antonuk, ETC

CONTROL OF DEVICES

As detailed above, at ACN's functional core are the virtual properties of each device, and each property represents some controllable or monitor-able aspect of a device. A property could be remote-controllable over the ACN network, or conversely, could represent a value maintained locally by the device (such as the current value of a local control slider that might be moved by an operator at any time). Properties can also be permanently fixed (i.e., a unit serial number) or held constant by a device, and they can be set to be read only, write only, or set for read/write access. A single, physically controlled unit can be made up of one or more virtual devices; for instance, a dimmer rack might have 96 ACN devices for level control of each dimmer, and also contain a 97th device related to the whole unit, which reports information such as the monitored temperature, fan speed, incoming voltage, current draw, and so on. Alternatively, something like an LED fixture might be represented as a single device, with three color intensity properties (RGB).

The Device Management Protocol (DMP) is the mechanism to control, configure, or monitor specific properties in a connected device, and DMP messages are sent over virtual device connections (created and handled, typically, by SDT). DMP only manages properties and their values it doesn't really care about or understand the meaning of any particular property (meaning is handled in the DDL). To accomplish all this sophisticated control, DMP uses only 17 messages, the most common of which are: `Get Property` and `Set Property`, `Subscribe` and `Unsubscribe`; `Get Property Reply`, `Get Property Fail`, `Set Property Fail`, `Subscribe Accept`, `Subscribe Reject`, and `Event`.

Let's look at a couple of these commands to get an idea of how they work. `Get Property`, for example, reads the value contained in a device property. The controller sends the `Get Property` command, and the device replies with the `Get Property Reply` command. For example, a controller might ask a moving light what its cumulative lamp hours are with a `Get Property` message, and the moving light device then replies with the `Get Property Reply` containing the number of hours. If, for some reason, the moving light can't respond with the lamp hours number it will issue a `Get Property Fail`, which contains error codes to pass along to the controller.

If allowed by the target device, `Set Property` will write to the device a new value for one of its properties, which it can then act on. For example, a controller might write a new color to an LED fixture's color property, and then the LED fixture would change to that property. If there is a problem, it replies with a `Set Property Fail` command.

Events and Subscriptions

There are many cases in entertainment control systems where devices may have data to send back to the controller, and that data may change completely, independently and asynchronously of the controller. For example, a household wall station in a venue might have a manual fader to bring up or down the houselights which could be changed at any time, while the actual level control might be accomplished through the control console. Or, an audio matrix might have a dB level that changes continually.

In simpler protocols (like RDM), this sort of communication is accomplished through a polling approach (see "Master-Slave/Command-Response," on page 81), where the master controller continuously asks each device if it has data to send. If it replies, the controller can stop and deal with that data. However, this approach wastes bandwidth, and can be slow (especially in large networks), since every device has to be polled whether it has data to send or not. So rather than using something like polling, ACN uses a "producer-consumer" model (page 83), and allows the controller to "subscribe" to "events" generated by a device.

An event is simply the change in the value of any property of a device, whether the change was caused internally or externally. Properties, of course, can take on a variety of forms. For instance, a video server might contain as a property the name of a clip to be played, which is commanded by the controller. Once that clip is rolled, the video server itself might maintain a running, constantly updating property containing the time code value as the clip plays. If a controller has subscribed to that time code property, changes in the property generate events, and, through DMP, properties can be continuously monitored by the controller. Additionally, of course, a device can offer subscriptions to multiple controllers, each of which may unsubscribe at any time. This is a very powerful and efficient way to manage the huge amount of data that could be found in sophisticated ACN networks.

ACN IMPLEMENTATIONS

It's important to understand that ACN is simply a way to manage properties and the functions of those properties. The ways the properties and related features are implemented are completely up to the manufacturer. For instance, in case of a failure needing a unit replacement, one manufacturer might require you to go in and repatch the new unit at the console, then manage the changes to the cues yourself. Another manufacturer could design a system in which if the console sees one unit go offline and another, similar unit come online, it might be able to automatically recognize and control the replacement unit.

Fixture "libraries" are already incredibly important in the world of DMX and RDM, and they will be even more important with ACN, since every unit can be controlled in a different way. Unit configuration and set up might be done through a console's fixture library, a network-based utility program, or a Web page served up by the unit itself over the network.

ANSI E1.31: STREAMING ACN (SACN)

"ANSI E1.31 – 2009, Entertainment Technology – Lightweight streaming protocol for transport of DMX512 using ACN" is a subset of "full" ACN specifically designed to carry DMX data. It has DMX's universes and slots, and because it only carries DMX frames, is covered in the DMX chapter on page 255.

ACN IN THE ENTERTAINMENT CONTROL MARKET

As of this writing, ACN has been in the market for more than five years, and unfortunately, "full" ACN has mostly found application only within the product lines of individual companies[3]. However, sACN has been successful and has found application in a variety of products, and sACN and RDMNet (page 268) have the potential to move more products into the native Ethernet world. It's my hope, in the long run, that sACN will lead the way to wider adoption of "full" ACN, because it could open up a whole new world of control possibilities. But for this to happen, it might take a new generation of lighting techs who are very comfortable with networking.

---

3. The lighting company ETC, and the microphone manufacturer Shure are the two most prominent as of this writing.

Chapter 22: Musical Instrument Digital Interface (MIDI)

The **Musical Instrument Digital Interface** (MIDI) was originally designed for synthesizer interconnection, and is now used widely for a variety of applications. MIDI's roots are in the early 1980s when, with the explosive growth of sophisticated keyboard synthesizers, musicians began to want their keyboards linked to simplify complex studio and live-performance setups. Several manufacturers had developed proprietary interfaces to connect their own equipment, but these systems would not work with gear from other manufacturers. In 1981, engineers at Sequential Circuits, a major manufacturer at the time, began talks with their counterparts at Roland and Yamaha in order to standardize inter-synthesizer connections. These talks eventually resulted in the formation of the MIDI Manufacturer's Association (MMA), and the release of the official MIDI specification in 1983.

BASIC STRUCTURE

In its most basic configuration, MIDI allows instruments to be connected in a simple master/slave relationship. No audio is transmitted over a MIDI line; only *control information* representing musical events is sent. So, if the MIDI output from synthesizer one (master) is connected to synthesizer two's (slave) MIDI input, pressing the third "A" note key on the master synthesizer simultaneously creates the electronic equivalent of an A3[1] key-press on the slave synthesizer. Synthesizer sound program changes

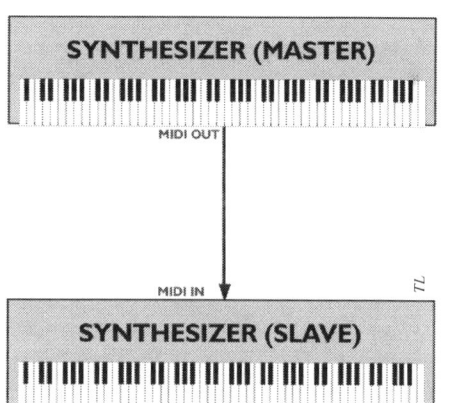

and other commands are transmitted in a similar manner: if a program is selected on the master synthesizer, the same program will also be selected on the slave. MIDI is unidirectional and open loop; so, in the configuration shown in the block diagram, nothing you do on the slave will have any effect on the master; in fact,

---

1. This may look like hex, but it's musical terminology meaning the A in the third octave.

the master has no way of knowing whether the slave is even connected or powered up.

PHYSICAL CONNECTIONS

MIDI was developed before layered, network control approaches became commonplace, and so it defines the interconnection hardware, the interface, and the command set. MIDI runs asynchronously at 31.25 kbits per second, which isn't very fast, but is actually fast enough to handle all but the most advanced musical applications with an acceptable latency. Systems have far more computer horsepower now than was available when MIDI was developed, so if more bandwidth is needed or latency is critical, you can always just send multiple parallel MIDI streams over multiple links.

MIDI is, by specification, opto-isolated (page 97): the inputs of a MIDI receiver connect directly to an opto-isolator chip. The communications circuit is a current-loop type (page 133); a current-on state is logical zero. Since MIDI was designed for simple synthesizer interconnection in a studio or on a stage, the maximum standard cable length was standardized at 50 feet; converters are available which convert MIDI into RS-422 and back to drive longer distances. However, because MIDI is based on a robust current loop interface that is highly immune to EMI, it may well work over significantly longer distances than 50'. For that reason, in my opinion, it's worth trying out your implementation (I've run MIDI hundreds of feet over microphone cable myself) before using converters/extenders, because they add additional failure points.

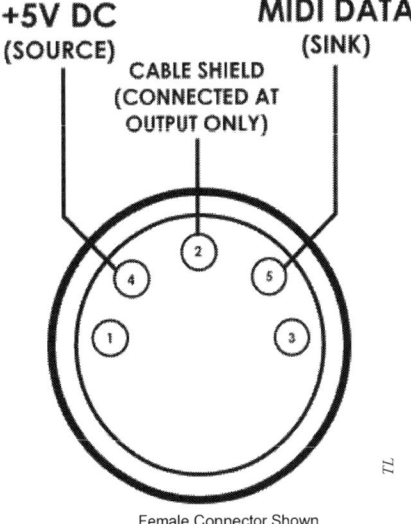

Female Connector Shown

MIDI transmits its data over standard shielded, twisted-pair cable, and uses five-pin "DIN 180°" connectors. Female connectors are used on equipment, with the shield connected only at the MIDI Out port. All cables are male-to-male, with the shield connected at both ends. The MIDI DIN connector has the control signals shown in the graphic at left. Since the standard is unidirectional, a MIDI instrument usually has both a receiver (MIDI In), a transmitter (MIDI Out), and sometimes a pass-through connector (MIDI Thru)—an opto-isolated copy of the data on the MIDI In port. This MIDI Thru port is used for daisy-chaining devices, although this kind of daisy chaining is not recommended for critical show purposes (see "Recommended MIDI Topologies," on page 294).

MIDI MESSAGES

MIDI's data is transmitted in a 10-bit word composed of a start bit, an eight-bit data octet, and a stop bit. Musical data is broken down into "messages," each composed of a "status" octet followed by zero or more "data" octets. A status octet always has its most-significant bit set to 1, i.e., `1xxxxxxx` (where the "x" bits are the control message and data); a data octet has its most-significant bit reset to 0, i.e., `0xxxxxxx`.

Pressing the second C note key on a MIDI keyboard generates a message in hex, such as `90 24 3F`, which breaks down as follows:

90	Status Octet	9 means Note On; 0 means channel 1
24	Data Octet 1	Note C2 (second C note from the left)
3F	Data Octet 2	Velocity of 63 (decimal)

Releasing the key generates the hex message `80 24 3F`, or Channel 1 Note *Off*, C2, Velocity[2] of 63 (alternatively, a note on with a velocity of zero could be used to trigger an "off" event).

There are two primary types of MIDI messages: Channel and System.

CHANNEL MESSAGES

Four bits of a Channel Message's status octet are used to denote the message's channel. This allows a controller to separately address, on a message by message basis, different devices. Four bits gives us 16 discrete control channels, and MIDI is a broadcast standard, so all messages are simply sent to all devices on a given MIDI link; each receiver is responsible for determining, based on its configuration, whether it has been set for that channel and should act on a particular message.

Within the channel classification, there are Channel Voice Messages, which control the instrument's voices, and Channel Mode Messages, which determine how a device will deal with the voice messages. These days, one physical MIDI-control-

---

2. Velocity indicates how quickly—that is, how hard—a key was pressed. Synthesizers and samplers can use this information to create different sounds for differing key-press intensities.

lable instrument or, more likely, a computer running a software synthesizer) will likely have many "virtual" instruments, each operating on a different channel.

The Channel Voice Messages are:

Function	Octets		Notes
	Binary	Hex	
Note Off	1000cccc 0nnnnnnn 0vvvvvvv	8c	cccc=Channel number1–16 (0–F_{16}) nnnnnnn=Note number, 0–127 (00–$7F_{16}$) cccc=Channel number, 1–16 (0–F_{16})
Note On (Velocity of 0=Note Off)	1001cccc 0nnnnnnn 0vvvvvvv	9c	cccc=Channel number, 1–16 (0–F_{16}) nnnnnnn=Note number, 0–127 (00–$7F_{16}$) vvvvvvv=Velocity, 0–127 (00–$7F_{16}$)
Poly Key Pressure	1010cccc 0nnnnnnn 0rrrrrrr	Ac	cccc=Channel number, 1–16 (0–F_{16}) nnnnnnn=Note number, 0–127 (00–$7F_{16}$) rrrrrrr=Pressure, 0–127 (00–$7F_{16}$)
Control Change	1011cccc 0xxxxxxx 0yyyyyyy	Bc	cccc=Channel number, 1–16 (0–F_{16}) xxxxxxx=Control number, 0–120 (00–78_{16}) yyyyyyy=Control Value, 0–127 (00–$7F_{16}$)
Program Change	1100cccc 0ppppppp	Cc	cccc=Channel number, 1–16 (0–F_{16}) ppppppp=Prog. number, 0–127 (00–$7F_{16}$)
Channel Pressure	1101cccc 0xxxxxxx	Dc	cccc=Channel number, 1–16 (0–F_{16}) xxxxxxx=Pressure value, 0–127 (00–$7F_{16}$)
Pitch Bend Change	1110cccc 01111111 0mmmmmmm	Ec	cccc=Channel number, 1–16 (0–F_{16}) 1111111=Least significant pitch octet mmmmmmm=Most significant pitch octet

I'm including the standard MIDI messages in this table for reference, but in entertainment control applications, you are most likely to encounter simple messages such as Note On, Note Off, Program Change, and Control Change. The Control Change message was defined as a sort of configurable generic control message—and there are a number of standardized controller types for the Control Change messages in the MIDI specification—for musical parameters like "breath Controller," "portamento time," and so on.

The control numbers $78-7F_{16}$ were reserved, and make up the Channel Mode messages:

Function	Octet		Notes
	Binary	Hex	
All Sound Off	1011cccc 01111000 00000000	Bc 78 00	cccc=Ch. number, 1–16 ($0-F_{16}$)
Reset All Controllers	1011cccc 01111001 00000000	Bc 79 00	cccc=Ch. number, 1–16 ($0-F_{16}$)
Local Control Off	1011cccc 01111010 00000000	Bc 7A 00	cccc=Ch. number, 1–16 ($0-F_{16}$)
Local Control On	1011cccc	Bc 7A 7F	cccc=Ch. number, 1–16 ($0-F_{16}$)
All Notes Off	1011cccc 01111011 00000000	Bc 7B 00	cccc=Ch. number, 1–16 ($0-F_{16}$)
Omni Mode Off	1011ccc 01111100 00000000	Bc 7C 00	cccc=Ch. number, 1–16 ($0-F_{16}$)
Omni Mode On	1011cccc 01111101 00000000	Bc 7D0 0	cccc=Ch. number, 1–16 ($0-F_{16}$)
Mono Mode On (Poly Mode Off)	1011cccc 01111110 0nnnnnnn	Bc 7E	cccc=Ch. number, 1–16 ($0-F_{16}$) nnnnnnn=Number of channels
Poly Mode On (Mono Mode Off)	1011cccc 01111111 00000000	Bc 7F 00	cccc=Ch. number, 1–16 ($0-F_{16}$)

SYSTEM MESSAGES

The other general category of messages is the channel-less group of System Messages. Within System Messages, there are three subcategories: System Common Messages, which are intended for all devices connected to the system; Real-Time Messages, used for timing and other functions; and System Exclusive Messages, designed so that manufacturers could send any kind of data over the MIDI line.

System Common messages include:

Function	Octet Binary	Hex	Where
MIDI Time Code Quarter Frame	11110001 0nnndddd	F1	nnn = Message type dddd = Values
Song Position Pointer	11110010 01111111 0mmmmmmm	F2	1111111 = Least significant octet mmmmmmm = Most significant octet
Song Select	11110011 0sssssss	F3	sssssss = Song number
Undefined	11110100	F4	
Undefined	11110101	F5	
Tune Request	11110110	F6	

MIDI Time Code is covered in the "MIDI Time Code" section, on page 330. Song Position Pointer (SPP) and Song Select are part of a MIDI synchronizer system (see "MIDI Sync," on page 287) used primarily in musical sequencers. Tune Request is used with an analog synthesizer (remember those?) to request that the synthesizer tune all its internal oscillators.

The System Real Time messages, which have only Status (no Data) octets, are:

Function	Status Octet Binary	Hex
Timing Clock	11111000	F8
Undefined	11111001	F9
Start	11111010	FA
Continue	11111011	FB
Stop	11111100	FC
Undefined	11111101	FD
Active Sensing	11111110	FE
System Reset	11111111	FF

The Timing Clock message, used for MIDI sync applications, is sent at a rate of 24 messages per quarter note. Start messages are sent when a play button is pressed on a master sequencer; Continue resumes play after a sequence has been stopped; Stop is the equivalent of pressing the stop button on a sequencer. See below for description of Active Sensing. System Reset, as you might expect, resets the entire system to the initial power-up state.

ACTIVE SENSING

MIDI includes a mode called Active Sensing, where devices repeatedly send an FE_{16} Active Sensing octet (a System Real-Time Message) every 300 ms. If a receiver does not pick up either a standard MIDI or Active Sensing message every 300 ms, it can assume that a problem has occurred upstream, and it will shut down all open voices and wait for the problem to be resolved. This is a very useful way to avoid the "stuck note" problem, where a Note On is received, then something goes wrong, and the Note Off is never received. Few entertainment control devices implement this feature, but when entertainment system engineers are looking at raw MIDI data, they may see this octet pulsing down the line if any musical equipment is connected.

MIDI SYNC

A MIDI Note On message is absolute regarding the note number, and so on, but relative regarding time; it contains no information about *where* in a musical composition that note belongs. A system based on Song Position Pointers (SPP) was implemented in the original MIDI spec to communicate this information for sequencing and tape-sync applications. The SPP system isn't really of much use in most entertainment control applications, but following is a brief introduction.

The SPP is a count of the number of 16th notes that have elapsed since the beginning of the song or since a recording deck's start button was pressed. If the MIDI system is locked to an audio or video deck player and the deck is fast-forwarded, the system sends SPP messages only before and after shuttling. When the deck is playing, the synchronizer device maintains its own SPP count, which is incremented every six Timing Clock messages (which are sent out at a rate of 24 per quarter note). The timing clock is based on tempo, not real time, so its frequency varies with the tempo of the song.

SYSTEM-EXCLUSIVE MESSAGES

The MIDI specification set aside two status octets—Start of System Exclusive (SysEx) and End of System Exclusive (EOX)—to allow manufacturers to send any kind of data down the MIDI line "exclusively" of the messages defined in the standard. This SysEx approach provided a way for manufacturers to send messages specifically to their own devices, and also a means for future expansion—a very wise move on the part of the specification authors. System-exclusive messages are not channel-specific; any device that receives the SysEx messages can act on them, assuming they are configured to do so and that they understand the messages.

SysEx messages are structured in this way:

Function	Octet		Where
	Binary	Hex	
Start of System Exclusive	11110000 0iiiiiii	F0	iiiiiii = ID Number of 0–124 (00–$7C_{16}$) If first octet is zero, the next two octets are extensions to the ID number
Data	0xxxxxxx ...		xxxxxxx = 7-bit data (or extended ID); any number of octets
End of System Exclusive	11110111	F7	

In a System Exclusive message, the $F0_{16}$ SysEx header octet is sent first, then comes a manufacturer's ID number, and then a number of data octets (of which only seven bits can be used, due to the MIDI structure), whose meaning is determined by the manufacturer. When complete, an EOX message is sent to return the system to normal operation.

Any manufacturer can register for a Manufacturer's ID number with the MMA, and if they make a product and publish the data within one year, the ID number becomes permanent. So many manufacturers around the world registered numbers that the MMA had to expand the ID number from one to three octets, but still keep the legacy numbers active. So they chose the unassigned 00_{16} ID for expansion; if a receiver sees the 00_{16} ID number, it should treat the next two octets as an extension to the ID number instead of as data.

Three of the 127 possible manufacturer's ID numbers are set aside for special types of messages. $7D_{16}$ is for research or academic development, and devices using this ID are not to be released to the public. $7E_{16}$ is used for "nonreal-time" messages, and $7F_{16}$ is used for "real-time" system-exclusive messages, very useful for our industry, as we will see later (e.g., "MIDI Show Control (MSC)," on page 301). The overall structure of messages using these special ID numbers is F0 $7F_{16}$ (or 7D or 7E), the "device ID," indicating to which device the message is addressed, two sub-ID octets, the data octets, and then the $F7_{16}$ EOX status octet to close out the message:

```
| F0 | 7F | Device ID | Sub-ID #1 | Sub-ID#2 | Data | F7 |
```

Since MIDI is a simple point-to-point interface, no priority levels can be assigned to particular messages. But the real-time and nonreal-time categories do, theoretically, allow a receiver to work with them in different ways.

The non real-time SysEx messages are:

Function	Sub-ID #1	Sub-ID #2
Unused	00	
Sample Dump Header	01	
Sample Data Packet	02	
Sample Dump Request	03	
MIDI Time Code Special	04	00
MIDI Time Code Punch In Points	04	01
MIDI Time Code Punch Out Points	04	02
MIDI Time Code Delete Punch In Point	04	03
MIDI Time Code Delete Punch Out Point	04	04
MIDI Time Code Event Start Point	04	05
MIDI Time Code Event Stop Point	04	06
MIDI Time Code Event Start Point with Additional Information	04	07
MIDI Time Code Event Stop Point with Additional Information	04	08
MIDI Time Code Delete Event Start Point	04	09
MIDI Time Code Delete Event Stop Point	04	0A
MIDI Time Code Cue Points	04	0B
MIDI Time Code Cue Points with Additional Information	04	0C
MIDI Time Code Delete Cue Point	04	0D
MIDI Time Code Event Name in Additional Information	04	0E
Sample Dump Multiple Loop Points	05	01
Sample Dump Loop Points Request	05	02
General Information Identity Request	06	01
General Information Identity Reply	06	02
File Dump Header	07	01
File Dump Data Packet	07	02
File Dump Request	07	03
MIDI Tuning Standard Bulk Dump Request	08	00

Function	Sub-ID #1	Sub-ID #2
MIDI Tuning Standard Request	08	01
General MIDI System On	09	00
General MIDI System Off	09	01
End of File	7B	
Wait	7C	
Cancel	7D	
NAK	7E	
ACK	7F	

Many of these non real-time messages are found primarily in musical MIDI applications, but several are worth mentioning here. General MIDI is discussed below. Wait, Cancel, NAK, and ACK are all part of the Sample Dump standard. Many of the MIDI Time Code (MTC, Chapter 25) commands listed here are part of the MIDI set up commands used to position sequencers and other devices in musical synchronization systems.

In addition to the non-real-time messages we've covered, there are also a number of globally defined, real-time SysEx messages, some of which are covered in their own chapters MIDI TIme Code (Chapter 25), MIDI Show Control (Chapter 23), and MIDI Machine Control (Chapter 24):

Function	Sub-ID#1 (hex)	Sub-ID #2 (hex)
Unused	00	
MIDI Time Code Full Message	01	01
MIDI Time Code User Bits	01	02
MIDI Show Control	02	
Notation-Bar Number	03	01
Notation-Time Signature Immediate)	03	02
Notation-Time Signature Delayed)	03	42
Device Control-Master Volume	04	01
Device Control-Master Balance	04	02
MIDI Time Code Real Time Cuing	05	
MIDI Machine Control Command	06	
MIDI Machine Control Response	07	
MIDI Tuning Standard	08	

MIDI RUNNING STATUS

With extremely complex, multichannel musical compositions, sometimes MIDI bandwidth limits will be approached, causing slight musical mistimings. To enable as much musical data as possible to be sent down the line using the fewest messages, the developers of MIDI created a feature called Running Status. Running Status allows one MIDI Status octet to be sent, followed by a group of data octets. The data octets in the group will then each be treated as the same type of message until another status octet is received. For instance, to trigger and release a chord of three notes using standard MIDI, we would need to send six messages and 18 octets (i.e., three Note On messages, and three Note Off messages).

Meaning	Normal (hex)
Note $3C_{16}$ On	90 3C 7F
Note 27_{16} On	90 27 62
Note 43_{16} On	90 43 80
Note $3C_{16}$ Off	80 3C 00
Note 27_{16} Off	80 27 00
Note 43_{16} Off	80 43 00

However, using MIDI Running Status, we would only have to send 14 octets:

Meaning	Running Status (hex)
Note $3C_{16}$ On	90 3C 7F
Note 27_{16} On	27 62
Note 43_{16} On	43 80
Note $3C_{16}$ Off	80 3C 00
Note 27_{16} Off	27 00
Note 43_{16} Off	43 00

The first 90_{16} status octet sets the receiver into Note On running status mode, covering the next two Note On messages. The 80_{16} status octet changes the running status into a Note Off condition, covering the next two Note Off messages. In our small example, a four-octet reduction is not all that significant, but you can see how this would scale up to significant savings over a larger time frame.

GENERAL MIDI

In some ways, "General MIDI" is a misnomer; it really doesn't affect MIDI at all—it specifies a standard way of *implementing* MIDI in musical instruments for simple musical applications. Nonetheless, it deserves a brief mention here.

Before General MIDI, musicians set up their systems in any way that was comfortable for them: channel 1 for this keyboard, channel 2 for that tone module, channel 10 for a drum machine, and so on. Similarly, each synthesizer used different program numbers for each of its possible sounds. Since each system was individually configured, a song composed on one system might come out completely wrong the notes meant for the piano voice might come out on the drum set voice) when played back on another system.

General MIDI standardizes setups so that you can build a song on one system, transport it to another system, play it back, and hear some semblance of what you created on the original system.

The highlights of General MIDI are:

- Middle C is MIDI Note 60 ($3C_{16}$).
- All voices must respond to velocity information.
- All synthesizers should have a minimum of 24-voice polyphony.
- All synthesizers must be capable of responding to all 16 MIDI channels.
- Percussion is on MIDI channel 10.
- The 128 program changes are standardized.
- Percussion sounds are mapped to specific notes and specific program changes.
- Each channel must respond to several continuous controllers in a specific way.

MIDI PROCESSORS/ROUTERS/INTERFACES

Our industry benefits from the enormous size of the musical market, since we can use the wide array of MIDI processors and interfaces that allow MIDI to be converted, changed, and transposed in a nearly infinite number of ways. A huge variety of devices are available, and brief summaries of a few of the main types are offered here.

For electrical reasons, MIDI cannot be "two-ferred"—a "Y" cable cannot be used. Instead, a splitter or "Thru" box is used, which takes a single input from one MIDI output and buffers and repeats the data for a number of outputs. MIDI splitters are very useful when a single MIDI device must connect to or drive a number of other devices.

A MIDI merger combines two MIDI signals, for cases when two outputs must be combined to drive one input. These devices buffer each MIDI stream to be merged, and then combine the data streams. The combined stream is then sent to the output.

MIDI patch bays (see photo) are generally combinations of mergers and splitters, with user-configurable patching capabilities.

Courtesy MIDIMan

A useful device for input/output with a MIDI system is the contact closure to MIDI interface. These devices can receive a contact closure input and send out a pre-programmed MIDI message.

Also useful is device is the MIDI to contact closure interface, also sometimes known as a MIDI Relay (see photo). This device has a relay which can be controlled by a user-programmable MIDI message.

Courtesy MIDI Solutions

One interesting MIDI device (or software component) is the pitch-to-MIDI converter, which translates audio pitch information into MIDI notes. In entertainment control, pitch-to-MIDI converters can be used for triggering devices from an audio signal; for instance, a gunshot could (with some MIDI filtering and processing) trigger a lighting effect.[3]

---

3. When I was working at Production Arts we did this for the movie *Carlito's Way*.

Chapter 22: Musical Instrument Digital Interface (MIDI) • **293**

RECOMMENDED MIDI TOPOLOGIES

In many musical MIDI installations, particularly in the recording or home studio environment, MIDI cables are simply connected from the Thru of one device to the In of the next to create a "daisy chain." However, since MIDI Thru ports actively repeat the signal, they need power at all times to pass data, and with a single loop, any break in the physical connections or a power failure in any device or cable in the chain will cause all downstream devices to lose data.

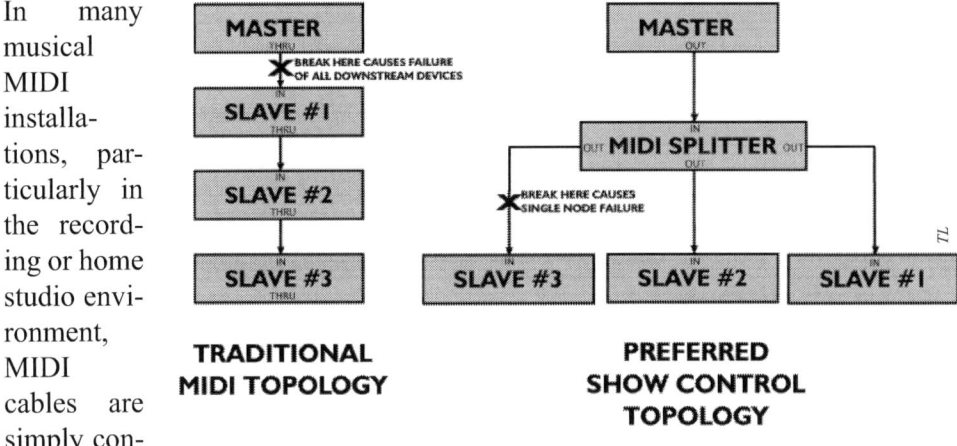

A better approach for critical entertainment control applications is to use a MIDI Splitter, or "Thru" box, which takes one MIDI input and creates multiple, redundant copies of that input. With such a device, a star, or home-run, topology (page 83) can be created. Of course, if the central splitter goes down, messages will not be passed to any downstream device, but splitters are very simple and generally reliable.

COMMON MIDI PROBLEMS

MIDI is an amazingly powerful system but, like any standard, it has ambiguities that become apparent in some applications; we will cover two common ones here.

The "Off by One" Problem

One common MIDI issue is the "Off by One" problem. This occurs because two companies have implemented a control range differently in their equipment: One company sets up its gear to work on program changes 0–127, and the other company designs its system to respond to 1–128:

Device 1 Program	Device 2 Program
0	1
1	2
2	3
3	4
4	5
5	6

This problem can also manifest itself with channel numbers 0–15 or 1–16. If you can't get two systems to talk, try adding or subtracting one from the parameter— you may be suffering the infamous "Off by One" problem. In either of these cases, the best way to look at the data is in hex—there is no ambiguity there.

The "Octave Problem"

Another similar issue you might encounter (especially with older equipment) is what I call the "Octave Problem." An octave is a doubling in frequency or, in typical western music scales, 12 half steps, which corresponds to 12 MIDI note numbers. In the original MIDI specification, MIDI note 60 was defined as middle C, but middle C was not rigidly specified until the release of General MIDI. So, one company may have implemented their device so that the octave containing middle C starts at some note number and another company implemented their system based on the same note, but in a different octave:

Device 1			Device 2	
Musical Note	MIDI Note (hex)	MIDI Note (decimal)	MIDI Note (hex)	MIDI Note (decimal)
C3	3C	60	48	72
C3#	3D	61	49	73
D3	3E	62	4A	74
D3#	3F	63	4B	75
E3	40	64	4C	76
F3	41	66	4D	77

So for example, if you're trying to trigger a sound effect using MIDI and you can't get it to work, try adding or subtracting 12 from the Note number—one of the systems may be off by an octave.

MIDI OVER NETWORKS

The seven-bit data set is a limitation for advanced musical applications, and with the rise of the network the physical, current-loop MIDI interface is increasingly becoming an anachronism. So in some cases, the MIDI command set has been separated from the older, legacy MIDI current-loop DIN cable hardware, and sent over networks.

Real Time Protocol (RTP) MIDI Payload

IETF RFC 4695, "RTP Payload Format for MIDI," completed in 2006 by John Lazzaro and John Wawrzynek of UC Berkeley, describes an openly standardized way to accomplish this goal. While in our market we are generally interested in transmitting MIDI over LANs, the approach documented in this RFC was designed to allow far-flung MIDI collaboration across the Internet, or even over wireless systems, which can have significant packet loss or delay. Since the protocol assumes that message delivery could be unreliable, it is highly complex and many of the standard details are something that only low-level developers (who should read the RFCs) need to understand. However, a basic overview is included here.

The RTP MIDI standard defines a payload to work in conjunction with **Real Time Protocol** (RTP, see RFC 3550), which was designed to aid the delivery of multimedia content over a network in a "real time" way, using a stream of data. The RTP MIDI payload can be transported using either TCP/IP or UDP/IP networks, in either unicast or multicast modes. The RTP MIDI payload is designed to carry any MIDI message covered in the MIDI 1.0 specification and this includes MIDI Show Control (MSC, Chapter 23), MIDI Machine Control (MMC, Chapter 24), and MIDI Time Code (MTC, Chapter 25)—all very useful for show applications.

Each RTP MIDI payload packet contains zero or more MIDI commands, a 16-bit sequence number, a 32-bit time stamp, other related information, and an optional "recovery journal." The recovery journal structure is a key part of the RTP MIDI Payload standard in that it can prevent "indefinite artifacts" (errors) of a MIDI stream. An indefinite artifact might be a sound that plays forever because a Note On message was received, but the related Note Off somehow got dropped or corrupted. RTP MIDI payloads do this by dynamically establishing a previous) "checkpoint" packet, and transmitting history information about the stream back to that previous packet. In this way, a receiving device can restore back through an error to this point if it detects that something has gone wrong (remember that computers can process a lot of information in the times between when typical MIDI messages might be received). This journal approach is very complicated, and how it works depends on what type of MIDI message is being handled.

RTP MIDI itself does not handle the session management duties, although it does suggest the use of IETF "session tools" like the Session Initiation Protocol (SIP, RFC 3261) and the Session Description Protocol (SDP, RFC 2327), but other session and connection management approaches are possible (see Apple Network MIDI below). Additionally, RTP MIDI doesn't provide guarantees of service, but it's possible to implement these using other network components.

Apple Network MIDI

Starting with its OS X operating system, Apple implemented a Network MIDI feature to transport MIDI over networks based, in part, on RTP. The approach uses two arbitrarily established UDP ports, which are assigned when the connection is set up. One is the "control" port; the other (the control port +1) is used to transmit the MIDI and RTP data. Using Apple's service discovery system **Bonjour**, the Apple Network MIDI system can work over networks without being aware of the IP addresses, and this accommodates IP address assignment by DHCP, link-local procedures, or fixed IP addresses.

Here's a screen capture showing cross-platform service discovery results using Bonjour:

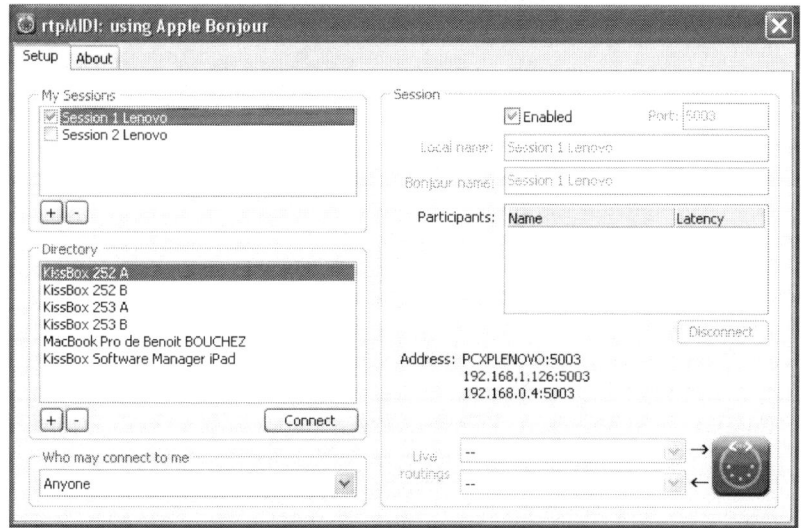

Courtesy Benoit Bouchez, Kiss Box

Apple's developers apparently considered that the session control protocols suggested in the RTP MIDI specification were too complex for typical LAN applications, and instead created their own simplified session protocol using a few custom commands to establish connections between an "initiator" and a "responder." To start, the initiator sends out an "IN" Invitation command once per second until a response is received, or until 12 Invitations have been sent. If an appropriate responder is present, it sends back an "OK" command.

To keep the systems in sync with each other, a synchronization procedure is started by the original initiator early in the life of the connection. Sync messages are exchanged and contain a 64-bit time quantity, in units of 100 microseconds, relative to some past point. Once the two stations have exchanged these clock messages, they can determine the offset (network travel and processing) time between the two. Apple claims that this approach implemented on a LAN should ensure synchronization within 1–2 milliseconds.

Apple also requires that the initiator start a new sync exchange at least once every 60 seconds. With this requirement, the sync exchange can fulfill a role similar to Active Sensing in MIDI (page 287), where if the receiver has not received a sync message from the other station within 60 seconds, it can terminate the connection, and also mute output audio or take other appropriate action.

When the session is established and with the stations synchronized, standard RTP MDI payload messages can be exchanged. Each message contains a timestamp, and if a station receives a message with a timestamp in the future, it should wait until that time to perform the requested answer.

Apple's simple, LAN-oriented approach to transmitting MIDI has been implemented in a number of products (see photo), and the Apple Network MIDI protocol is also available on Windows platforms, thanks to a German developer, Tobias Erichsen, who created software[4] that allows Windows machines to send MIDI message to or from other Windows, Mac OS X, iOS, and embedded computers.

Other Approaches
While, as of this writing, RTP-MIDI seems to be the most commonly used approach to sending MIDI over an Ethernet network, there are several other options, including ipMIDI, a proprietary approach for Ethernet LANs made by a German company called Nerds.[5]

---

4. http://www.tobias-erichsen.de/rtpMIDI.html
5. http://www.nerds.de/en/ipmidi.html

MIDI IN THE ENTERTAINMENT CONTROL MARKET

MIDI was, of course, designed purely for the musical instrument market, but it found its way into a wide variety of entertainment control applications. You might see it running animatronic axes, triggering sound effects, turning on and off simple (and safe) mechanical devices, controlling a light board using MIDI Show Control (MSC, Chapter 23), controlling a multitrack digital audio hard disk playback system using MIDI Machine Control (MMC, Chapter 24), or time synchronizing a variety of equipment using MIDI Time Code (MTC, Chapter 25).

Chapter 23: MIDI Show Control (MSC)

MIDI Show Control (MSC), released by the MIDI Manufacturer's Association (MMA) in 1991, is an open, standardized set of extensions to "musical" MIDI (Chapter 22) for peer-to-peer show control applications. As such, MSC does not replace any controller-to-device standard, such as DMX512-A, or even musical MIDI messages; in fact, it works well in conjunction with most other standards and protocols.

MSC COMMAND STRUCTURE

MIDI Show Control is one of the few open standards to contain show-related commands such as "Light Cue 34 go." The messages are sent over the MIDI SysEx structure, using the Universal Real-Time System Exclusive ID, with a reserved Sub-ID of 02_{16} assigned by the MMA. Commands can vary in length, up to a maximum of 128 octets.

MSC Message Format
The general MSC command format (in hex) is:

| F0 7F | Device ID | 02 | Command Format | Command | Data | F7 |

MIDI Show Control messages all start with an $F0_{16}$ octet to indicate that the message is Universal System Exclusive. The next octet is a $7F_{16}$, which denotes Real Time (all MSC messages are "real-time"; see page 287 for more info); then a Device ID is sent. This Device ID determines to what address a message is intended, although, since MIDI is a broadcast standard, all messages go to all connected devices on a link. The MSC Sub-ID, 02_{16}, is sent next. This second Sub-ID octet may seem to actually belong immediately after the $7F_{16}$ octet, but it is placed here so that the messages conform with the Universal System Exclusive standard. After the Sub-ID, the Command Format is sent, which tells the receiver what type of message to expect—lighting, sound, machinery, or myriad other types (see table below). Next, the actual command is sent. These commands include simple messages such as Go and Stop, as well as more esoteric commands. Finally, any associated data, such as the cue number, is sent. The message is closed out with an $F7_{16}$ End of SysEx octet.

Device ID

The Device ID, as just detailed, is used to indicate the intended receiver of a particular message. Every device must respond to at least one of the 112 individual device IDs ($00-6F_{16}$) and the $7F_{16}$ All Call number, which is used to transmit global messages to all receivers in the system. Many devices allow the user to configure the Device ID, although this flexibility is not specifically called for in the standard.

Additionally, a receiver can respond to commands issued to one of 15 groups ($70-7E_{16}$), which allow groups of devices to be addressed simultaneously, although this feature is not required by the standard. Keep in mind that for every separate Command Format, you can have a complete set of Device IDs.

A Bit of MIDI Show Control History

MIDI was adopted by the professional sound industry soon after its introduction in 1983, since the sound and music businesses have always been intertwined. Many musical MIDI messages are well suited to sound control—a program change command applies as well to an audio processor as it does a musical synthesizer. The entertainment lighting industry jumped on the MIDI bandwagon in the late 1980s, with musical-equipment manufacturers such as Sunn, which made club and small touring lighting systems, leading the way. Other companies such as Electronic Theatre Controls, Strand, and Vari*Lite, soon caught on. However, standard MIDI messages don't translate easily to disciplines like lighting control: What musical MIDI message do you use to initiate a cue? How do you control a submaster? How do you fire a macro? How do you set a grand master level?

Some lighting companies created proprietary implementations where standard MIDI Program Change messages triggered cues; moving light companies often based their implementations on note commands, where arbitrary notes were assigned to various buttons on a proprietary console. These note messages could be recorded into some sort of MIDI sequencer, and played back later to initiate complex cue-based operations. These varying implementations led to a situation very similar to the state of the music industry in the days before MIDI, or the lighting industry before DMX—similar devices had similar control capabilities but spoke different languages. In this case, it was more like different dialects because, though all the controllers spoke MIDI, few could understand each other.

Talk of creating a standard MIDI implementation for show applications culminated in 1989 at the "MIDI Mania" panel discussion at the Lighting Dimensions International (LDI) trade show in Nashville. Andy Meldrum, a programmer for Vari*Lite, put forth a proposal for an open, standard implementation of System Exclusive messages that would allow controllers from any manufacturer to be connected and controlled. Participants were excited about the possibilities of a standard protocol and, in December 1989, Charlie Richmond of Canada's Richmond Sound Design organized a Theatre Messages working group within the MIDI Manufacturer's Association (MMA) and set up a forum on the USITT's Callboard electronic bulletin-board system. Through contributions on Callboard and a great deal of footwork on Richmond's part, a formal standard was developed and sent to the MIDI Manufacturer's Association for an acceptance vote in 1991. The standard was approved by the MMA and sent to the Japanese MIDI Association for final, international approval. MIDI Show Control Version 1.0 became a standard in the summer of 1991.

Command Format

In general terms, the Command Format octet indicates what type of equipment is intended to be addressed by a particular message. Basic commands such as Go and Stop have been globally defined (more on that below), and room was left in the standard so that other commands, specific to types of equipment, could be created. A large variety of Command Formats[1] were defined in the standard; each type of controller has a General Category number and specific subsections. The existence of a Command Format here, of course, does not mean that MSC-controllable equipment exists; it simply means that the designers of the specification left room for the possibility of that type of equipment (and, of course, many of these equipment types are now obsolete).

Hex	Command
00	Reserved for extensions
01	Lighting—General
02	Moving Lights
03	Color Changers
04	Strobes
05	Lasers
06	Chasers
10	Sound—General
11	Music
12	CD Players
13	EPROM Playback
14	Audio Tape Machines
15	Intercoms
16	Amplifiers
17	Audio Effects Devices
18	Equalizers
20	Machinery—General
21	Rigging
22	Flies
23	Lifts
24	Turntables
25	Trusses
26	Robots

---

1. Ironically, no Show Control command format was designated.

Hex	Command
27	Animation
28	Floats
29	Breakaways
2A	Barges
30	Video—General
31	Video Tape Machines
32	Video Cassette Machines
33	Videodisc Players
34	Video Switchers
35	Video Effects
36	Video Character Generators
37	Video Still Stores
38	Video Monitors
40	Projection—General
41	Film Projectors
42	Slide Projectors
43	Video Projectors
44	Dissolvers
45	Shutter Controls
50	Process Control—General
51	Hydraulic Oil
52	H_2O
53	CO_2
54	Compressed Air
55	Natural Gas
56	Fog
57	Smoke
58	Cracked Haze
60	Pyro—General
61	Fireworks
62	Explosions
63	Flame
64	Smokepots
7F	All-Types

Commands

The heart of a MIDI Show Control message is the single octet Command that follows the Command Format octet. There are 127 possible commands for each Command Format; see the detailed descriptions below for more information. Each command can contain up to 128 octets of related data, and two types of data are globally defined in the standard: cue numbers and time numbers.

Cue Numbers

The cue number is a vital part of MIDI Show Control. Cues are addressed via Cue Number and optional Cue List and Cue Path. Cue Path indicates from which of the available media the Cue List should be pulled; Cue List indicates to the receiver in which of the currently "open" Cue Lists the Cue Number resides. These meanings aren't set in stone, though; lighting equipment, for example, might use the Cue List number to indicate which playback control or fader is being used to execute a particular cue. For instance, if a lighting console had eight faders, Cue List value of 1 could be used to tell the console to execute the cue on fader 1; Cue List 2 could denote fader 2, and so on.

The actual numeric cue data is encoded using ASCII, with the $2E_{16}$ ASCII decimal-point character used for its intended purpose. The Cue Number/List/Path data is sent sequentially with 00_{16} delimiters. If the Cue Path or List is not needed, it can simply be omitted; however, a Cue Path cannot be sent without a Cue List. If a receiving device is not able to deal with Cue Paths or Lists, it simply discards that data.

So, to indicate that a particular device should select Cue 47.3 on Cue List 2 from Cue Path 37, the following octets would be sent (shown in hex):

```
34  37  2E  33  00  32  00  33  37
 4   7   .   3       2       3   7
```

More than one decimal point may be used: numbers like 67.3.4 are valid. If a particular piece of equipment cannot support the multiple decimal points, subsequent sub-cue numbering is discarded. Some equipment also accepts non-numeric characters.

Time-Code Numbers

Time is also globally defined in the MSC specification, and the coding method is the same as that used in MIDI Time Code (see page 330) and MIDI Machine Control (Chapter 24), with the same units of hours, minutes, seconds, frames, fractional frames: `hr mn sc fr ff`.

RECOMMENDED MINIMUM SETS

While manufacturers may decide which of the commands defined in the spec to implement, there are logical groupings that help indicate to an end user how thoroughly a manufacturer has implemented MSC. The commands are grouped into three recommended minimum sets, designed to help equipment of various sophistication be compatible. Recommended minimum set 1 contains only three commands: Go, Stop, and Resume; set 2 has full data capability but does not contain any time-code commands; set 3 contains all the commands defined in the specification.

MSC COMMANDS

Let's go through each of the MSC commands. As shown above, the general MSC command structure (in hex) is:

```
| F0 7F | Device ID | 02 | Command Format | Command | Data | F7 |
```

Only the Command and Data octets are detailed below, and data in angle brackets <> is optional. To make a complete message, you would have to know the target Device ID, Command Format, Command, and related data (cue number, etc.).

Messages for All Types of Equipment

The first group of MIDI Show Control commands is common to nearly any cue-based system—lighting controllers, sound playback systems, or anything else that can be controlled using cues.

Go

```
01 <Q_number> <00> <Q_list> <00> <Q_path>
```

This command's function is fairly self-explanatory: it starts a transition to a preset cue state. If a cue number is not specified, the controlled device simply executes the next cue in its cue list. The transition time is stored in the controlled device; if you want to send transition time as well, you should use the Timed Go command below.

Stop

```
02 <Q_number> <00> <Q_list> <00> <Q_path>
```

This command is also fairly self-explanatory: it simply stops the specified transition. If no cue is specified, it stops all cues currently in progress.

Resume

`03 <Q_number> <00> <Q_list> <00> <Q_path>`

Resumes the specified cue, or resumes all stopped transitions if no Q number is specified.

Timed_Go

`04 hr mn sc fr ff <Q_number> <00> <Q_list> <00> <Q_path>`

The time communicated in this Go message is generally either the cue transition time or the time at which the cue should be executed, depending on the implementation. If a receiving device cannot deal with a Timed Go command, it should simply execute the cue (Go) with its prerecorded time. `hr mn sc fr ff` is the time, down to fractional frames.

Load

`05 <Q_number> <00> <Q_list> <00> <Q_path>`

This command might also be known as Stand By. The cue specified in the message is loaded and placed in a ready mode for execution.

Set

`cc cc vv vv <hr mn Sc fr ff>`

This command defines the value of a generic control. This definition is intentionally vague because of the command's potentially broad application. "`cc`" in the above composition denotes a two-octet Generic Control Number, which might be thought of as a sort of sub-Device ID, sent least-significant octet first. "`vv`" denotes a Generic Control Value.

Fire

`07 mm`

This command is primarily designed to fire console-based macros, as might be found on a lighting controller. "`mm`" here denotes a seven-bit (128 values) Macro Number.

All_Off

`08`

Restore

`09`

Reset

`0A`

All Off kills all outputs but leaves the controllers in a restorable condition. This command might be useful when a stage manager or director is yelling, "Hold please!" or in emergency situations. The `09` Restore command restores the controlled devices to the conditions existing before an All_Off command was issued; Reset terminates all running cues and resets addressed controlled devices to a starting condition. Controlled devices are reset to a "top-of-show" condition.

Go_Off

`0B <Q_number> <00> <Q_list> <00> <Q_path>`

This command sends a selected cue to an off condition. If no cue is specified, all currently running cues are terminated. Go_Off would be useful in a situation where a cue contains a loop that runs until a certain event takes place; once that event takes place, you want to shut the cue down (fade-out the sound, dim the lights, and so on) in a controlled fashion.

"Sound" Commands

A second set of commands, known as "sound" commands, is defined in the specification. At the time of the standard's creation, Charlie Richmond, chief architect of MIDI Show Control, was one of the only manufacturers involved with the standard who was making open show-oriented controllers that functioned in a cue-based mode, and few of the lighting manufacturers involved at the time could see the use for these commands. For these reasons, many of these "sound" commands are derived from the operation of Mr. Richmond's systems, which have one more internal clock (typically one per cue list), which can be set or reset through a command. This clock can run on its own, be synced to external sources such as MIDI Time Code, be reset for each cue, or be ignored altogether.

Go/Jam Clock

`10 <Q_number> <00> <Q_list> <00> <Q_path>`

This command executes a cue and synchronizes, or "jams," a controlled device's internal clock to the cue's time number, which is stored in the controlled device.

Standby_+

`11 <Q_List>`

`Standby_+` is similar to the Load command detailed above, but it sets the next cue in a cue list to a ready mode, awaiting execution.

Standby_-

`12 <Q_List>`

`Standby_-` works the same way as `Standby_+`, except that it loads the previous cue in the list.

Sequence_+

`13 <Q_List>`

If you have multiple cues with the same base number, such as 1, 1.25, and 1.3, the 1 component of these cues is called the "parent" cue. `Sequence_+` loads the next parent cue to standby operation. Let's say you have a cue series of 2.0, 2.5, 2.6, 3.0, and 3.8. You're in scene 2 (cue 2.5), a performer forgets his blocking, and you need to jump ahead to scene 3 (cue 3.0) without speeding through the cue actions in cue 2.6. Sending a `Sequence_+` command will send the system to cue 3.0. The `Q_List` selection optionally selects which cue list you want to act on.

Sequence_-

`14 <Q_List>`

`Sequence_-` functions the same as `Sequence_+`, except that it places into standby mode the parent cue previous to the current cue. If you have cues 2.3, 2.5, 3.0, 3.1, and 4.8 and are currently executing cue 3.1, executing `Sequence_-` would call up cue 2.3.

Start_Clock

`15 <Q_List>`

`Start_Clock` starts a controlled device's internal clock. If the clock is already running, this command is ignored. In some controlled devices, multiple cue lists can each have their own independent clock; the optional cue list parameter of this command selects which cue list to act on.

Stop_Clock

`16 <Q_List>`

`Stop_Clock` stops a controlled device's internal clock, holding the current time value.

Zero_Clock

`17 <Q_List>`

`Zero_Clock` resets a cue list's clock to a value of `00:00:00:00.00` (the last set of zeros denotes fractional frames). If the clock was running when it received the command, it will continue running from the zero value; if it was stopped, it will remain stopped but be reset.

Set_Clock

`18 hr mn sc fr ff <Q_List>`

`Set_Clock` sets a controlled device's internal clock to a particular value. Optionally, this action takes place only on a specific cue list's clock. Like the `Zero_Clock` command, `Set_Clock` works whether the clock is currently running or stopped.

MTC_Chase_On

`19 <Q_List>`

`MTC_Chase_On` instructs a controlled device's internal clock to follow an incoming MIDI Time Code (or other) time data stream. By specifying an optional cue list, this command affects only a specific cue list's clock. If no MTC is present in the MIDI stream when the command is received, the internal clock is unaffected; however, once MTC appears, the internal clock must follow the time code.

MTC_Chase_Off

1A <Q_List>

MTC_Chase_Off stops following the incoming time code, and then returns the clock to its previous mode with the last MTC value received.

Open_Cue_List

1B Q_List

Close_Cue_List

1C Q_List

These commands "open" or "close" a cue list, making it available for operation.

Open_Cue_Path

1D Q_Path

Close_Cue_Path

1E Q_Path

These commands "open" or "close" a cue path for use by the system.

Sample Messages

To show you how all this goes together, let's put together some sample messages. Remember, the general MSC command format (in hex) is:

| F0 7F | DevID | 02 | Cmd Fmt | Cmd | Data | F7 |

First, let's look at a message that tells a lighting console, configured to respond to messages on Device ID #1, to execute cue 18.6 on fader 2 (many light boards use the cue list octets to indicate fader numbers):

- The Device ID is simply 01_{16}.
- The command format is also 01_{16}, which means lighting.
- The value for Go is simply 01_{16}.

Cue 18.6 in ASCII/hex is **31 38 2E 36** (**2E** is the ASCII decimal point character), and the cue list is **32**. A **00** delimiter is placed between the cue number and cue list.

Adding the "overhead" octets F0, 7F, 02, and F7 we get:

F0 7F	DevID	02	Cmd Fmt	Cmd	Data	F7
F0 7F	01	02	01	01	31 38 2E 36 00 32	F7

To tell the same lighting console to run cue 25 on the fader 1, we change only the cue and cue list octets:

F0 7F	DevID	02	Cmd Fmt	Cmd	Data	F7
F0 7F	01	02	01	01	32 35 00 31	F7

To tell the same lighting console to stop cue 25 on fader 1, we change only the command octet:

F0 7F	DevID	02	Cmd Fmt	Cmd	Data	F7
F0 7F	01	02	01	02	32 35 00 31	F7

This is easy, right? Once you get the basic structure set up and working (it's always worth testing your syntax before going further), it's easy to make minor changes.

Let's look at one more example. Let's tell a Sound Controller on Device ID 6 to execute a Go on cue 11, cue list 30, and cue path 63.

F0 7F	DevID	02	Cmd Fmt	Cmd	Data	F7
F0 7F	06	02	10	01	31 31 00 33 30 00 36 33	F7

LIMITATIONS OF MIDI SHOW CONTROL

The MIDI Show Control standard was certainly never a panacea; it is, however, a tremendously versatile protocol, which has been accepted for a range of different applications: controlling lighting consoles, triggering computerized sound systems, and so on. But, like any protocol that attempts such wide applications, it does have some limitations.

Command Response Time

Some criticized MSC for being designed on the MIDI platform; they were concerned that MIDI response time may not be adequate. While all MSC commands are designated "real time," there is no reason that some data-hogging device on a MIDI Link couldn't bog down the system enough to cause important messages to be delayed. However, this is a problem easily avoided by simply putting each controlled device on its own data link (something that was cost prohibitive at the time of the standard's development). In addition, the MSC messages, even with the tremendous overhead, are pretty fast. If a "Light Cue 362 Go" command is given to Device 1, the following octets (in hex) are sent: `F0 7F 01 02 01 01 33 36 32 F7`.

This is 10 octets, each eight bits with start and stop bits, for a total of 100 bits. At 31,250 bits per second (MIDI's data rate), this message would occupy the link for 100/31,250 seconds, or just over 3 milliseconds. This means that 100 such messages could be sent in 300 milliseconds, which is probably about how long it takes the average board operator to press a go button after hearing a human stage manager's "Go" command. In any case, if your application is more time-critical than a few milliseconds, you should probably be using another standard. And, of course, we can now send the MIDI commands over Ethernet at 100 Mbits per second or faster—see the "MIDI over Networks" section, on page 296.

Open Loop
Standard MIDI Show Control commands are completely open loop: no feedback or confirmation of any kind is required for the completion of any action. When a controller sends a message out, it has no idea if the target device even exists (although in most shows someone should eventually notice such a problem, or confirmation systems can be designed for critical purposes or for start-of-day check ups).

This open-loop approach was chosen for a number of reasons, first and foremost of which was to keep the standard simple to use, easy to implement, and fast—an open-loop approach keeps traffic on a MIDI link to a minimum. In addition, with an open-loop system the failure of any one device in a show network should not affect any other system on the network—even if the system is waiting for a response from a broken lighting controller, there is no reason to prevent a properly functioning sound computer from executing its cue on time. As for MSC applications that could be potentially life threatening, the philosophy of the standard is clearly spelled out (in all capital letters) in the specification itself:

IN NO WAY IS THIS SPECIFICATION INTENDED TO REPLACE ANY ASPECT OF NORMAL PERFORMANCE SAFETY WHICH IS EITHER REQUIRED OR MAKES GOOD SENSE WHEN DANGEROUS EQUIPMENT IS IN USE.... MIDI SHOW CONTROL IS NOT INTENDED TO

TELL DANGEROUS EQUIPMENT WHEN IT IS SAFE TO GO: IT IS ONLY INTENDED TO SIGNAL WHAT IS DESIRED IF ALL CONDITIONS ARE ACCEPTABLE AND IDEAL FOR SAFE PERFORMANCE. ONLY PROPERLY DESIGNED SAFETY SYSTEMS AND TRAINED SAFETY PERSONNEL CAN ESTABLISH IF CONDITIONS ARE ACCEPTABLE AND IDEAL AT ANY TIME.[2]

Version 1.1/Two-Phase Commit

To address safety and other issues, MIDI Show Control version 1.1 was developed, which added some additional commands to allow a new type of operation: Two-Phase Commit (2PC). This 2PC approach was primarily developed by Ralph Weber, then of Digital Equipment Corporation, with assistance from Charlie Richmond. 2PC is a communications approach that allows systems to proceed with absolute certainty; in 2PC, all actions are acknowledged before execution and after completion. The two phases are Standing By, when devices are readied for action, and Go, when the actions are executed. These are the same two phases used in communications between stage managers and system operators in the theatrical world. The standard, released in 1995, was (as far as I know) never implemented in any system or product. Theme park system designers considered it simultaneously too complex and not sophisticated enough, and theme parks were the key market for the standard. It is now considered obsolete, although you will still find it in the official MIDI standard.

MSC IN THE ENTERTAINMENT CONTROL MARKET

MSC was designed to be a universal standard, with the ability to control everything and anything on a show. This universal command approach, in my opinion (which changed over time—see my "Limitations of Standards" section on page 7) was more desirable in the past, when we did not have the universal data *transport* facilities offered by Ethernet, and IP's common interconnection. With exceptionally powerful computer hardware, and nearly every controller device including an Ethernet port, it's far easier than ever to simply talk to any device with whatever protocol the device creator dreams up, proprietary or not.

Of course, MSC, like DMX, musical MIDI, and other older protocols, isn't going anywhere and will be around for years to come, and it has found wide use for things like triggering lighting cues.

---

2. MIDI Show Control 1.1 Standard, p. 8.

Chapter 24: MIDI Machine Control (MMC)

Developed in the late 1980s, **MIDI Machine Control** (MMC) was adopted by the MMA in 1992 as a standard way to give music sequencers control of SMPTE Time Code–based tape machines. Of course, tape is now obsolete, but the standard works just as well with digital audio or video playback devices. MMC was based on the broadcast control standard ES-Bus, and is very complex—the MMC part of the MIDI standard is over a hundred pages long. However, the standard is very useful for simple A/V control, so an introduction is included here.

MMC SYSTEMS

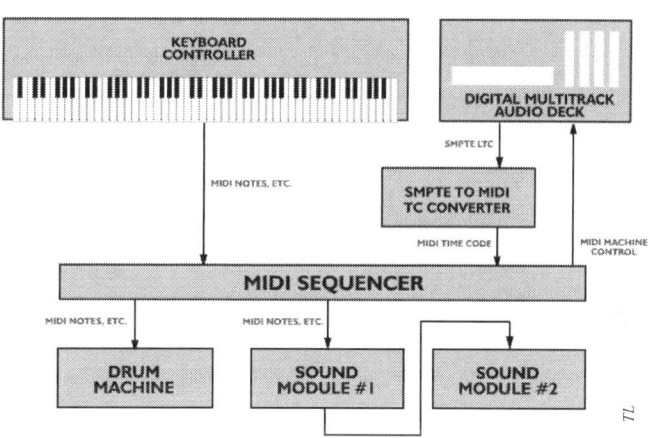

Systems built using MIDI Machine Control can take many forms. The basic structure, however, is fairly simple: A master control computer running sequencer software (or entertainment control software in entertainment control applications) sends MMC commands to a playback deck, which can generate Time Code as it plays. If the deck is generating SMPTE TC, it is converted to MIDI Time Code and sent back from the deck to the control computer, over an optional closed-loop MIDI line. In this way, the master controller can determine the time code location of the track, and make decisions based on that and other information. MMC can operate in either open- or closed-loop modes; because MIDI is unidirectional, a return MIDI line must be included in any system where closed-loop operation is desired.

COMMAND/RESPONSE STRUCTURE OF MMC

MIDI Machine Control messages fall into three categories: commands, responses, and information fields. Master controllers send commands; slave devices reply with responses (see page 81). Information fields are a sort of hybrid; they are registers, maintained by controlled devices that can report their contents when polled by the master. Some of these registers are read-only and send their data back to the master on request; others can be written to by the master. Data contained in the registers include the current time code address of the controlled media, execution status of recently received commands, and so on.

MMC MOTION CONTROL

MIDI Machine Control commands fall into several groupings. Two of the more interesting groups for show control applications are Motion-Control States and Motion-Control Processes.

Motion-Control States

An audio or video deck can do only one type of thing exclusively at a time (e.g., it can't fast-forward while in rewind mode). For this reason, some MMC commands are gathered into a mutually exclusive group; these commands are known as the Motion-Control State (MCS) commands. Executing an MCS command cancels any other currently running motion-control mode, and starts the new action as soon as possible. In this way, the master controller doesn't have to know about the current state of the controlled device; it simply commands it to the new state. However, if the master wants to know the current state of the machine, it can find out by querying the Motion Control Tally information field. The basic MCS commands are Stop, Pause, Play, Fast Forward, Rewind, Search, and Eject (see details below).

Motion-Control Processes

Two other mutually exclusive commands are the Motion Control Processes (MCP)—Locate and Chase—that cause the controlled device to go into a special mode, capable of issuing its own Motion Control State commands. Locate causes the deck to move to a specific time-code frame on the media; Chase causes the machine to lock itself to a time code source. When in either of these modes, the controlled device may issue any number of Motion Control State commands internally to position the media.

MMC MESSAGE STRUCTURE

Like MIDI Show Control (Chapter 23), MMC is made up of Universal Real-Time System Exclusive MIDI messages, and two Sub-IDs have been designated for MMC: 06_{16} and 07_{16}. Messages with an 06_{16} ID are commands sent from a controller to a controlled device; the 07_{16} ID denotes responses sent from slave devices back to the master.

MMC commands use the real-time command structure:

```
| F0 | 7F | Destination ID | 06 | Command String | F7 |
```

MMC responses are structured as follows:

```
| F0 | 7F | Source ID | 07 | Response String | F7 |
```

Note that the meaning of the ID is dependent on its context. The ID refers to the address of the target controlled device if sent in a command by a master controller; in a response from a controlled device, the device ID is the address of the sender—the target is assumed to be the master controller.

Command/Response Lengths

The number of data octets in an MMC message is variable and is either implied by the command number, or designated within the message itself through the use of a "count" octet. Count is simply a count of the subsequent data octets (in hex), not including the count octet itself or the F7 end of exclusive. For instance, a Return to Zero command to an audio or video machine would include a count octet:

```
F0 7F 01 06 [Locate] [Count = 06] [Target] 60 00 00 00 00 F7
```

MMC also allows more than one command or response to be sent within a single 48-octet SysEx MIDI message or, conversely, long single messages to be sent across multiple SysEx messages. This is known as "segmentation," and two special messages (Command Segment and Response Segment) are included for this purpose.

Guideline Minimum Sets

As in MIDI Show Control, no commands are required for implementation in an MMC-capable device; however, MMC has four "Guideline Minimum Sets." Implementing a particular commands set implies that a device is generally capable of a certain level of operation. The Guideline Minimum Sets are as follows:

1. Simple Transport—open loop, no TC reader.
2. Basic Transport—closed loop possible, no TC reader.
3. Advanced Transport—closed loop possible, TC reader, event triggering control, track-by-track record control.
4. Basic Synchronizer—closed loop possible.

COMMON MMC COMMANDS

In this section, we'll go through some of the MMC commands and responses most useful for live entertainment applications. The overhead (F0 7F, etc.) octets are not included.

Stop

```
01
```

Play

```
02
```

Fast Forward

```
04
```

Rewind

```
05
```

Eject

```
0A
```

These Motion-Control State commands are self-explanatory.

Chase

```
0B
```

This Motion-Control Process causes the selected device to chase an incoming time-code stream. Chase mode is canceled by the receipt of an MCS or MCP command.

Pause

```
09
```

Places the controlled deck in pause mode, where the device stops as soon as possible and can restart as fast as possible. On video decks, this command causes the device to stop "with picture."

Locate

```
44 [Count] [Sub-Command] [Information Field/Target
Time Code]
```

When the subcommand octet is 00_{16}, Locate causes the deck to move the media to the time-code position indicated in the specified information field. If a subcommand of 01_{16} is sent, the target time-code address is specified within the Locate command.

MMC IN THE ENTERTAINMENT CONTROL MARKET

MIDI Machine Control was designed primarily for the musical sequencing market, but has found application in our market for the control of linear playback devices.

Chapter 25: SMPTE and MIDI Time Code (MTC)

Time code is one of the most widely used synchronization tools in show control, since it is easy to use, allows very precise synchronization between multiple systems, accurate repeatability, and makes for straightforward programming of time-based systems. **Society of Motion Picture and Television Engineers Time Code** (SMPTE TC) was created first, and then **MIDI Time Code** (MTC) was developed to transport SMPTE time code over a MIDI link. While SMPTE (pronounced "sim-tee") TC is (at its lowest level) analog and MTC is digital, they share many common aspects, so we will cover the basics of time code first, and then get into specifics of each type.

BACKGROUND

As of this writing, the latest version of the SMPTE standard is in two parts, last updated in 2008: SMPTE 12M-1 and SMPTE 12M-2, entitled "SMPTE Standard for Television, Audio, and Film—Time and Control Code"; MTC was released as a standard in 1986.

TIME CODE ADDRESSES

TC has its origins in videotape editing, so it breaks time down into hours, minutes, seconds, and frames; up to 24 hours worth of frames can be encoded. One discrete TC address exists for every frame on an encoded media—a typical address might be `05:20:31:18`, or five hours, 20 minutes, 31 seconds, and 18 frames. Time code can be recorded on a wide variety of media and systems, and used for a wide variety of synchronization applications. TC is absolute: If a show starts at `05:00:00:00` and a time code of `05:20:31:18` is read, the system can determine that this frame is 20 minutes, 31 seconds, and 18 frames from the beginning of the show.

TRADITIONAL AUDIO/VISUAL SYNCHRONIZATION

Because TC has its origins in the film and video world, a brief introduction to a common time-code application will be helpful here, before jumping into time code for live entertainment applications.

Since there were few ways to conveniently record high-quality audio on film, film production traditionally used a double system: images were recorded on film and sounds are recorded separately onto an audio recorder. A "slate," on which is written shooting information such as the scene and take, is used to sync the two systems. When shooting, the film camera and audio recorder are both rolled, and once they are both up to "speed," the clapper on the slate is quickly closed to make the familiar "clap" sound. This system provides the editors in post-production a visual image of the clapper closing, and the sound of the clapper on the audio recording. They can then manually line up the audio sound track with the images to this point, and mechanically link them to run together as needed.

With time code, this process can be simplified. If the camera is capable of generating and recording time code, the TC is sent to the audio recorder and recorded along with the audio. Whenever the camera and the audio deck roll, they both record the exact same time code, frame for frame. The inverse can also be done: The audio deck can generate the TC, which can be sent directly to the camera or to a special "Time Code Slate" with an electronic TC display, which is then shot for a few seconds each time the camera is rolled. These approaches give the editors encoded time code or a visual image of the time code and the time code signal on the audio recording.

In post-production, the editors could use the recorded time codes for easy electronic synchronization, by connecting video and audio machines together (of course this is usually now done all in one computer running editing software), making them operate as one virtual machine, so that the audio will run in precise

A Bit of SMPTE Time Code History

The development of SMPTE Time Code goes back to 1967, when, to facilitate electronic videotape editing, the California-based company EECO developed a method of encoding video frames with an absolute time and frame number, based on a method used by NASA to synchronize telemetry tapes from its Apollo missions.[1] The EECO code was successful in the marketplace, and several other manufacturers developed their own proprietary time-coding methods. As with so many other prestandard-market-situations already discussed, this situation became problematic, as equipment from one manufacturer would not work with gear from another. In 1969, the SMPTE formed a committee to create a standard time-coding scheme, and the result was eventually adopted by the American National Standards Institute (ANSI) as a national standard.

In the early 1980s, personal computer-based audio and music systems started to proliferate, and the only way to get these systems to receive the analog SMPTE time code signal was to get an expensive interface card which could convert it into a digital form easily readable by the computer. MIDI (Chapter 22) offered an inexpensive digital interface widely used in the musical and sound markets, so the logical next step became creating a way to transmit SMPTE data over MIDI. So, in 1986, MIDI Time Code (MTC) was developed by Evan Brooks and Chris Meyer, of the companies Digidesign and Sequential Circuits, respectively.

---

1. According to EECO, Inc., *The Time Code Book*, 1983.

synchronization with the image at all times. To do so, a cable is run from the video deck to the audio deck, so that the video deck will "master" the audio deck.

With the machines in the right modes, anytime the video deck is rolled, the audio machine compares the incoming time code from the video player with the time code position of its own media, and controls and continuously varies its transport speed ("vari-speeds") to lock the two machines precisely together. For example, if "Play" is pressed on the video machine, a time code is generated and sent to the audio deck. The audio machine will rapidly start falling behind, and then will start moving forward quickly to catch up; it will ramp up and slow down its speed until audio and video are perfectly synchronized, frame for frame. So, in other words, the audio machine "chases" the video deck, and after a bit of "pre-roll" time (time code that rolls before the start of the media to allow the machines to synchronize), the audio tracks can be used in perfect sync with the video image.

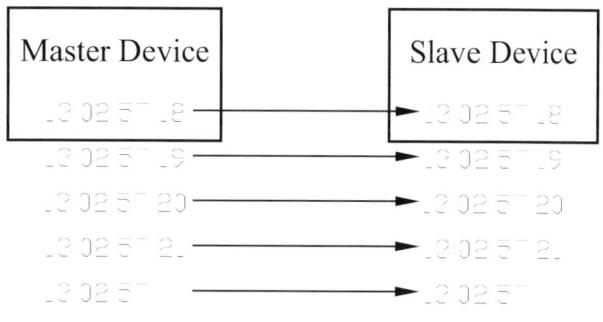

LIVE ENTERTAINMENT TIME CODE APPLICATIONS

While the example above comes from the post-production world, we oftentimes want to do very similar things in the live entertainment industry. We may want an audio multitrack machine to chase a video presentation; or have four video decks run separate images to separate screens, all in perfect synchronization with an audio deck; or synchronize two audio decks to get more tracks; or have an animatronic character synchronize exactly to a soundtrack.

In addition, we also use time code to trigger event-based systems. For instance, if we need to synchronize light cues (events) to a sound track, we could record TC along with the audio. Then, whenever the audio deck is rolled, time code would be played back and routed to the lighting console. Professional lighting consoles generally accept time code directly, and each cue can be programmed to be triggered at a precise time: cue 18 might be programmed to execute at 07:02:59:23. Whenever the audio track rolls and the pre-programmed time code address comes along, the appropriate light cues will trigger. In a similar fashion, many preset-based audio mixers, video routers, pyro systems, playback devices, or other systems can also be triggered. Entire attractions are run from time code.

TIME-CODE TYPES

Since different media and countries have different framing rates, there are different types of time code. Film is typically shot and projected at 24 frames per second (fps), so there is one type of TC for film—24-frame time code. Monochrome American video shared the 60-Hz frequency of North American AC power mains, and because it was interlaced,[1] the framing rate is exactly 30 fps. This is another type of TC, known as 30-frame time code. (Other time code frame rates in the SMPTE standard include 60, 59.94, 50, 25 (for Europe), and 23.98; with digital media, you might encounter any of these frame rates, or others not included here).

Drop-Frame Time Code

Color video in America was standardized by the National Television System Committee (NTSC) in the 1950s (and used for over-the-air broadcast until 2009 in the US), and was designed to be backwards compatible with monochrome TV. For reasons outside the scope of this book, it runs at a frame rate of approximately 29.97 fps. If regular 30-fps TC is used to count video frames running at a rate of 29.97 fps, the time-code second will gain 0.03 frames per second in relation to a true second; a cumulative gain of 108 frames, or 3.6 seconds per hour of running time, will result.

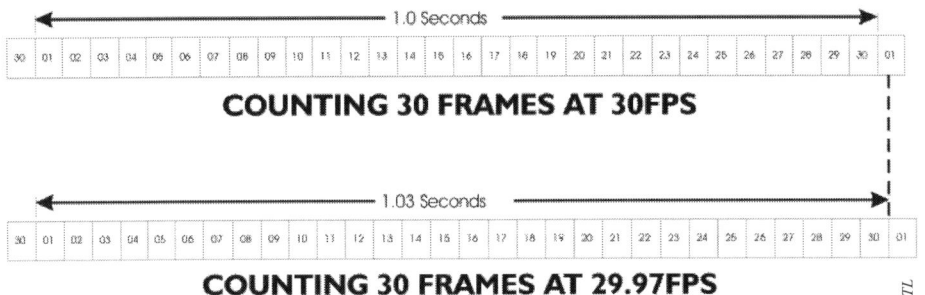

This error may be a bit difficult to comprehend, but the key is that time code receivers don't (generally) compare the frame counts of the incoming time code signal with an external clock, because the incoming TC *is a clock*. So, if we're using 30 fps TC to count NTSC 29.97 fps frames, it will take 1.03 actual seconds to reach a count of 30, or 1 time-code second. While an error of +0.03 real seconds per time-code second doesn't seem like much, this accumulates to 3.6 seconds per hour, or 1 minute and 26.4 seconds per day. It would certainly be a

---
1. Every other line is scanned first, and then the remaining lines are scanned on alternate "fields." Two fields make up one frame.

problem if a light cue was off by 3.6 seconds at the end of an hour-long presentation.

For this reason, another type of code was developed, known as Drop-Frame-NTSC Time Compensated Time Code (drop-frame TC), and bits in the time code signal are used to indicate when drop-frame compensation is applied, so that receivers can decode the signal properly. Drop-frame TC is so named because the extra 108 frame counts that would accumulate over the course of an hour are simply omitted. Conceptually, this is similar to a sort of reverse "leap-year," although here we're dropping frames every minute instead of adding a day every four years.

Since it is impossible to drop fractional frames, the first two frame numbers of every minute, with the exception of minutes 0, 10, 20, 30, 40, and 50 are dropped:

```
60 minutes - 6 not dropped = 54 minutes

54 minutes @ 2 frames dropped/minute = 108 frames
```

Even this method still only approximates real time; drop-frame TC still has an error of –3.6 milliseconds every hour. Even though this type of time code has its roots going back more than half a century, these odd frame rates may still be found today.

PRACTICAL TIME CODE FOR LIVE SHOWS

There are a number of issues to keep in mind when dealing with time code on live shows.

Hours

Shows that run 24 hours are rare,[2] so it is often more convenient to use the time code hours for other purposes. For instance, the time code hour 1 might be used for show one, hour 2 for show two, and so on. Keep in mind that it's always a good idea to leave some time before the show start for preroll, to make sure everything is synchronized; this is why shows often start at `01:00:00:00` instead of `00:00:00:00`.

---

2. Although there was at least one such show in Times Square at the 1999/2000 New Year's Eve celebration. For that show, they had to figure out how the machines dealt with rolling over from `23:59:59:29` to `00:00:00:00`.

Frame Rate

The most important thing in any time code application is to be sure that the framing rate is chosen and agreed upon well in advance, everything is double-checked, and time is left for testing. For many straightforward live shows, a simple 30 frame per second, "non-drop" time code is used, simplifying everything. In other cases, where video media is used, the chosen frame rate may be dictated by the media used.

If you see an error in a system close to the 3.6 seconds per hour described above, then it's likely that either the transmitter is sending drop-frame, and the receiver is not properly configured for it, or vice versa. Time code test equipment (such as the "Distripalyzer," on page 329) is invaluable for solving these problems.

Offset

It is often desirable to be able to "offset" time code, adding or subtracting time. This might be useful when you want to transfer one show to another, or when you have latency problems somewhere in a system.

Chasing

Another feature of well-designed equipment that handles time code is the ability to somehow *disable*, or not listen to, the incoming time code stream (and ideally, this chase on/off should be remote controllable). This is extremely useful when you are trying to work on a system and someone else is doing something that causes the time code to roll—it is really annoying when you are deep into programming a cue and the console starts executing cues based on incoming TC!

Dropouts and Freewheeling

Occasionally, due to tape wear (in the old days), connection difficulties, or other problems, time code "drop outs" may occur, where the signal momentarily disappears or becomes too corrupted to decode. This is often a symptom of a serious problem, but many devices are capable of "freewheeling" through these dropouts by simply rolling on with their own time bases until the code reappears or can be decoded properly again. On any well-designed piece of equipment incorporating this feature, freewheeling should be able to be enabled or disabled, and the number of freewheeling frames should be configurable.

Time-Based Applications without Time Code

Time code was developed in the days of all analog equipment, which could drift out of sync easily. Most computers have incredibly accurate time bases, so it's now possible to simply start two digital devices at the same time and have them run "wild" together.

SMPTE TIME-CODE

There are two types of SMPTE time code: Linear Time Code (LTC) and Vertical Interval Time Code (VITC—pronounced "vitcee").

Linear Time Code

Linear Time Code (LTC) is so named because it was designed to be recorded on an audio track linearly along the edge of a videotape, not helically like the video signal. LTC can be distributed as and works similarly to an audio signal, but actually uses a type of bi-phase modulation, that is similar, conceptually, to Manchester encoding (page 141). In bi-phase modulation,

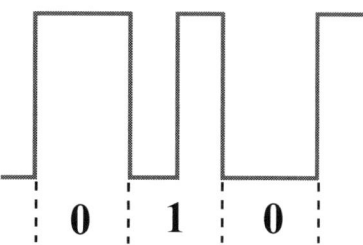

there is a transition from high to low or low to high at the start (and because bits are trailed together, at the end) of every bit. A bit with no internal transitions is a zero; a transition within a bit time denotes a 1 (see graphic). This bi-phase coding scheme makes the polarity of LTC irrelevant; it also allows the code to be read backwards as well as forwards, and allows LTC to be recorded on a wide variety of audio media.

Each LTC frame is composed of 80 bits, numbered 0–79; and because bit timing in the reading of LTC is derived from the actual code read (and not by comparison to an external clock), LTC can be decoded in either direction and at speeds well above (typically up to 100 times) or below (typically down to 1/50) its normal speed. At the end of each LTC frame, there is a 16-bit sync word—0011111111111101—which is a series that cannot be repeated anywhere else in the time-code word with any combination of time or user data. When the decoding system encounters this sync word, it knows that this point in the code is the end of one frame and the start of another; the word is also used to derive playback direction.

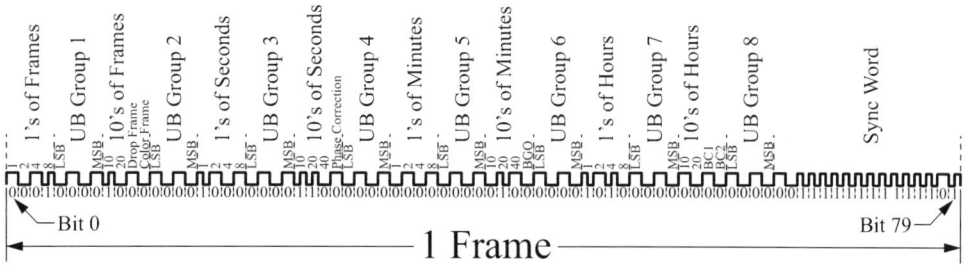

The standard specifies a "preferred connector" for balanced outputs: a 3-pin XLR. Males are to be used for outputs, females for inputs.[3] The pin-out follows the AES XLR standard: pin 1 is shield, and pins 2 and 3 are used to carry the differential audio signal (remember, polarity is irrelevant in this bi-phase scheme). The preferred connector for single-ended time code transmission is the BNC.

LTC is simple to use and gained broad acceptance, but it has a number of limitations for some applications. If media containing LTC is moving extremely slowly or is stopped, the code can be misinterpreted or not read, because the *transitions* in the signal carry the information; with no movement in the LTC signal, there are no transitions. While LTC can be read at a varying speed, at high transport rates the frequency[4] of the signal can exceed the frequency response capabilities of standard audio circuitry, and become distorted. In addition, when using (obsolete) tape-based recording, LTC generally becomes so distorted and noisy after two generations of (nondigital) rerecording that it must be "jam-synced," a process in which the TC is read, regenerated, and rerecorded. LTC is generally recorded at a high level in an audio track, and because it is a square wave, crosstalk with adjacent tracks or even low-level signals in adjacent cables is possible. There are ways to deal with all of these problems, and LTC is used reliably in many applications. For video applications, there is another type of time code: Vertical Interval Time Code.

Vertical Interval Time Code

Vertical Interval Time Code (VITC) was developed in the late 1970s after videotape recorders capable of playing back high-quality still images were developed. As just described, LTC becomes unreadable when the media is not moving. In VITC, time-code information is inserted as digital data into each frame's "vertical blanking interval," the time during which the electron beam was "blanked" as it rastered back to the top of a cathode-ray tube (CRT) after scanning a video field. Because VITC is part of the video signal itself, it can be read and decoded even in extreme-slow-motion or still-frame applications. Additionally, in video-sync applications, VITC does not use up one of the videotape's precious audio tracks.

VITC uses 90 bits to encode the time code instead of 80 as with LTC; most of these additional bits are used for cyclic redundancy checking (CRC, see page 132). Because VITC is encoded within the video signal itself, it can be used only

---

3. This may seem backwards if you're not an audio person, but it is the standard sex arrangement for audio connections (because of phantom power emanating from mixers).
4. At normal speed, the bit rate of 80 bits × 30 frames equals 2,400 bit/s, or a fundamental audio frequency of about 1,200 Hz.

in video applications, and cannot be changed without rerecording the video. Since video equipment is needed to read VITC, implementations of LTC are generally less expensive for nonvideo applications as generally found in live entertainment.

SMPTE TIME CODE HARDWARE

Because of the extensive broadcast equipment market, and the wide acceptance of the SMPTE Time Code standard, there is a large variety of specialized time code generation, processing, distribution, and interfacing hardware.

Time Code Generator

A time code generator is used in many applications to create a master time code source. Many live shows do not need a generator, but might use one in the studio to create the code recorded on the audio or video media; other shows use an externally controllable generator as a master, with all other devices slaving to it.

Distripalyzer

One unique and invaluable device for any time code application is the Distripalyzer, made by Brainstorm Electronics. In addition to performing standard functions such as jam-syncing and time code distribution, this unit features some built-in test equipment: the Distripalyzer can analyze an incoming time code stream and measure and display the frame rate, the time-code format, drop frame status, and a number of other parameters.

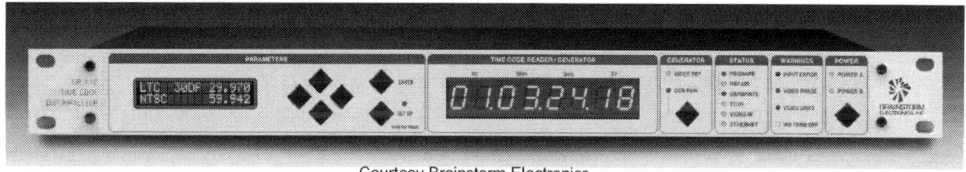
Courtesy Brainstorm Electronics

SMPTE Display Units

Another invaluable device is a SMPTE display, which simply shows the current time code addresses as they roll by. This is a very useful device for troubleshooting and for system monitoring.

SMPTE to MTC

Many devices take only MIDI Time Code (described below). There are a variety of converters that will generate MTC messages from an incoming SMPTE stream.

SMPTE LINEAR TIME CODE IN THE ENTERTAINMENT CONTROL MARKET

The recommendations here apply to LTC, which you are far more likely to encounter in a live show than VITC.

Regeneration

As mentioned above, LTC degenerates each time it is recorded or rerecorded, especially with (obsolete) analog systems. So, it is recommended that any time you re-record LTC you "regenerate" or "jam-sync" it, cleaning up bit transitions through reshaping and, in some systems, totally recreating the signal to cleanup jitter (timing) problems. Some devices incorporate this regeneration as a feature, but stand-alone regeneration units are also available. Most digital decks simply generate the LTC signal digitally without going through an analog or A-D recording process, so this obviates the need for regeneration.

Distribution

To be safe, you should generally not "two-fer" LTC signals; instead, you should use some sort of audio distribution amp (DA). This approach solves potential impedance problems, isolates each output from others on the same link, and ensures that each device receives a clean signal.

MIDI TIME CODE

Since SMPTE TC is an analog audio signal, machines that have only digital inputs (computers) must sample and convert the analog signal somehow in order to extract the time code data. Alternatively, systems can use MIDI Time Code (MTC), which breaks the analog SMPTE TC frames into MIDI messages and digitally transmits them directly down a MIDI line.

MTC Messages

MIDI Time Code is transmitted in two formats: full messages and quarter-frame messages. Full messages contain a complete time-code address and are sent whenever a system starts or stops, or when the system moves in or out of various transport modes (i.e., from Fast Forward to Play). Quarter-frame messages transmit pieces, or "nibbles", of the current TC address, and as the name implies, they are sent out at the rate of four per TC frame. Eight quarter-frame messages are used to communicate one full TC address, so the time-code count can be fully updated only every other frame. However, a receiving system can still *sync* four times per TC frame, by examining the timing of the quarter-frame messages. In addition, the receiver can tell the TC direction from the received order of the quarter-frame messages.

The 11110001 ($F1_{16}$) status octet is used as the MTC quarter-frame system-common octet. After this MTC status octet is sent, a data octet is sent, which can contain a nibble for the message number, a nibble representing one of the TC digits, or a nibble containing other data.

Function	Binary	Hex	Where
Full Frame Message	11110000	F0	System Exclusive
	01111111	7F	Real-Time Sys-Ex
	01111111	7F	Device ID: Entire System
	00000001	01	MIDI Time Code Sub ID
	00000001	01	Full Frame Message
	0tthhhhh		tt = TC Type (00 = 24, 01 = 25, 10 = 30 DF, 11 = 30 NonDrop)
			hhhhh = Hours, 0–23
	00mmmmmm		mmmmmm = Minutes, 0–59
	00ssssss		ssssss = Seconds, 0–59
	000fffff		fffff = Frames, 0–29
	11110111	F7	End of System Exclusive
Quarter Frame Frames LS Nibble	11110001	F1	
	0000dddd	0d	dddd = Frames LS Nibble
Quarter Frame Frames MS Nibble	11110001	F1	
	0001uuud	1d	uuu = Undefined; set to 0
			d = Frames MS nibble
Quarter Frame Seconds LS Nibble	11110001	F1	
	0010dddd	2d	dddd = Seconds LS nibble
Quarter Frame Seconds MS Nibble	11110001	F1	
	0011uudd	3d	dd = Seconds MS nibble
Quarter Frame Minutes LS Nibble	11110001	F1	
	0100dddd	4d	dddd = Minutes LS nibble
Quarter Frame Minutes MS Nibble	11110001	F1	
	0101uudd	5d	dd = Minutes MS nibble
Quarter Frame Hours LS Nibble	11110001	F1	
	0110dddd	6d	dddd = Hours LS nibble
Quarter Frame Hours MS Nibble/TC Type	11110001	F1	
	0111ttdd	7d	tt = TC type (same as above)
			dd = Hours MS nibble

So for example, a full-frame MTC message containing the time code address 05:20:31:18 (sent in 30 nondrop) breaks down in hex as follows:

F0   7F   7F   01   01   65   14   1F   12   F7

SMPTE User Bits in MIDI Time Code

The MTC standard provides a method for encoding SMPTE TC user bits in MTC, as shown.

Function	Octet Binary	Hex	Where
User Bits Message	11110000	F0	System Exclusive
	01111111	7F	Real-Time Sys-Ex
	01111111	7F	Device ID: Entire System
	00000001	01	MIDI Time Code Sub ID#1
	00000010	02	User Bits Message
	0000aaaa		aaaa = First User Bit Group
	0000bbbb		bbbb = Second User Bit Group
	0000cccc		cccc = Third User Bit Group
	0000dddd		dddd = Fourth User BitGroup
	0000eeee		eeee = Fifth User Bit Group
	0000ffff		ffff = Sixth User Bit Group
	0000gggg		gggg = Seventh User Bit Group
	0000hhhh		hhhh = Eight User Bit Group
	000000ii		ii = Group Flag Bits
	11110111	F7	End of System Exclusive

MIDI TIME CODE IN THE ENTERTAINMENT CONTROL MARKET

On a show, you might find LTC, MTC, or a combination. When using MTC, there's a few things to consider.

Timing Accuracy

MIDI Time code is actually capable of *higher* resolution that SMPTE, because it uses four quarter-frame messages per single SMPTE frame. At a normal framing rate of 30 fps, 120 MTC messages would be sent out each second. However, some MTC messages might be delayed in a MIDI merger or other processing device due to heavy traffic. So, if timing is critical in a particular application, a dedicated MIDI link should be run directly from the generating device to the receiver, with no other MIDI traffic allowed on that segment. If the MIDI link is simply a piece of cable with no other traffic, it will have a predictable latency.

Distribution

MTC messages are simply standard MIDI messages, and the distribution topologies and other recommendations found in the "Recommended MIDI Topologies" section, on page 294, apply.

OTHER TIME CODES

There are several other time codes you might find in show applications.

Computer Time Codes

In addition to the widely used SMPTE Time Code, there are a number of other "time codes." For instance, many show programs create an internal time code clock based on the internal clock of the computer. To export time code outside the computer, you would likely see MTC or SMPTE LTC (which could be sent over the network using RTP, see the "MIDI over Networks," on page 296), or you might see proprietary, networked approaches.

Sync or Tach Pulses/Pilot Tones

In the past, for simple, relative synchronization, a "sync" or "tach" pulse was sometimes used. For example, a camera might send out a square wave pulse when it exposes each frame, or a motorized system may send out a pulse train that varies in frequency as it runs. With either of these approaches, other devices that receive the information can synchronize to the pulses.

Another, analog approach found in older film production equipment is the use of "pilot tone." With such a system, a tone is recorded on a media's track. As the device moves, a tone of a specific frequency is generated; as the device speeds up, the tone's frequency will increase.

Video/Audio Sync

In video and audio studios, video frames from all equipment must lock together exactly in time; audio samples, correspondingly, must also be synchronized. See "A Quick Note on Video Sync," on page 38, and "A Quick Note on Word Clock," on page 32, for more information. Many time code devices can accept one of these sync signals to synchronize the time code with the gen lock or word clock signals.

Chapter 26: Open Sound Control (OSC)

Open Sound Control (OSC) is an open protocol for message-based communication between computers, synthesizers, samplers, gestural controllers, and other similar devices. It was developed in the late 1990s at UC Berkeley's Center for New Music and Audio Technologies (CNMAT) by Adrian Freed and Matt Wright, and Version 1.0 was published online in 2002. The low-level details of OSC are dense and based on intense jargon that only a propeller-head programmer could love but, fortunately, users of OSC generally only have to map parameters on their controlled systems to the control parameters. Still, users should always have an idea of how a protocol works because this defines the protocol's strengths and limitations. So, we'll start here with an overview before moving on to an example protocol.

OSC OVERVIEW

One of the more confusing aspects of OSC for many users is that it doesn't actually contain definitions of any sort of functionality that a user might use—unlike MIDI, there are no "note on" or "note off" kind of commands. Instead, OSC defines a structure for an "address space" and then lets the system or product designers decide on all the functionality that makes up that address space. In some ways, OSC is sort of analogous to a Microsoft® Excel spreadsheet, in that it doesn't describe or standardize a user's data, but instead gives system designers a structured way in which to exchange that data.

Clients and Servers

OSC was designed to run on any OSI compatible network which, these days means UDP/IP (or TCP) over Ethernet, and also via USB, especially for peripheral devices. Any device in OSC that sends OSC packets is a client; a device that receives packages is a server. Clients in OSC can only connect to one server at a time, but servers can receive packets from many clients. For example, sound gen-

> #### A Bit of OSC History
> OSC was based on previous work by a group including Wright and Freed on the Zeta Instrument Processor Interface (ZIPI), which was designed as a successor to MIDI but never gained widespread use and was never built into any products. OSC is a sort of descendant from ZIPI's "Music Parameter Description Language," which is also like the Device Description Language in ACN, and the now defunct AES-24 standard (more on AES-24 on page 342).

eration software running on a computer (server) could be controlled simultaneously by two different kinds of controllers (clients):

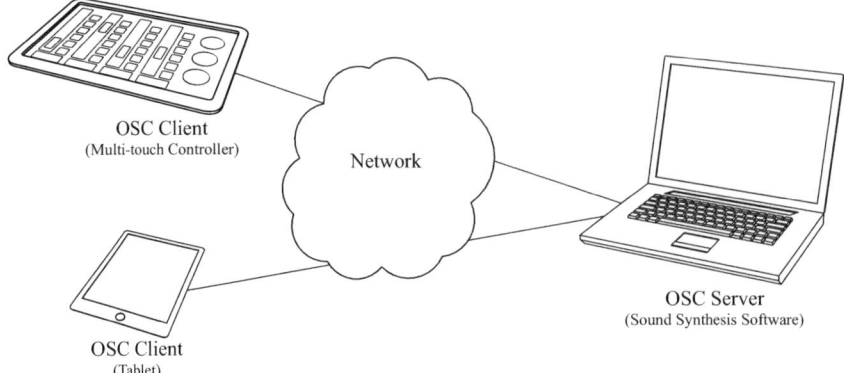

Addresses

The OSC server has an address space (not to be confused with the IP address), that gives a tree structure to a number of "containers," which contain "methods," that actually accept and act on the "arguments," or data:

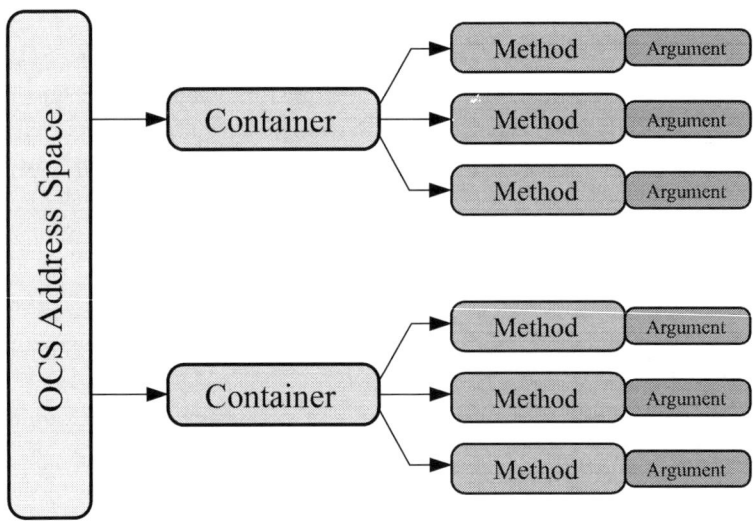

Each method and container has a name, which is simply a string of ASCII characters. For example, a mixer (OSC server) could have several channels, each with a volume control method called "Volume," another method called "Pan," and another for "mute." It might accept a level as its volume method an argument for level "50" at the address /Mixer/Channel_2/Volume:

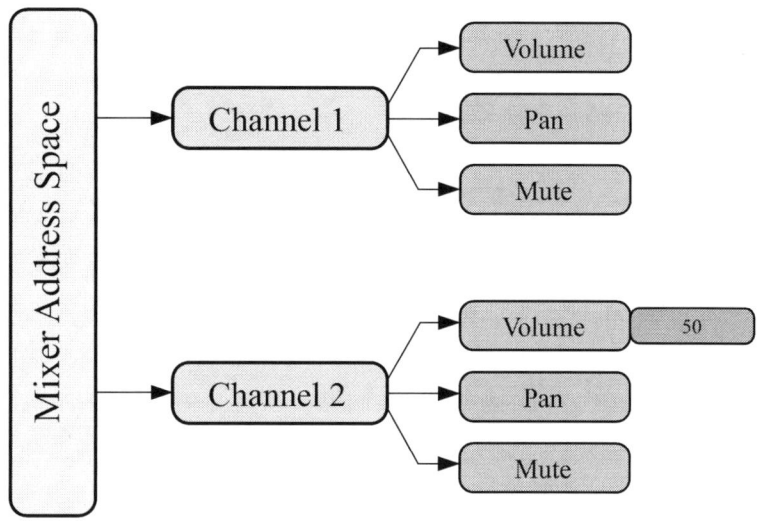

Packet Structure

The actual OSC packet is made up of four "atomic" data types, with each data block designed to always be a multiple of 32 bits for easy parsing:

Data Type	Description
int32	32-bit Integer (numbers such as 1, 2, 3)
OSC-timetag	64-bit count of seconds and fractions of a second since January 1, 1990; same as used in Network Time Protocol (NTP)
float32	32-bit Floating Point Number (a way of storing numbers such as 1.23 in a computer)
OSC-string	A string of ASCII Characters, followed by 0-3 null characters (00_{16}) is appended to ensure the block size is 32 bits.
OSC-blob	A variably sized "blob" of any kind of 8-bit data, preceded by an integer count of the number of octets. A pad of 0-3 null characters (00_{16}) are appended to ensure that the data is some multiple of 32 bits.

Each OSC message starts with an address pattern, which is a string beginning with a slash /, continues with some information about the data to be sent, and then finishes with the "arguments," which are the commands being sent. The messages can be variously sized, so the length, or "size," is also included in the message structure. Messages can either be sent individually, or in a "bundle," which contains a block of messages. Here's the packet structure for a basic message, with the OSC terminology:

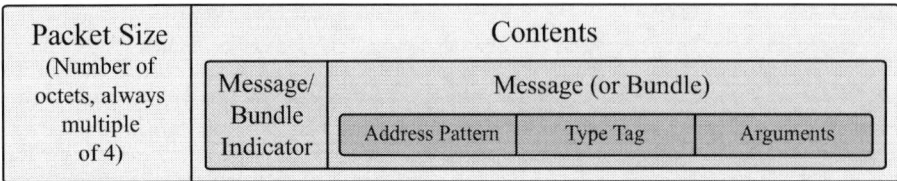

The Packet Size, Message/Bundle indicator is usually handled by the user's software, which would create a message something like this:

`/channel/1--,si-Guitar--000F`

(Note: The "-" here represents required padding added to make the message block into a multiple of four octets).

So this message would break into an Address Pattern of `channel/1`, followed by the argument Type Tags of `s` and `i`, which mean "string" and then "integer." The "`Guitar`" is the string argument for the address pattern, and the `000F` is the value of "15" as an integer, representing the value of guitar to be set.

OSC IN THE ENTERTAINMENT CONTROL MARKET

Fortunately, in most implementations in our market, the user is isolated from the low-level details of OSC; the user simply has to make sure the two devices can communicate (this might involve something like setting IP addresses or some other configuration), and then make sure the right OSC messages are assigned to the right parameters of their controlling device and target.

For example, to control a Meyer Sound D-Mitri processor using OSC, you would send messages on port 18033 via UDP datagrams, or TCP connections. To fire a "next cue Go" you would send something like:

`/go`

To set the level on a matrix cross point, you might send something like this:

`/set,Bus 13 Output 58 Level-90.0f`

Since OSC specifies only a structure for data and not actually operational commands or meaning, if you're going to use it make sure that the devices you select have excellent documentation and support.

Chapter 27: Other Control Protocols

In this section, we will cover a few miscellaneous protocols that may be useful in entertainment and show control systems, but aren't widely used enough to warrant their own chapter.

OPEN CONTROL ALLIANCE (OCA)

As of this writing, the **Open Control Alliance** (OCA) is developing a new, interoperable audio system control protocol. OCA is based in part on work done 20 years ago within the Audio Engineering Society (AES) on the AES-24 standard (which never was commercialized and was eventually withdrawn—see note below).

OCA is being designed to run on a variety of transports and infrastructures, although the current focus is, of course, on Ethernet and IP (and possibly USB). It's being designed to work well with various streaming solutions like Dante (page 33), AVB (page 32), RTP (see page 296), and others like Cobranet (page 33); Bosch has announced that their OMNEO system is basically AVB combined with OCA.

As of this writing, OCA has the following design features/goals:

1. Scalability: 2 to 10,000 nodes, localized or wide-area.
2. Reliability: Loss-free exchange of control and monitoring information, and prompt detection of malfunctions.
3. Growth Potential: Expandable definition to support emergent products while still maintaining legacy compatibility.
4. Multi-vendor Flexibility: Rich support for proprietary extensions that do not compromise core functions.
5. Versatility: OCA supports media devices at all levels of complexity, with all kinds of functions: Re-configurable processors are fully supported.
6. Security: The OCA security option offers controller authentication and full encryption of control and monitoring traffic. This option uses standard algorithms that are available worldwide.

7. Reliable Remote Firmware Update: OCA supports an option for accurate firmware updating over the network from a central point.
8. Full Discovery Features: OCA allows controllers to discover what OCA-compliant devices are on the network and what elements are inside each one. Controllers are notified when devices enter and leave the network.
9. Platform Independence: OCA relies on no manufacturer-specific protocols, platforms, or programming environments. System requirements are minimal.
10. Efficiency: Although OCA is not a minimalist architecture, it uses network bandwidth and node processing power conservatively.

OCA is designed as an "architecture, not a protocol" and is currently made up of three main parts:

- Framework (OCF): Device model, Functional mechanisms.
- Class Tree (OCC) Object-oriented definition of control & monitoring functional repertoire.
- Protocol Implementations. OCA is planned to be a family of protocols for different contexts; OCP.1 is for TCP/IP networks; others will accommodate other networks.

OCA's Roots in AES-24

Here's an excerpt from the second edition of this book, released in 2000:

When the first edition of this book [1994] was released, work was under way on a radically new sound communications network standard—Audio Engineering Society (AES) 24. This was a very ambitious project, since the group was attempting to design an object-oriented and network independent system from the ground up—not standardize or modify existing systems. After many years of on and off again work, AES-24-1-1999, part one of the proposed standard was published in the April 1999 issue of the *Journal of the Audio Engineering Society*. While this part of the standard has been reviewed and issued, it is unlikely that the remaining (and key) pieces of the puzzle will ever be completed and so, sadly, the standard is basically dead.

Two primary factors led to the failure of this standard effort. The first is that no clear market demand or commercial pressure ever developed for the creation of a unified audio control standard, so key manufacturers, many of whom had proprietary systems already on the market, supported the group's efforts only in token, and not substantive, ways. The second factor is that with the growth and power of DSP technologies, audio systems are becoming increasingly centralized, with more power housed in less physical devices, so there are less devices and types of devices that need to be connected and controlled ... The AES-24 standard was eventually officially withdrawn in 2004. While it is unfortunate that AES-24 failed, the effort was not wasted; the list of those involved reads like a "who's who" of the industry, and those people have incorporated what they learned and developed into many products and systems on the market..

OCA does include specific control definitions for elements such as level, EQ compression, etc., but allows manufacturers to extend this to include proprietary control approaches and features. That makes OCA very different from OSC (see page 335) or ACN (page 341) (although sACN does define functionality), which only supply device description frameworks and methods to control them.

OCA in The Entertainment Market

In the short term at least, in the sound market I know best—live sound—end users aren't screaming for a universal audio control technology, and haven't been since the days when AES-24 was developed and we routinely used racks and racks of individual analog processors from different manufacturers. But Bosch, the main driver behind this effort, owns the audio brands DYNACORD, Electro-Voice, RTS, and Telex, and they do have a clear market need, in large systems for facilities like airports and installed PA systems. These kinds of systems have security and life safety issues that we don't (yet, anyway) face in live sound, and wider adoption of the standard is of course in their interest. Time, of course, will tell how it all plays out!

MIDI VISUAL CONTROL (MVC)

In 2011, Roland, a manufacturer of musical and video equipment, spearheaded an effort to create a "MIDI Visual Control (MVC)" standard based on its Roland's V-LINK products. The protocol is a mixture of standard musical MIDI messages, and new Universal System Exclusive messages with a structure fairly similar to MIDI Show Control (Chapter 23), including a Device ID, sub ID, etc. More information is available from the MIDI Manufacturers Association at `http://www.midi.org/`.

NETWORK TIME PROTOCOL (NTP)

The Network Time Protocol (NTP) was developed in the mid-1980s to synchronize machines across a network. The Protocol uses messages over UDP port 123, and can synchronize free-running, local computer times to within less than 128 ms, and typically 5-10ms.[1] The operational details of NTP are beyond the scope of this book, but a brief overview is included here.

To achieve this accuracy, NTP rates clock accuracy according to "stratums." Stratum 0 includes devices such as atomic or GPS clocks, and are of the highest possible accuracy. These clocks are not attached directly to the network, but are generally attached to local machines directly using a serial or other deterministic

---

1. `http://www.ntp.org/ntpfaq/NTP-s-algo.htm#Q-ACCURATE-CLOCK`

connection. Stratum 1 clocks are the computers that receive time signals from the Stratum 0 devices, and these computers are called time servers in NTP. Stratum 2 computers can request time information from one or more Stratum 1 time servers, and will ignore any time data that seems too far off to be reasonable. Stratum 2 servers can also work in conjunction with other Stratum 2 machines to increase the accuracy for all connected machines of the same Stratum. Stratum 3 and greater machines get timing information from Stratum 2 machines, and can also work together with Stratum 3 machines.

The actual timestamps exchanged are 64-bit messages, with 32 bits allocated to a cumulative count of seconds and 32 more for fractional seconds. The seconds count is incremental from January 1, 1900—because of this, NTP will effectively "roll over" in 2036.

NTP would be very useful on show networks for keeping local clocks on many machines tightly synchronized. However, it's not as useful as a "show clock" as something like SMPTE or MIDI Time Code (see Chapter 25, on page 321), which can be started relative to the beginning of a show. It is, however, of course, to dynamically schedule cues on multiple machines and then use NTP to keep the local clocks synchronized.

SIMPLE NETWORK MANAGEMENT PROTOCOL (SNMP)

The **Simple Network Management Protocol** (SNMP, RFC 1157 and others) was developed for—not surprisingly—network management. With the increase in network use in entertainment, SNMP has found a place in configuring entertainment devices (particularly for some audio gear) as well as IT equipment (routers, switches, etc.) on shows, so I'm including a brief overview here.

SNMP communications take place between a network-management system, or "manager", and a device element, or "agent," which is traditionally a router, hub, network printer, and so on. SNMP runs over UDP on port 161 and 162, and most of the communications involve the setting or polling of simple variables, as defined by the manufacturer of a device element in a Management Information Base (MIB), using commands like "GetRequest." Parameters are written using commands like "SetRequest."

VIRTUAL NETWORK CONTROL (VNC)

One of the other great benefits of networking a control system is that you can use something very helpful like **Virtual Network Control** (VNC). VNC is an open source, free, desktop sharing system that allows a computer to be controlled from a remote location, just as if the user were sitting in front of the machine. To make the system work, run VNC server software on the machine you want to control, and then run VNC client software from any other machine reachable on the network. Using TCP messages, the user screen of the controlled machine is sent out, and mouse and keyboard controls work from the client. The system works over any sort of network, including wireless, and is really useful for programming of show systems that run on general purpose computers. Obviously, security for important systems when remoted becomes an issue and must be addressed.

INDUSTRIAL I/O SYSTEMS

There are a number of Input/Output (I/O) systems designed for the larger world of industrial controls, that have entertainment applications. These (mostly) proprietary systems may be used as part of a motion control system, or might be part of a show control I/O system. Since they are proprietary, we will just go through a brief overview here (listed alphabetically), and I have provided Web links for more information.

AS-Interface (AS-i)

The AS-Interface (AS-i) is a simple, master/slave, two-wire industrial control protocol often used for simple machinery I/O applications. Interfacing equipment based on the open system is available from a wide variety of manufacturers, and the system is often used with PLCs. The system is designed primarily for simple sensors and actuators, and uses a simple cabling system running up to 100 meters, carrying both data and power. 62 slaves can be connected to a single master. More information is available on http://www.as-interface.com/.

Controller Area Network (CAN)

The Controller Area Network (CAN) is the foundation of DeviceNet (see below), and was originally designed by the German company Bosch for the automotive market. CAN is a broadcast, differentially-transmitted serial network, which was designed specifically for difficult electromagnetic environments, like automobiles, where it has found widespread use. The system features CRC error detection, and it runs at rates of up to 1 Mbit/s. Like Ethernet, CAN has a MAC layer and a LLC layer. Messages on CAN are small (limited to eight data octets) and a prioritization system is included in the network to ensure that high priority messages will be delivered first. More information is available on http://www.can-cia.org/.

DeviceNet

DeviceNet was originally developed by Allen-Bradley (which became Rockwell Automation), but is now maintained by the Open Device Net Vendors Association (ODVA). DeviceNet is based on the CAN network (see above), and is a multidrop system with as many as 64 nodes on a single logical device. Speeds can run up to 500 kbit/s, and the system was engineered to operate in high-noise electrical environments. The network can cover distances as far as 500m, depending on the data rate used (higher rates mean shorter distances). There is also a method to sending DeviceNet information over Ethernet, which is confusingly named EtherNet/IP™ (IP here stands for Industrial Protocol, not Internet Protocol). EtherNet/IP uses the Common Industrial Protocol™ (CIP) for upper layer operation. More information is available on http://www.odva.org/.

Echelon LonWorks®

LonWorks was developed by Echelon Corporation in the 1990s as a proprietary, OEM networking system for distributed-control system applications using twisted-pair cables. In the 2000s, the LonWorks communications protocol was adopted by ANSI and IEEE for several kinds of applications, like in-train controls, and new power-line distributed versions became available. Several entertainment control companies (most notably Meyer Sound Labs and Renkus-Heinz, and others in the audio industry) adopted the system for monitoring and control for their products. However, even though all these systems use LonWorks technology, none of them are interoperable, so there's not much to cover here; more information on http://www.echelon.com/technology/lonworks/.

EtherCAT

EtherCAT stands for "Ethernet for Control Automation Technology" and this high performance, Ethernet-based I/O system has been gaining wider acceptance as of this writing. It is designed to send small control messages very quickly, and uses a "processing on the fly" approach, with a single Ethernet frame carry data for multiple nodes—possibly even the entire control network. Nodes pass through the data "telegram" while processing the data addressed to it. More information on http://www.ethercat.org/.

Modbus

Modbus was originally published by the Programmable Logic Controller (PLC, see page 45) manufacturer Modicon in 1979, and the simple, easy-to-use protocol eventually came into widespread use, becoming a de facto standard for industrial automation interconnection. Modbus was originally designed using RS-232, then moved on to multidrop RS-485, and then eventually on to Ethernet with Modbus/TCP. Modbus/TCP is now a fully open standard, maintained by Modbus-IDA, an independent trade association.

Two varieties of Modbus are available: ASCII and RTU. The ASCII version uses simple two-character ASCII messages; messages in the RTU version contain two

4-bit hex digits. Both contain a CRC check. The system works in a command/response mode: a master makes a query of a slave device, which then responds. Every query message has a destination address, which indicates which "coil" (as in relay coil—remember, this protocol has its roots in PLCs) is to be controlled or monitored. The first coil (1) is addressed as coil 0. Typical commands include "Read Coil Status," "Read Input Status," "Force Single Coil" and "Force Multiple Coils." More information is available on http://www.modbus.org/.

Profibus

The ProcessField Bus (Profibus) is a popular fieldbus (especially in Europe) for connecting distributed I/O systems. Standard Profibus comes in five different versions, ranging from RS-485 over twisted-pair to fiber, with transmission speeds of up to 12 Mbit/s. Versions are also available for the transmission of time-critical data, and 32 nodes can be connected in one segment. Profinet is the Ethernet-based network version of Profibus, running over TCP, UDP, and IP, with additional support for real-time operation. More information is available on http://www.profibus.com/.

LEGACY VIDEO CONNECTION STANDARDS

You may still encounter these older, serial-based protocols, so we will cover them briefly here.

Pioneer LDP/DVD Control Protocol

Pioneer® made a line of popular industrial DVD players, and was one of the dominant manufacturers of industrial videodisc (laser disc) players. Their command set for these devices has been used by many other companies in a variety of devices, so it can now be considered a de facto standard. A few basic commands are included here; check Pioneer's website if you need more details, or need the complete command set.

Physical Connections/Data Specs

The system is based on RS-232, and uses transmit (TxD), receive (RxD), Data Terminal Ready (DTR), and ground. The DVD-V7200, one of the more popular Pioneer DVD models, uses a 15-pin connector. Data is sent in eight-bit words with no start bits, one stop bit, and no parity at either 4,800 bit/s or 9,600 bit/s, which is configurable in the DVD.

Command Structure

The command protocol is simple and straightforward and, therefore, very easy to use and reliable. ASCII characters are sent from a controller to the unit over a bidirectional serial link. To tell the player to play, for example, the controller would send the ASCII characters "P" and "L," followed by a carriage return. The player would then acknowledge successful completion of this command by returning an "R" and a carriage return. If an error occurs or the target device doesn't understand a command sent to it, it returns an error code. This simple approach is very powerful and allows very sophisticated operation, and the fact that it is standard RS-232 and ASCII means that a wide variety of computers or entertainment controllers can be used to send the commands. You can see an actual control exchange with one of these machines in "Serial Connection Example," on page 152.

Commands

Here are some of the more commonly used commands. The symbol << means Carriage Return (ASCII character $0D_{16}$), and anything in <> brackets is optional.

Start

`SA<<`

Starts disc rotation. Good for "spinning up" the disc at the beginning of the day.

Search

`<FR> nn SE <<`

`<TM> nn SE <<`

`<CH> nn SE <<`

Causes the unit to search for an address specified by `nn`. The address can be a frame number, time, chapter, or a variety of other search points, depending on either the mode set in the unit or the `<optional>` prefix added to this command. For example, `FR3928SE<<` searches the unit to frame 3928. Typically, when the unit reaches the specified address, it enters Pause mode.

Play

`<Address> PL<<`

Causes a disc to play. Putting an optional address in front of the play command will cause the unit to play until that address (frame, chapter, etc., depending on which mode is set) is reached, at which point an "auto stop" occurs.

Pause

PA<<

The disc stops temporarily, and the image may or may not stay on the screen, depending on other mode settings of the player.

Still

ST<<

The disc stops and the image freezes on whatever was on screen when the command was received.

Scan

NF<<

NR<<

NS<<

Like fast forward or reverse; the image is displayed as the scan takes place. NF causes forward scan, NR causes reverse, and NS returns the unit to normal playback.

Reject

RJ<<

Stops disc rotation; good for shutting things down at the end of the night.

Chapter Request

?C<<

It is often useful for a controlled device to "poll" a target device in order to find out its current status. The Chapter Request command asks the player for its current chapter, and the player returns two ASCII characters representing a two-digit chapter number.

Frame Number Request

?F<<

Frame Number Request asks the player to return its current frame number. The player responds with seven ASCII characters, representing the frame number value.

Time Code Request

?T<<

Time Code Request asks the player for its current time code. The player responds with five ASCII characters: the first three representing the number of minutes and the last two representing seconds. Other numbering schemes are used for disks other than DVD.

P-Block Number Request

?A<<

The P-Block Number Request polls all of the following at once: Title number, chapter number, current minutes, and current seconds. The player responds with 10 ASCII characters: the first two for title number, the second two for chapter number, the next four for minutes, and the last two for seconds. Other numbering schemes are used for disks other than DVD.

Sony 9-Pin Protocol

Sony® is one of the largest corporations in the world, and is a major manufacturer of broadcast and professional video equipment. Sony developed a (now obsolete) serial control protocol for use in their equipment, which became a de facto standard commonly referred to as "Sony 9-Pin." Because Sony never formally issued this protocol as a standard, finding accurate information was difficult; relevant commands for a particular type of equipment are typically contained in that piece of gear's manual.

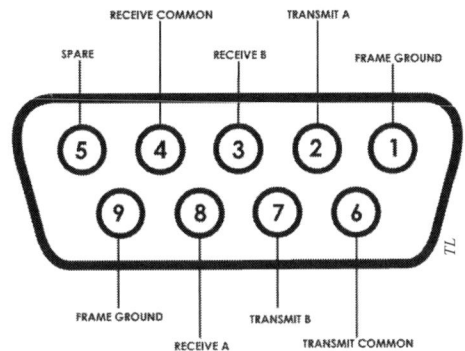

Physical Connections/Data Specs

The implementation uses RS-422 on a nine-pin connector, and uses four wires. Data is sent at 38.4 kbit/s in eight-bit words with one start bit, one stop bit, and an odd parity bit.

Command Structure

The command protocol is fairly complex for a serial, point-to-point connection method—it includes length information and a checksum. While these features require the transmitter and receiver to have added intelligence, they also make the standard very robust.

Each message consists of from three to eighteen octets in this order:

`Command1/DataCount`	Four bits of command, four bits of octet count
`Command2`	Command
`<Data>`	Optional data
`Checksum`	Checksum of previous octets

The first octet is laid out as `ccccbbbb`, where `cccc` is the Command Group (see below), and `bbbb` is a four-bit count of the number of data octets in that message, ranging from 0–15 ($\mathtt{0000\text{-}1111}_2$). If the data octet count is zero, that message has no data octets; any number from 1–15 ($1\text{-}F_{16}$) indicates that the command has that number of data octets, located between the Command2 octet and the checksum. The `cccc` nibble of the Command1 octet indicates the message function as follows:

Command1	Function	Message Direction
0	System control	Controller to controlled device
1	System control return	Controlled device to controller
2	Transport control	Controller to controlled device
4	Preset and select control	Controller to controlled device
6	Sense request	Controller to controlled device
7	Sense return	Controlled device to controller

The function of the Command2 and data octets varies with the command (see below). The checksum is the sum of the *numeric value* of the octets from Command1 through the last data octet before the checksum.

To send a command such as Play, the following octets in hex would be sent:

```
20 01 21
```

In the first octet, 20_{16}, the 2 means that the command is a transport control message; the 0 means that there are no data octets. The second octet is the Command 2 octet—01 here—indicating the Play command. The 21_{16} octet is the checksum (20+01).

If a controlled device correctly receives and processes a control message with no data, it responds with an acknowledgment "Ack" message. If the command message contained data, the receiver responds with a copy of the command and the associated data. If an error occurs, it responds with negative acknowledgement ("Nak") and returns the associated error data.

Commands

Here are a few of the more commonly used commands in hex:

Stop

```
20 00 20
```

Play

```
20 01 21
```

Eject

```
20 0F 2F
```

Fast Forward

```
20 10 30
```

Rewind

```
20 20 40
```

The 2 in the first octet of these commands means "Transport"; the zero means there are no data octets in this command. The second octet is the actual command and the third octet is the checksum.

Cue Up With Data

```
24 31 ff ss mm hh cc
```

This command cues up a deck to a frame number specified in the four data octets. The 24_{16} octet means that this is a transport command, with four data octets. The 31_{16} octet means "Cue Up with Data." The four data octets are binary-coded decimal numbers representing the desired frame, seconds, minutes, and hours, with the most significant nibble of each octet representing the tens unit and the least significant nibble representing ones. The `cc` above represents the checksum. So, to tell a target device to cue up to `12:23:39:05` (12 hours, 23 minutes, 39 seconds, and five frames), we would send a hex message like this:

```
24 31 05 39 23 12 C8
```

where the $C8_{16}$ is the checksum of all preceding octets.

Current Time Sense

```
61 0C dd cc
```

It is often desirable to "poll" a target device, asking it to report its current media time. This is the function of the Current Time Sense command, and what exactly the target device should return is indicated by the octet shown as `dd`. `dd` can have several different values, depending on which bits are turned on in the octet. The function of the bits are:

Bit	Function
0	LTC time
1	VITC time
2	Timer 1
3	
4	LTC UB
5	VITC UB
6	
7	

LTC and VITC are different varieties of SMPTE time code (see Chapter 25, on page 321). UB stands for "User Bits," which can be encoded in the time code. There are many permutations to this table, and each one gets a different response

Chapter 27: Other Control Protocols • 353

from the target device. Let's look at a common data octet, requesting the time in LTC format. dd would be set to 00_{16}:

```
61 0C 01 DE
```

Because of the 01_{16} data octet, the receiver would respond with an LTC Time Data message.

LTC Time Data

```
74 04 ff ss mm hh cc
```

The target device would then send its data back in the same format as previously specified.

Part 5: **Show Control**

In this final part of the book, we'll bring together everything we've covered: entertainment disciplines, entertainment control systems, data communications and networking, and discuss the basics of show control. Then, we'll examine a few practical (albeit hypothetical) show control problems and solutions.

Please note that I made the examples, systems, and approaches to illustrate specific points, so please take these systems only as *possible* solutions to the design challenges presented, since one of the most fascinating (and simultaneously daunting) aspects of entertainment control is that there are a hundred different ways to achieve any particular goal.

Although most of the names have been changed to protect the innocent (and the guilty), the components, devices, and systems in these hypothetical case studies are generally based on real products and applications.

Chapter 28: Show Control

Show control (as we discussed way back on page 2) simply means connecting together more than one entertainment control system, and this simple idea can bring amazing sophistication and precision to a wide variety of types of performance. While many of the concepts we've discussed throughout the book are related to connecting entertainment control systems together, here we will discuss general and design issues related specifically to show control. First, though, let's discuss a bit about how we got here.

EVOLUTION OF SHOW CONTROL

You could probably trace the development of electronic[1] show control back to the Disney's *Great Moments with Mr. Lincoln* animatronic show at the 1964 World's Fair; a version of this attraction also opened at Disneyland 1965. That show marked one of the first instances where show producers started integrating then new automation technologies (many analog) into their attractions on a large scale, and these (mostly) time-based technologies brought an unprecedented level of sophistication and repeatability to shows that reached a huge audience through dozens or hundreds of performances a day. These centralized, "canned," time-based show control techniques and technologies reached another milestone in terms of scale with the opening of Disney's EPCOT® in 1982, and this kind of time-based, "canned" show control is still an important part of shows in theme parks, concert tours, museums, corporate events, and many other venues.

In 1994, when the first edition of this book was released, an exciting trend in show control had begun, where increased computer horsepower was allowing show control systems to become far more interactive and flexible than the older canned, fixed-time applications. One of the best known landmarks of this development was the performer–interactive *Indiana Jones™ Epic Stunt Spectacular!*, which opened at Disney-MGM Studios (now Disney's Hollywood Studios®) in 1989. Another landmark show with sophisticated, performer-driven interaction was 1993's *George Lucas Super Live Adventure* (which was produced in the United States, but unfortunately toured only in Japan) and that was followed by *EFX* in

---

1. I'm saying electronic here to keep from including automata, which are very cool but really a separate area of interest.

Las Vegas in 1995. These live shows used show control not to reduce labor costs, but instead to allow performer interaction with show elements, thereby providing cueing sophistication and precision in ways that would otherwise be impossible.

Successors to shows like *EFX* include the amazing *KÀ*™ from Cirque du Soleil™, which opened in Las Vegas in 2005 (and is still running as of this writing). *KÀ* represents a key milestone on an exciting evolutionary branch of the show control state-of-the-art, since it uses sophisticated peer-to-peer connections across departments to allow performers themselves guide the creation of some stunning moments on stage.

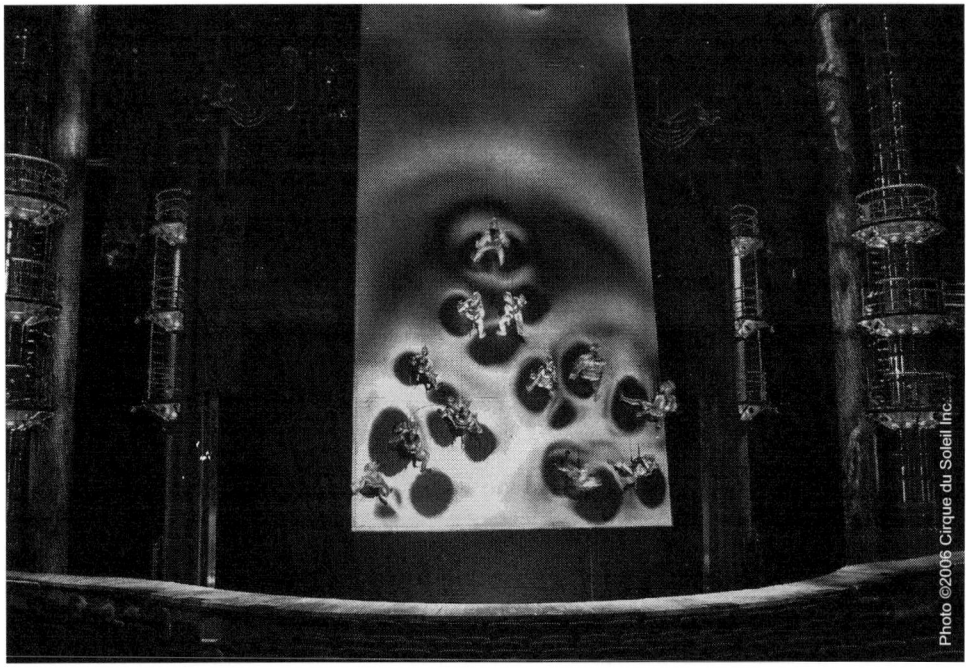

For example, in one scene (shown in the photo), performers wirelessly operate their own hoist winches, while their positions are sensed in near real time and sent, along with the positions of incredibly complex moving stage machinery, out on the show networks. Sound effects are triggered based on this data and a sophisticated graphics system creates interactive imagery in real time, based on the position of the performers. These graphics are projected back onto massive moving platforms, immersing the performers in a beautiful, interactive, real-time performance environment. Robert Lepage directed this show, and he has used similar techniques on subsequent shows, like *La Damnation de Faust* at the Metropolitan Opera, which led to his 2011 Ring Cycle at the Met, which again used interactivity.

As computer systems have grown more powerful and cost effective, the potential for both audience members and performers to interact with show systems has increased exponentially. With the tools available now, there is really no technical reason not to free (at least to some extent) the performers from the constraints of rigid time-based or "canned" systems and, instead, give them as much freedom as a show's storyline allows. This is a trend exemplified for me by the Australian dance troupe Chunky Move, whose 2006 production *Glow* built a foundation that was extended into an amazing 2008 production of *Mortal Engine*.

This beautiful "dance-video-music-laser performance" took place on a blank, white raked stage, and the dancer's movements were used to drive the creation of the audio, video, and laser environment—all in real time. This kind of interactivity had been done by others for years (with companies like Troika Ranch), but *Mortal Engine* for me marked a maturing of these concepts and techniques into an amazing piece of art; it was first time I've seen a dance audience crowding around the control position in the back of the house after a show!

We've arrived at a point where pretty much anything's possible in terms of show-control systems. Performer-interactive show control, as we've seen in *Indiana Jones*, *KÀ*, *Mortal Engine*, and many other shows, is of particular interest to me but many, many shows are run very successfully off of fixed, time-based systems.

So let's take a step back and ask: Why show control in the first place?

WHY SHOW CONTROL?

There are myriad reasons to connect systems together, but the key reasons for me are:

- To increase cueing and synchronization precision beyond human capability.
- To allow the performers to interact with the technical elements.
- To maximize the productivity of available labor.

KÀ's Connections

I would speculate that shows such as *KÀ* are possible not just because we now have so much computer horsepower (we had less, but still enough in the early 1990s, as evidenced by *Superlive*) but, even more importantly, because a critical mass of artists and technologists have worked with show control technologies long enough to become comfortable with them, and now use them to get rid of the rigid clock and give flexibility back to the performers. J.T. Tomlinson gives some interesting history and supports my premise: "Show control for *GLSLA*," Tomlinson told me in 2007, "was a Golder Group system, specified by myself, then designed and built by Damon Wootten, with George Kindler as consultant. Damon previously worked for Charlie Richmond, who recommended him to me for this show. We used show control to trigger multiple systems on commands from the stage manager, and also to sync various systems via time code from the LCS/Doremi sound system (LCS serial number one!) or 35 mm film projectors. Commanded systems included lighting, automated scenery, projection, lasers, fog effects, and pyro. One real challenge during production was getting SMPTE from an optical film sound track! *GLSLA* and *KÀ* are also similar in that regardless of the amount of computerized controls, neither show is/was automated in the theme park or cruise ship sense; i.e., all or mostly to time code. *GLSLA* and *KÀ* are both cued in the classical theatre fashion: a calling Stage Manager maintains the pacing. We had good help on *Superlive*, including the fine fellows mentioned above, and these current Las Vegas players: JT Tomlinson (now Head of Automation at *KÀ*) was Technical Director, Todd Toresdahl (now Head of Automation at *Mystere*) was Head Carpenter, Bill Wendlandt (now Assistant Head of Automation at *KÀ*) was Assistant Carpenter, Keith Bennett (now Production Manager at *Mama Mia*) was Head Electrician, JO Henderson (now Lead Automation Operator at *O*) was the Show Control operator, and Jonathan Deans (now Cirque Sound Designer) was Sound Designer." I am sure all would concur that *GLSLA* was ahead of its time.

WHAT IS A SHOW CONTROLLER?

For many years, if you wanted to connect systems together, you would need a custom-engineered, turn-key system programmed by some sort of "guru." But it's now possible to do very sophisticated show control by simply connecting a number of peer systems on a network (without a centralized "show controller" device), simply using one or more of the various protocols and techniques already discussed in this book.

However, there are many times when, for a variety of reasons, simple peer-to-peer interconnection is not enough: perhaps one of the devices you want to link has only limited conditional control capability; perhaps there are political reasons why it makes more sense to use an intermediary system to react to changes as they happen; or perhaps one of the systems doesn't allow easy, time based editing, doesn't have a time of day clock, and so forth. In these cases, you still need a show controller.

For me, a true show controller must have a few key characteristics:

- The system must be able to communicate bidirectionally using a variety of protocols (ASCII, MIDI, DMX, Modbus, etc.) over Ethernet/IP and a variety of serial and other links.
- The system should be able to output and receive contact closures and other similar I/O methods (if the system can speak IP, we have many ways to accomplish this goal).
- The system must be able to work in any "interrupt-driven" (page 82) way, meaning that it can act on any information coming from anywhere at any time.
- The system's program must be able to be edited while the show is running.
- The system must be able to manage multiple asynchronous timelines, and sync to multiple timing sources, external and internal.
- The system must be capable of stand-alone, automated operation (e.g., working off a time of day clock, etc.) without any human intervention (for fixed installations).
- The system should be able to present a customizeable user interface.

MY SHOW CONTROL DESIGN PROCESS

Show control is not necessarily accomplished through a particular piece of hardware; it is more a design approach that may or may not incorporate a show controller. Here are the questions I ask when first approaching a show control application:

1. What are the safety considerations?
2. What type of show is it?
3. Is the show event-based, time based, or a hybrid?
4. What is the control information source?
5. What is the type of user interface required?
6. What devices must be controlled/connected?

Let's go through each of these questions in some detail; afterwards, we will apply this process to some case studies to show you what kind of specific approaches are commonly used in some typical situations.

QUESTION 1: WHAT ARE THE SAFETY CONSIDERATIONS?

Safety, of course, must be the utmost consideration in any situation—in entertainment technology and elsewhere. Many entertainment control system applications are basically safe for audience and crew in all but extreme circumstances: lighting, sound, video, and so forth.[2] Other systems, such as scenic automation, pyro,

---

A Note On Media Control Systems

For real show applications, be wary of the "media control" touch-screen based systems used in many corporate boardrooms, restaurants, and home theaters. For simple, operator-controlled applications, these systems are powerful and cost effective, and for that reason, specifiers are often tempted to use them for show applications. However, you should resist that temptation; these media control systems (while sometimes used in shows for user interface purposes), rarely offer the timing accuracy/repeatability or the flexibility inherent in true show control systems (as of this writing, none of the commonly available media control systems meet my criteria detailed above). In addition, these systems are usually programmed in a compiled, general-purpose programming language, meaning that programmers must create from scratch much of the functionality of real show controllers. More importantly, media control systems typically do not offer the programming flexibility and speed needed to survive the "edit on the fly" live show environment.

I once had a programmer of media control systems call me; he had been asked to provide an estimate for the programming of a show in a museum. He said, "How long does it take to program XYZ show control system." I asked, "How much time do you have?" He was baffled. I told him, "If you have a day, it's a day. If you have a month, it's a month." He usually based his programming estimates for media control systems on the number of buttons he needed to create on the touch screen—one hour for this type, two hours for this, and so on. Eventually, after much explaining, he got it—this is a show, not a boardroom. He declined to bid the show project.

and flame systems, are very dangerous and must be approached with extreme caution—you should get expert advice before integrating with them. But with proper safety precautions and procedures in place, these systems can and often are integrated using show control to provide highly precise timing accuracy, enhanced safety, or other functionality not possible with human operators. We covered this in detail in the "Principle 1: Ensure Safety" section, on page 111.

QUESTION 2: WHAT TYPE OF SHOW IS IT?

For our purposes here, we will break shows down into linear and nonlinear. Each type presents different challenges and limitations; see the "Show Types" section, on page 67, for more information.

QUESTION 3: IS THE SHOW EVENT-BASED, TIME-BASED, OR A HYBRID?

The three basic types are event-based, time-based, or hybrid event/time based; see the "Cueing Methods" section, on page 69, for more information.

QUESTION 4: WHAT IS THE CONTROL INFORMATION SOURCE?

This may sound complex, but what we're looking for is simple: we need a way to get information about the show into the control systems. Is a skilled technician operating the system? Is a nontechnical person running the system? Are the audience members themselves running the system? Is the show run from a time-of-day clock? Does the show start at sunset and run through sunrise? Determining this and how you can get this information into some machine-readable form will be a big factor in deciding which system or technology is best for your application.

QUESTION 5: WHAT IS THE TYPE OF USER INTERFACE REQUIRED?

If the system is run by skilled technicians comfortable with cues, protocols, and timing issues, they may want as much user interface, feedback, and status information as they can get. In other applications that are simpler or more routine, there may be only unskilled operators who might be confused by too much detailed feedback and status information. For example, in a retail environment, someone who knows a lot about shoes but nothing of show technology may be starting and/or selecting the shows, and much of the editing functions may need to be completely locked down. Each of these applications needs a different kind of user interface, and you need to determine this in figuring out your approach.

---

2. Of course, you can electrocute, deafen or blind people, but you get the idea.

Human Interface Guidelines

Show control user interfaces most often involve a customized computer-based user screen, perhaps used in conjunction with a hardware panel of operator controls, or a touchscreen. A number of **Human Interface Guidelines** (HIG) have been developed to help offer guidance to designers and developers. Most of these HIGs are proprietary (Apple, Android, Windows, etc. all have them), so I have based my common sense, entertainment control-oriented list on the open "usability principles" of the GNOME Human Interface Guidelines.[3]

Design for People

This may seem obvious, but you have to keep in mind exactly who your users are when designing any user interface. Are they experienced technicians? Are they managers with some familiarity with the show but not with networks? Are they performers? The audience? Also, user interfaces are designed in most cases for someone other than you, so it's best to get feedback from other people before locking in on anything.

Don't Limit Your User Base

Will your user interface be used by people who don't speak your language? Will people with some form of disability need access to your user interface? Graphics and other metaphors can communicate a lot of information directly, without the heavy usage of text. On the other hand, I often find it confusing, for example, if graphical icons are the only explanation for buttons.

Create a Match Between Your Application and the Real World

To quote the GNOME guidelines, "Always use words, phrases, and concepts that are familiar to the user rather than terms from the underlying system."[4] I couldn't have said it better myself. Remember that we are telling stories and doing shows, and your operators may not really care if you are using a TCP/IP IO Interface Unit. Aren't they really monitoring some sensor, or show condition? Don't get caught up in unnecessary jargon.

Make Your Application Consistent

Give every screen presented to the user a consistent "look and feel." Don't make the user relearn the interface simply because they have now entered some diagnostic mode.

---

3. http://www.gnome.org/
4. http://developer.gnome.org/hig-book/3.0/principles-match.html.en

Keep the User Informed

Again, from the GNOME guidelines,[5] "Always let the user know what is happening in your application by using appropriate feedback at an appropriate time." I am big on putting a lot of information into my user interfaces, but at the same time, I try not to make them too cluttered. That status or diagnostic information may not even be used under normal circumstances, but when something goes wrong, that one extra indicator or status update may give the operator the information they need to fix a problem quickly (or work around it and get through the show).

Keep It Simple and Pretty

This is pretty self-explanatory. Put complex diagnostic screens on another page, which, while still available, doesn't bother the operator when not needed.

Put the User in Control

Keep in mind that the entire point of an entertainment control system is to allow us to do a show. The computer, while it may be very sophisticated and do things we can't do, is still there to serve our needs.

Forgive the User

Remember that in our industry, operators are often working very quickly and under enormous pressure. It doesn't take long to add "error trapping" to your system, so that if a user enters something that doesn't make sense, the system won't do something stupid. Add undo if you can, and most of all, give the user a way to stop any process already under way.

Provide Direct Manipulation

Wherever possible, allow the user to manipulate data, control information, and so on directly, rather than having to open some arcane window or dialog box. Exploit the metaphor of the touch screen or the mouse or whatever you've got.

QUESTION 6: WHAT DEVICES MUST BE CONTROLLED/CONNECTED?

Since show control, by definition, connects systems together, it's not possible to complete the system design until you know about all the systems that are being linked together. Once this is known, you can make a detailed list of each of the devices connected to the system. Such devices might include video servers, A/V

---

5. http://developer.gnome.org/hig-book/3.0/principles-feed-back.html.en

routing devices, lighting control consoles, programmable audio mixers and playback devices, animatronic characters, special effects systems, park-wide monitoring and control systems—you name it.

In addition, each controllable device or system will accept one or more control protocols/standards over one or more connection methods. Selecting the right control approach for each device (or, selecting the right device based on how it can be controlled) takes some thought about the design. It's often best to make a comparison chart with the benefits and drawbacks of each approach and select the best approach based on that. I always start this process by typing up a spreadsheet showing each device and system, including what control ports each device has (leave plenty of room in the spreadsheet for IP addresses, port numbers, etc.) and what protocols can be sent over that control port.

OTHER CONCERNS

I left budget off the list above because I firmly believe the most effective design process is to first figure out what is truly needed by the story you are trying tell, and see what that costs. If you can't afford to do it right, then cut something. Extensive "value engineering" to meet inaccurate budget estimate often results in a system that appears to work on paper, but fails to meet the requirements of the show. If a client is insisting that you compromise your system to the point of ineffectiveness in order to meet a budget goal, you probably won't regret (in the long run, anyway) walking away from the project.

In addition, one major item to consider in your budget, of course, is time. Show control offers enormous benefits, such as increased precision and repeatability, and at best, freedom and flexibility for the performers. However, those enormous benefits come at the cost of a large amount of programming, testing, and tech time. Be sure to include the impact of that time in your budget.

That said, show control is increasingly affordable, and show control functionality now often comes "for free" with other devices. For instance, in the 1990s lighting consoles rarely had direct time code input, so a show controller was often necessary to read the clock from a master source (video, time of day, etc.) and trigger light cues. Many consoles (and, of course, many other kinds of devices) are able to accept (and/or generate) time code and many other types of control inputs directly. Also, Ethernet and TCP/UDP/IP have created enormous potential connectivity at very low cost.

FINALLY

Finally, you should consider one additional question: Do you really need show control for this show? As I've said, I believe that the simplest solution that gives the desired results is best ("Principle 3: Simpler Is Always Better," on page 117). Is show control adding unnecessary complexity, or worse, convolution? ("Principle 5: Complexity Is Inevitable, Convolution Is Not," on page 118) If it is, you should rethink whether or not there is a simpler way to get the job done.

Done right, show control is capable of offering a level of sophistication in your show not otherwise possible. But if you don't have the necessary time and other resources available on a particular show, don't even attempt show control. Try some experiments in a no-pressure situation first to see if it will work for you.

MOVING ON TO SOME EXAMPLES

Now, to bring it all together, let's take a look at some practical examples of show control. One note: in other parts of the book, I have included lots of cross references to relevant sections. To keep these examples more readable, I have not included those cross references going forward. If there's a protocol that you haven't heard of, check the Index (page 453); if there's a term that doesn't make sense, check the glossary (page 435).

Chapter 29: A Theatrical Thunderstorm

A huge flash of lightning illuminates the stage and, two seconds later, "Kaboom!" the audience's seats shake as a huge thunderclap rolls through the venue. Ephraim Cubit, his three sons, and the girl next door are all huddled under the kitchen table as the storm continues. From the ever-shortening time between the lightning flashes and thunderclaps, we can tell that the storm is fast approaching. "I sure wish I had cut down that big, dead oak tree next to the house!" shouts Ephraim. As the word "house" leaves his lips, there is a blinding flash and simultaneous thunderclap. Seconds later, an oak branch smashes through the window upstage. As the sound of the giant thunderclap decays, Ephraim suffers one final indignity: his lights go out, leaving the stage in darkness.

THE MISSION

This musical version of *Desire Under the Oaks* is being produced at a major regional theatre. The director will "spare no expense" to ensure that this sequence comes off perfectly night after night, although she has also insisted that the sequence not be "canned." We must keep in mind that "sparing no expense" in the regional theatre world is equivalent to sparing every expense at a theme park; controls will be selected with an eye on keeping the cost as low as possible while still achieving the director's goal.

DESIGN CONSIDERATIONS

Let's evaluate this application using the design process I laid out in "My Show Control Design Process," on page 362. Of course, we will also incorporate my general principles from the "System Design Principles" section, on page 111.

Question 1: What are the safety considerations?
The large, scenic tree branch effect could be potentially life-threatening, and care should be taken to ensure that the effect cannot be triggered accidentally.

Question 2: What type of show is it?
Because it's a theatrical production performed from a script that doesn't change from show to show, it's linear.

Question 3: Is the show event-based, time-based, or a hybrid?

Given the fact that the director doesn't want anything "canned" and that there are live actors involved, the system should be mostly event-based. However, the delays between lightning flashes and thunderclaps add a time-based element.

Question 4: What is the control information source?

The stage manager is in charge of the run of the show, and will call all the cues, including those in the thunderstorm sequence. It's possible, however, that we might link the systems together, and take control information from the lighting or sound department, or use a separate master clock for the time-based part.

Question 5: What type of user interface is required?

Skilled operators are present, and they will want as much user interface and feedback as possible. However, it's a relatively straightforward show, so it's likely that we can accomplish the goals with off-the-shelf systems, and it's not likely that custom user interfaces will be required.

Question 6: What devices must be controlled/connected?

Let's go through each one in detail.

THE SYSTEMS

The show will run for four weeks. Lighting, sound, and props run crews are already on contract to the theatre, and most systems are already "in house." A show control system will be used only to ensure accurate timing for the tree branch effect; the rest of the show will be run manually by the operators on cues from the stage manager.

Lighting

For the production, the master electrician has rented a DMX-controllable lightning-effects generator and, of course, the theatre already has a lighting console that outputs DMX. The lightning generator is connected to the console's DMX512-A output through an opto-isolator/splitter, which is even more important in this application because of the high voltage in the lightning unit and the potential for electrical damage in case of a failure.

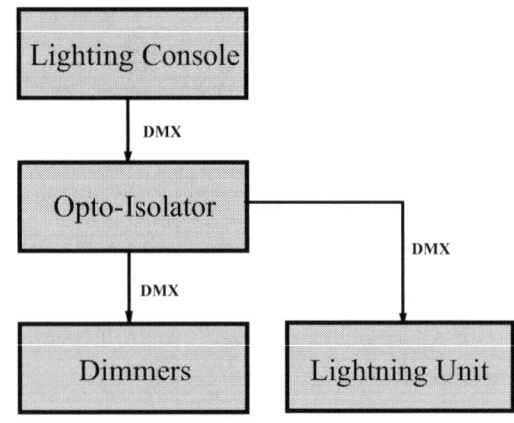

The console, which is backstage, is capable of firing cues from MIDI Show Control commands, and has been configured to receive those commands as MSC device #1. In addition to receiving MSC commands, the lighting controller also generates "musical" MIDI messages when its front-panel buttons are pressed. This could be useful for our thunderstorm, since MIDI Note On/Off messages are generated when the submaster bump buttons are pressed and released.

The following MSC Lighting cues have been programmed into the board:

Cue	Effect	MIDI Command
218	Lightning strike 1	MSC Cue 218 Go
219	Lightning strike 2	MSC Cue 219 Go
220	Lightning strike 3	MSC Cue 220 Go
220.5	Big lightning strike	MSC Cue 220.5 Go
221	All lights out	MSC Cue 221 Go

The lightning strike "looks" from cues 218 to 220.5 have also been loaded into submasters 1–4; when the submaster bump buttons are pressed on this console, the appropriate lightning effect is triggered and a corresponding MIDI Note is sent out via the console MIDI port. The submasters are programmed so that the lightning flashes last as long as the operator holds down the bump button.

Sound

In addition to a traditional musical-reinforcement sound system, the theatre has a MIDI-controllable sound playback system, which can be triggered either by standard MIDI Show Control commands or by MIDI Note commands assigned to each cue. The sound FX playback system has been set to receive MIDI Show Control Sound commands on Device ID #1, and the following cues have been programmed:

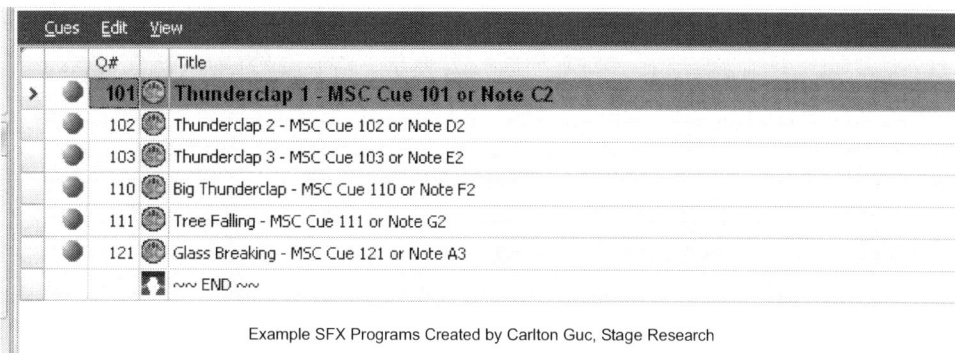

Example SFX Programs Created by Carlton Guc, Stage Research

In addition to the thunderclaps, the sound designer added two cues. As the branch effect falls, it doesn't make realistic tree crashing sounds, so its fall must be augmented with a "tree falling" sound effect. For safety reasons, stage glass is used in the window through which the branch crashes, and since this fake glass doesn't make much sound when it breaks, a "glass breaking" sound effect is also necessary.

The audio outputs of the sound FX system are connected to inputs on the main audio mixer, so the operator must remember to bring up those faders to the prescribed levels for the cue sequence. Since this is a live show, the sound mix position is in the back of the orchestra seats.

Props

The theatre's technical director has devised a simple but effective way of electrically releasing the tree branch. The base of a scenic branch is mounted on a pivot and weighted, so that, without restraint, it will fall through the window. An electrical solenoid is mounted so that, without power, the branch is locked in the "out" position; when power is applied to the solenoid, the solenoid is activated and the branch falls (this is a kind of fail safe design—a power failure will not cause the tree to fall). Through experimentation, the TD has determined that the solenoid must be energized for at least one second to give the branch time to fall past the solenoid's shaft. In addition, a manual safety release pin has been installed parallel to the solenoid shaft, so that unless the pin is manually released, the tree cannot fall. Once the actors are safely onstage and under the table, the pin is manually released by the prop crew, "arming" the effect.

The solenoid is controlled by the light-board operator, using a push-button station positioned next to the light board. The master electrician had considered firing the effect using a dimmer controlled by the console, but all the dimmers in the theatre were already in use for the show lighting.

SHOW CONTROL SCRIPT

To clear up any confusion, the stage manager precisely scripts out the whole sequence, working closely with the lighting and sound designers:

Actions	*Trigger*	*Control Messages*
First lightning strike	Stage manager	Light Cue 218
First thunderclap	Two second delay	Sound Cue 101
Verify safety of tree	Stage manager	Verbal Headset Confirmation

Actions	Trigger	Control Messages
Second lighting strike	Stage manager	Light Cue 219
Second thunderclap	One second delay	Sound Cue 102
Third lightning strike	Stage manager	Light Cue 220
Third thunderclap	1/2 second delay	Sound Cue 103
Big thunderclap/ Lightning strike	End of "house" line	Light Cue 220.5 Sound Cue 110
Tree falling/sound FX	Stage manager	Solenoid Go Sound Cue 111
Glass breaking	Tree through window	Sound Cue 121
Lights out	Stage manager	Lights 221

APPROACH 1

The design team wisely decides to test the entire sequence well in advance of technical rehearsals, using the equipment already on hand. The light board will be used as a master MIDI source, triggering the sound FX system; the tree will be triggered manually by the electrician using the push-button controller on command from the stage manager. All system interconnection is done via MIDI, and since the light and sound boards are farther apart than can easily be reached with a standard, off-the-shelf MIDI cable, the sound department makes some adaptors from MIDI DIN to XLR, so that the MIDI can be run down the audio snake to the FOH console. Since this run is longer than the MIDI spec-defined length of 50 feet,[1] the sound department tests the signal at the end of the snake, and it works just fine.

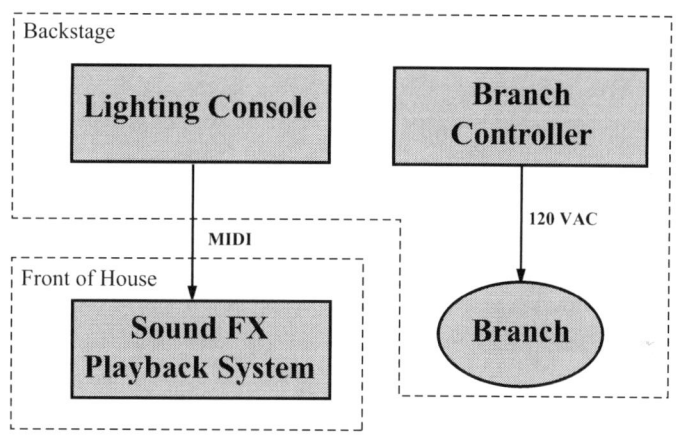

For the test, the four lightning/thunderclap effects are triggered by the electrician, using the console's submaster buttons, on cues from the stage manager. The MIDI

---

1. As we discussed on page 282, even though MIDI specifies a maximum distance of 50', it's possible to send it a lot longer. Always test to make sure!

output from the lighting console is routed to the MIDI In of the sound playback system, so that pressing a submaster button triggers both the lightning effect and the appropriate sound. So that the delay between lightning flash and thunderclap will be consistent night after night, the sound designer programs a "Wait" cue into the sound playback system, which, when triggered, automatically delays the thunderclap sounds by the appropriate amount. Once the stage manager confirms that the tree effect has been armed, he calls the cue for the falling branch over the headset to the electrician; the sound operator is cued by a cue light for the falling-branch sound. The breaking-glass sound is taken as a visual cue by the sound operator as he sees the tree poke through the window.

The Results

The director and design team are delighted with the results of the test, but there are a few problems. The electrician is not able to consistently control the duration of the lightning flashes to the director's and lighting designer's satisfaction, and the director now feels that the timing between the first three thunderclaps should be exactly the same every night, regardless of what the actors are doing (in other words, she wants to "can" the effect, which is exactly what she said she did not want to do at the outset). In addition, during the test, the big thunderclap was so loud that the electrician never heard the "Tree Release Go" command and missed the cue. Finally, the director thought that the lightning for the big blast was not bright enough, so an additional lightning machine is rented and connected to DMX from the lighting console.

APPROACH 2

The sound designer figures out a way to solve many of the problems with the first approach. The sound FX playback system can, in addition to generating sound, also generate any sort of MIDI message. So, she programs in MIDI Show Control messages for the light board. She also now removes the wait time from the start of each thunderclap cue and sets the delay between the lightning flashes and the thunderclaps using a time clock available in the sound FX playback system. She reverses the MIDI lines

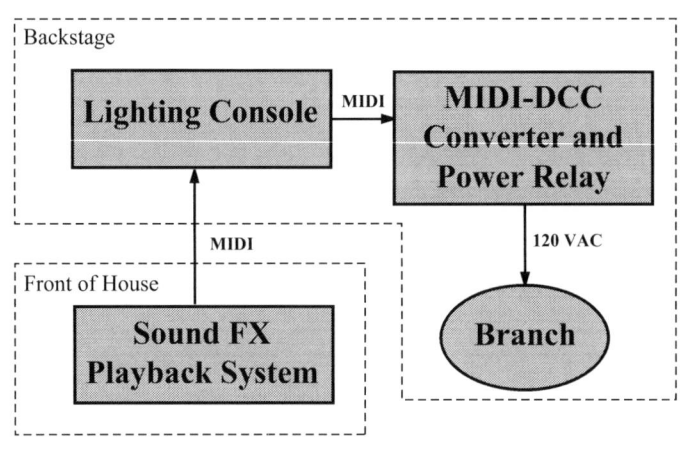

already in place, sending data from the sound FX system at the sound position to the lighting console backstage

To fire the tree effect reliably, the electrician buys an inexpensive MIDI-to-dry-contact-closure (DCC) relay interface, which can fire the solenoid using a MIDI Show Control "Set" command in the Machinery command format on MSC device 1. The relay interface will be controlled by the sound FX system, and since MIDI is already run from the sound FX system to the light board backstage, the relay interface is simply placed next to the console and given a MIDI input from the MIDI Thru jack on the light board (since this project is still in the testing phase, daisy chaining from the thru of the lighting console is just fine).

The sound designer programs two sequences into the sound FX playback system that the sound operator will execute with single button presses.

Sequence 1 contains the messages for the first three lightning strikes (Note: the "timecode" column is the system's internal clock):

Sequence 2 consists of the simultaneous lightning effect and huge thunderclap, followed by the release command for the tree and the falling-tree sound effect:

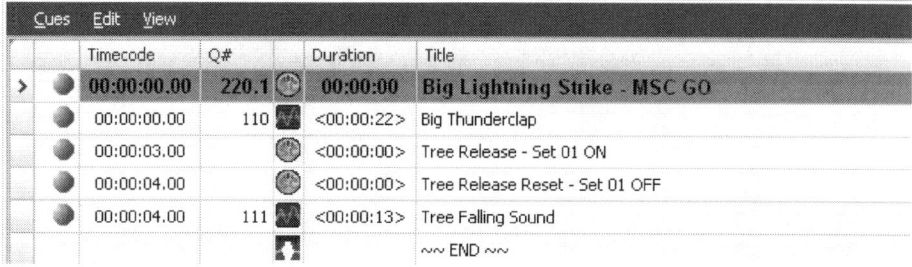

These two sequences allow the entire thunderstorm to be run with two simple Go commands. Since the rate at which the tree branch falls is not predictable, the

breaking-glass sound cue is still taken visually by the sound operator. The final "lights out" that ends the act is called by the stage manager and taken manually by the electrician.

The Results

This new approach works well, but the additional lightning machine generates a huge static blast through the sound system (the new fixture was mistakenly plugged in the same power circuit as the mixer), and the sound operator was so busy pulling down channels on the board to kill the static that he was late in executing the second, tree-falling sequence. Flustered, he missed the breaking-glass sound altogether.

APPROACH 3

To overcome all these problems, the team decides to use even more of the show control functionality of the sound FX playback system and use the movement of the tree itself to trigger its sound effect. Since it seems that this approach will be the final one used for the show, a MIDI splitter (and backup) is bought, and placed backstage by the lighting console. One MIDI line from the splitter is run to the lighting console; another runs to the MIDI-DCC converter.

This leaves only the problem of triggering the breaking-glass sound effect. Since the sound operator is no longer running the rest of the sequence, he is free to run the cue manually, but could be tied up at any point with unanticipated live sound reinforcement problems. So, the sound designer decides to trigger the sound automatically.

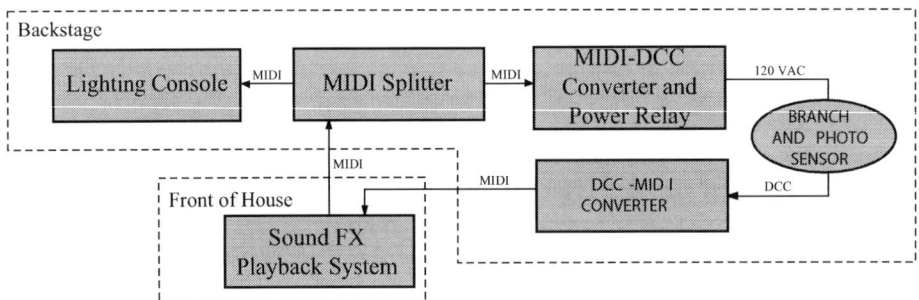

The tree does not fall at a predictable rate, so the sound department installs a photo-electric sensor across the window frame; when the falling tree breaks the beam of light, the photo sensor generates a contact closure. A DCC-MIDI interface is programmed so that a MIDI Show Control command is generated when the beam of light is interrupted (another is generated when the beam is cleared). An additional snake line is used to run the MIDI signal from the DCC converter back up to the sound FX playback system. The Sound Designer updates the second sequence:

The Results

This approach works beautifully, the director is ecstatic, the show gets great reviews, and sells out.

Chapter 30: Put on a Happy Face

"Put on a happy face," booms the voice from the screen. "Xylenol is the one for you!" Xylenol's trademark yellow smiley face appears on the screen and suddenly two spotlights follow a loud swoosh sound, sweeping from behind the audience to the screen. The spots highlight the main screen, which is rotating to reveal a pile of old car batteries. The houselights come up, and the room full of doctors—some humming the catchy Xylenol theme music—line up for free Xylenol merchandise handed out by scantily clad models.

This is how Michael, marketing VP for pharmaceutical giant Klaxo, wants to end the pitch for Xylenol, the antidepressant drug manufactured from used car batteries. Klaxo's chief stockholder, Mr. Schumaker, is flying in from Germany to approve this presentation, which will be used to pitch the drug to doctors at an upcoming trade show. Michael, understandably, is very anxious for the show to come off perfectly.

THE MISSION

Michael wants this show to be different from others on the trade show floor—he wants it to have a more "theatrical" feel, although he doesn't have the budget for a large crew. Ralf, presentation manager for Klaxo, has been assigned the task of making Michael's dreams a reality. Ralf is well versed in many computer programs and boardroom control systems, but he finds controlling equipment beyond the realm of the computer screen to be a bit of a mystery, so he has hired a show control consultant.

DESIGN CONSIDERATIONS

Let's evaluate this application using the design process I laid out in "My Show Control Design Process," on page 362. Of course, we will also incorporate the principles I explained in the "System Design Principles" section, on page 111.

Question 1: What are the safety considerations?
The turntable is potentially dangerous, and appropriate safety measures must be taken. The old car batteries will be just a scenic piece, so there are no acid dangers.

Question 2: What type of show is it?
The show is linear, and is the same every time.

Question 3: Is the show event-based, time-based, or a hybrid?
Some sort of time code will likely be needed for the media, and probably a few event-based triggers will be needed as well.

Question 4: What is the control information source?
A booth attendant will be available to start every show; the show will then likely be run from time code.

Question 5: What is the type of user interface required?
The attendant will not be a skilled technician, so the interface should be as simple as possible.

Question 6: What devices must be controlled/connected?
Let's look at each in detail.

THE SYSTEMS

Klaxo has contracted with an exhibit firm to make the booth for the show, but is producing the media in-house, and will rent any necessary control, video, or audio gear from a staging company.

Video
Video is a key element of the show. Klaxo's corporate communications department has contracted an experienced video producer to produce the content, which will play, synchronized, on three screens across the front of the theatre. The imagery will be played back from a computerized image presentation system that allows authoring and changes to be made on site; this system uses one computer for control and one for each of the presentation screens. The image presentation system can accept event-based control signals over an Ethernet/IP link, and also export time code back over the network.

Audio
The video servers will act as the primary audio source for the show—each of the three networked image computer servers can output two-channel stereo sound, giving six channels of synchronized audio. A MIDI controllable mixer is incorporated into the system as well:

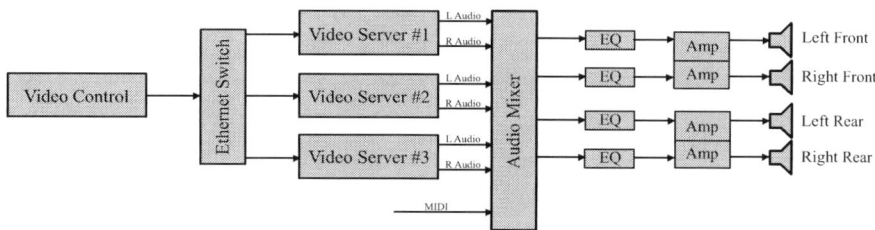

Lighting

Klaxo has rented four simple moving lights: two for the "swoosh" effect and two others for house and ambient lighting. The fixtures can be controlled using DMX512-A. Although the show control consultant recommends renting a moving light console, Ralf decides there is not enough money in the budget to do so, the lights will be run directly from the show controller, and he will program the lighting cues himself. They decide on the following channel assignments:

DMX Slot	Fixture	Function
1	1	Dimmer
2	1	Pan
3	1	Tilt
4	1	Color
5	1	Gobo
6	2	Dimmer
7	2	Pan
8	2	Tilt
9	2	Color
10	2	Gobo
11	3	Dimmer
12	3	Pan
13	3	Tilt
14	3	Color
15	3	Gobo
16	4	Dimmer
17	4	Pan
18	4	Tilt
19	4	Color
20	4	Gobo

Machinery

The scene shop that built the turntable has supplied a system that can accept two simple ASCII serial commands: "C1" and "C2." When the ASCII text "C1" is followed by a $0D_{16}$ Carriage Return and a $0A_{16}$ Line Feed, the turntable rotates to the postshow "battery" position; C2 resets the turntable to the preshow screen position. The 10-foot-diameter turntable rotates 180° in about five seconds. The shop also provides a manual control console.

On the show control consultant's advice, Ralf has the turntable control system built with a simple manual "enable" control, which will be operated in a "dead man switch" mode by the attendant responsible for loading the doctors into the room. The attendant must be holding down the button for the turntable to operate; if any drunken doctors get up and try to ride or grab the turntable while it's in motion, the attendant will simply release the button and the turntable will stop. For safety reasons, the turntable is manually reset at the end of the show.

To ensure that the attendant knows her cue, Ralf decides to have an indicator light come on just before the effect needs to be armed and go out when it is safe to release the button and allow the show to complete normally. Rather than constructing anything, Ralf finds a small, DMX-controllable LED fixture to use for the indicator; this will be controlled through the show controller via DMX.

Show Control

Since there is a video system that needs tight integration and it's a time-based show, the show control consultant chose a show control system that has direct support for the image presentation system and good timeline functionality. It can also, with various interfaces, take as input or output a number of other parameters, such as DMX.

APPROACH 1

Ralf takes the consultant's recommendation to set up a full-scale test at corporate headquarters two weeks before the trade show. The control computer is located backstage right next to the turntable controllers and the video and audio racks

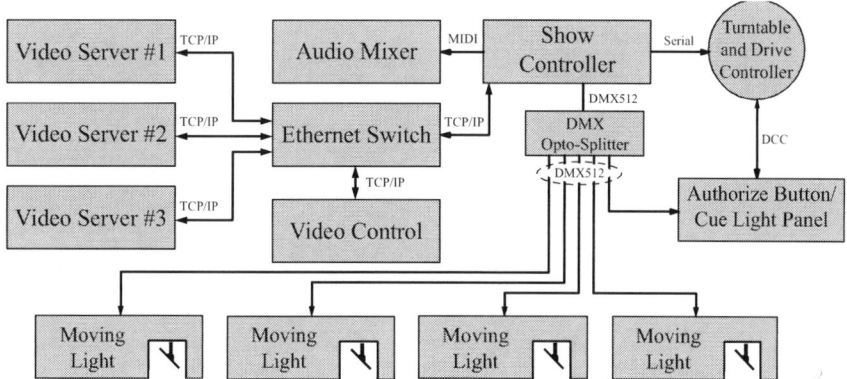

The show controller instructs the image presentation system to start and then receives time code back from the presentation over the network, to which the show controller synchronizes an internal timeline, which triggers light cues, and so on. This way, the show is always synchronized to the visual media, even if there is a slight delay or some other issue.

The consultant creates a user screen:

Example Medialon Manager User Screen Created by Alan Anderson, Medialon

Chapter 30: Put on a Happy Face • 383

Here are the last few cues of a simplified show control script:

Time Code	Action	Messages/Operations
00:03:20:18	"Put on a happy …"	(soundtrack)
00:03:21:00	Turntable warning light on	DMX Out 21 to 100%
00:03:21:20	Smiley face on screen	(video)
00:03:25:00	Swoosh sound starts	(soundtrack)
00:03:25:01	Lights 3 and 4 dimmers to full	DMX outs 11 and 16 to 100%
00:03:25:15	Turntable rotate to battery position *(if enabled)*	Serial "C1<<"
00:03:26:00	Lights 3, 4 to intermediate position 1	DMX Outs 12/17 to 50/50%
00:03:26:00	Lights 3, 4 to intermediate position 2	DMX Outs 13/18 to 50/50%
00:03:27:00	Lights 3, 4 to final position	DMX Outs 13/18 to 10/90%
00:03:30:00	Turntable warning light off	DMX Out 21 to 00%
00:03:58:00	Houselights (spots 1&2) to Full	DMX Outs 1/6 to 100%
00:05:00:00	Reset to top of show Wait for "Go"	Image Presentation System Pause

The Results

The presentation goes very well; Mr. Schumaker from Germany and Michael from marketing are both dutifully impressed. Ralf barely hears the compliments, because he was awake all night getting the moving lights programmed. So he gives in to the show control consultant's recommendation and rents a moving-light console that can accept MIDI Show Control messages. It is sent overnight and is programmed before the exhibit leaves headquarters.

APPROACH 2

The system layout for the trade show is the same as what was built up in testing, with a few exceptions. Since the moving-light controller is so easy to program, Ralf adds a number of cues to the presentation, each triggered using a MIDI Show

Control lighting message. The console is configured as MSC device 1, and the moving lights are connected through a DMX opto-splitter.

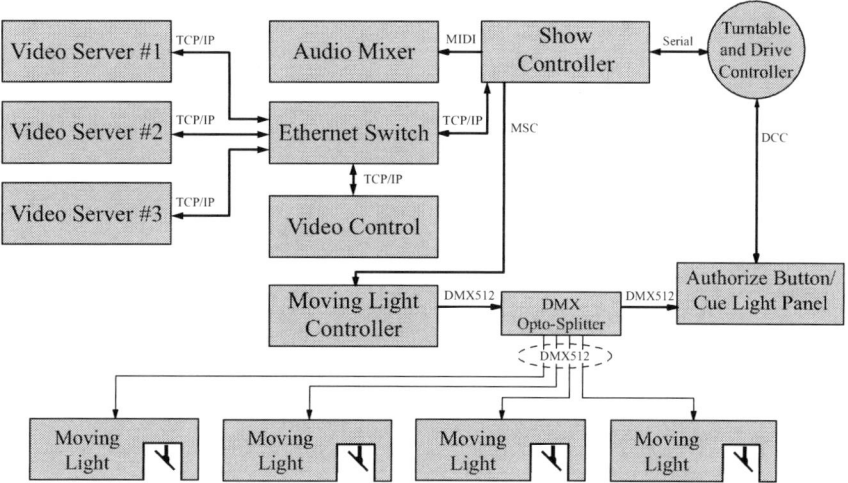

The turntable sometimes takes just a bit more than five seconds to rotate, so if the attendant released the button as soon as the indicator went out, the turntable would not fully complete its half rotation. A second is added to the warning light off time command.

The Results

The presentation is the hit of the show. Prescriptions for Xylenol increase 50% in subsequent weeks, so Ralf gets a big raise. Ralf also contracts the show control consultant a year in advance for the next show—Michael wants the models next year to be part of the presentation, interacting with it!

Chapter 31: **Ten-Pin Alley**

The animatronic bowling ball beckons the crowd, "Hey kids, come over here and watch an amazing show with me, Bowly!" The house lights dim and spectacular moving lights fill the space with color. "You, too, grown ups," says Bowly. A catchy bowling tune replaces the store's background music and sound effects, and a huge screen starts lowering into place. "Here we go!" shouts Bowly, and video images of dancing bowling pins fill the screen, cut precisely to the music beat. At the end of the five-minute show, the house lights and regular background music restore, the screen retracts, and Bowly says, "Hey, thanks for coming, the next show will be in a few minutes. In the meantime, why not come inside and take a look around?" Video monitors around the floor, which had been relaying a feed from the show, spring back to life and display live bowling highlights from around the world, intercut with real-time score updates from the adjacent bowling alley.

"Big Pete," president of 300 Score Entertainment (3SE), wants this sequence to take place every 30 minutes at StrikeTown bowling superstores/alleys in malls around the world. The flashy shows are designed to attract customers into the store, get them to stay a little longer, and maybe buy some StrikeTown sportswear or a new bowling ball, and later bowl a round, or dine and dance in the attached bar/restaurant complex.

THE MISSION

Big Pete has hired a design firm to produce the shows, and the firm will be producing the video and audio media. The design firm's producer/director is working with Big Pete's architect for scenic elements, and has hired big-name Las Vegas lighting and sound designers to create the lighting and sound in the store. Freelance lighting, sound, animatronic and show control programmers are hired to "change out" the media, and they will return two to three times a year and put in updated media overnight.

Big Pete has hired a general contractor (GC) to build the first StrikeTown. The GC has hired an entertainment lighting contractor to handle entertainment lighting systems; a scenery automation company to handle the large screens; an A/V contractor to handle audio, video, and control; and an animatronic character company to provide Bowly and his control system. On the A/V contractor's recommendation, Big Pete has hired a show control consultant to coordinate the control of all the systems.

Shows will run automatically from a centralized "Scheduler" computer, which can be updated from corporate headquarters in Milwaukee over the corporate intranet, or from the store manager's desktop. Shows can also be run manually from a touch screen controller for special events, such as weddings, or appearances by famous bowlers with StrikeTown endorsements.

3SE bean-counters will allow only one maintenance person for all the technical systems in the store and the bowling alley, and this person will be responsible for everything from the show systems, to a fancy glass elevator to bring bowling balls up from the basement warehouse area, to the pin reracking equipment. Management decides that each system should run a full self-test every morning and also respond to a simple "query" just before the start of each show, so that maintenance personnel can be alerted and shows canceled if a system is not working properly. The system will have both a "day" mode when the store is open and a "night" mode when the store is closed and show equipment is shut down.

DESIGN CONSIDERATIONS

Let's evaluate this application using the design process I laid out in "My Show Control Design Process," on page 362. Of course, we will also incorporate the principles I explained in the "System Design Principles" section, on page 111.

Question 1: What are the safety considerations?
The large screen and the animatronic character "Bowly" are potentially dangerous, but the audience is physically partitioned away from each and interlocks can be placed on the access points.

Question 2: What type of show is it?
It's a linear show, since it's all prerecorded.

Question 3: Is the show event-based, time-based, or a hybrid?
The show, since it's all synchronous, will use some sort of time code.

Question 4: What is the control information source?
The Scheduler will start each show, and will be responsible for scheduling day and night modes. In addition, the store manager will be able to override certain shows or force a "party" mode.

Question 5: What is the type of user interface required?

There is only one technician on the premises, so the system should present him with a detailed interface, but should also present a simple system to the manager for overrides, parties, and so on.

Question 6: What devices must be controlled/connected?

Let's go through each system.

THE SYSTEMS

A wide variety of equipment needs to be connected and controlled for this system.

Scheduler

The Scheduler is going to be created, installed, and maintained by 3SE's Information Technologies (IT) department, since it is going to run on the corporate intranet. The IT department originally wanted to oversee all the control aspects of the show and store, but after hearing some war stories from the show control consultant, they realized that they would be getting in way over their heads. So a separate system, specified by the show control consultant, will run the shows and be triggered at the appropriate time over the network by the Scheduler. The Scheduler will maintain a Web page on the corporate intranet for remote monitoring, and will offer web-cam views of the store and backstage equipment room.

The Scheduler will send a variety of ASCII commands via IP to the Show Controller. Five minutes before show time, the Scheduler will send a "Warning Show #n" command and later a "Go Show #n" message to start the show. Each command will be acknowledged by the Show Controller, which will control everything associated with the show. Both the Scheduler and Show Controller will be able to communicate with various store systems.

Show Video

Video is a key element of the shows. To ensure maximum flexibility and to allow shows to be easily updated, video will be run off a hard-disk based video server, which can have its contents updated over the corporate intranet. The control connection to the server is via Ethernet, using proprietary ASCII commands. The server also has a balanced XLR time code input and output, and can chase or generate time code and send it out via MIDI Time Code using RTP MIDI over the Ethernet connection.

Each video segment is assigned a different time code hour, and five minutes of video black is included at the head of the segment before the actual show starts so that "Bowly" can start and run and other systems can roll up. One minute of video black fills out the end of each segment. Here are the details of two typical video segments:

Time Code	Event
11:00:00:00	5 minute pre-roll video black
11:05:00:00	Start of "Bowling's Greatest Bloopers" segment
11:09:04:25	End of bowling segment
11:10:04:25	End of video black
12:00:00:00	5 minute pre-roll video black
12:05:00:00	Start of "King Pin" segment
12:10:01:00	End of "King Pin" segment
12:11:01:00	End of video black

The show video projectors also need some control so that they can be configured, put into or taken out of "standby" mode at the start and end of each day, and queried before shows. The projectors have Ethernet ports and accept and respond with proprietary ASCII commands.

To test the entire video system—from server to projector—each morning, a special test pattern is rolled from the video server and projected on the screen. If the projector is on and working, the projector's light will activate a special photo sensor placed in the screen area. For the query check before each show, the video server will be sent a simple status message, which it will acknowledge, and then the projectors will be queried for lamp hours and on/off status.

Store Video
The store video systems have been contracted separately, and video for the 100 or so monitors around the facility is sent through a routing switcher, which is controllable via Ethernet and ASCII commands. The switcher can route any of a number of satellite-received sports channels to the various monitors, put up logos, take graphics feeds for messages or score highlights from the alley, or take the video feed from the shows. No morning self-test is designed for the store video systems, but the switcher can respond with simple "ack"-like messages during the preshow checks.

Show Audio
Audio will be played off a multitrack hard-drive playback unit, which can chase SMPTE or MIDI Time Code or accept MIDI Machine Control commands, or accept both over Ethernet using RTP MIDI.

The audio mix for the store will be accomplished through a computerized mixing matrix system, which can accept either Open Sound Control (OSC). All the amplifiers for the store are connected, monitored, and controlled from a control system made by the amplifier manufacturer, which also can talk via a proprietary link over the network.

For the full morning audio system test, a cue is fired in the matrix system playing a test tone to the speakers, and a small microphone in the store near the main speaker position is routed to a device that detects that specific tone and closes a relay. The contact closure from the relay is sent back to the Show Controller. For the preshow query, the audio matrix is sent a message to which it responds. The amplifier control system takes a message and responds with a status update.

Store Audio
The contract for store audio is separate from the show audio, and while similar equipment (amplifiers, etc.) is used in some parts of the system, a matrix system designed for permanent installations (rather than shows) is used. The store audio matrix system can call up presets or respond to simple queries over a network link or via SNMP commands over Ethernet. No morning self-test is designed for this system.

Animatronics
"Bowly" has his own control system, which can chase SMPTE or MIDI time code, or be triggered by RS-232 serial commands or contact closures. The consultant gets the manufacturer to add Ethernet support to keep the whole system consistent. A programming panel is provided for inputting Bowly's moves, and a small audio deck with time code output capability is also provided in the animatronic control rack, so that Bowly can be programmed "offline" separately from the show and then integrated when his programming is complete.

For the morning self-test, a heavy-duty limit switch is discreetly positioned near Bowly, and when the test program is executed, Bowly simply hits the switch. For normal preshow checks, Bowly's control system responds with a simple "Ack" command over the network.

Stage Machinery
The projection screen motion-control system takes commands over Modbus TCP. The machinery system, which is based on a Programmable Logic Controller (PLC) has several virtual "coils" that can be "written" using standard Modbus commands. One coil is to set the target position to up; another is for down, and a third is the "Go." Since the communications take place over Ethernet, their integ-

rity is protected by Ethernet's built in CRC error detection. A manual control panel with auto/manual lockout is provided, as is a complete E-Stop system.

The screen is housed in a vertical area closed to the public, accessible only via a locked door. For safety reasons, a sensor on the door sends a signal to both the motion-control system and to the Show Controller each time the door is opened. When the sensor is tripped, the screen, if moving, will stop, and the system will have to be manually reset using a key-switch near the door before the screen will be allowed to move again.

The motion control system is closed loop, with encoders mounted on the screen drum itself. For the morning self-test, to be sure that the screen hasn't jammed or fallen off the drum, photo-sensors are positioned near the bottom of the screen's travel and near the top. During the morning check, the screens are run out (the screens come in during "night" mode to keep them stretched out), and each sensor is examined by the Show Controller.

For the quick preshow self-test, the motion controller moves the screen about an inch, looks for motion on its encoder, and then reports its status back to the Show Controller.

Show Lighting

The show lighting system consists only of moving lights and LED fixtures, so there are no dimmers, and a controller is selected that is available in both a console version with full controls and a rack-mount version. A full console is rented for show change-outs, and the show is then loaded into the rack-mount system for daily operation. The controller can chase SMPTE or MIDI Time Code, or accept proprietary ASCII commands over Ethernet.

For the system self-test, the same photo-cell sensors designed for testing the projectors are used. One at a time, cues are executed, directing each moving light to shine its beam on one of the photo sensors. For pre-show tests, the console can be queried via an ASCII command.

Store Lighting

The contract for store lighting has been let separately from show lighting. The store system is fairly simple, with two general "looks": normal and show. The store controller can accept ASCII commands over the network, and also acknowledge via a simple "Ack" for the pre-show test. No morning self-test is designed for this system.

THE APPROACH

The show control consultant picks a Show Controller that can generate time code; chase it; send serial, MIDI or other messages; send and receive any sort of messages over Ethernet; and send or receive contact closures. The show control consultant decides as much interconnection as possible should be done via Ethernet, and he lays out the following fixed IP addresses in consultation with the company's IT department (the show network is separate, but by coordinating IP addresses, there's room for the future and there won't be a problem if the systems are accidentally connected). He puts each department on a separate VLAN and subnet, and uses inter-VLAN routing for communications between the show controller and the other systems when necessary (only key control devices shown):

Device	IP Address	VLAN
Show Controller	10.111.101.01	1
Scheduler	10.111.101.11	1
Video Server	10.111.102.11	2
Video Matrix	10.111.102.12	2
Show Audio Matrix	10.111.103.11	3
Store Audio Matrix	10.111.104.12	4
Animatronic Controller	10.111.105.11	5
Stage Machinery Controller	10.111.106.11	6
Show Lighting System	10.111.107.11	7
Store Lighting System	10.111.108.11	8

At the five-minute warning, the Scheduler sends its control message via TCP/IP to the Show Controller. The Show Controller then checks each system, sending various commands to each of the subsystems, and waits for replies from each. If all the critical systems check OK, the controller sends a message back to the Scheduler saying that the show is standing by, and this is logged on the scheduler's Web page. If the systems do not check out OK, the Show Controller sends a message to the Scheduler detailing the problem, and the Scheduler can put this information on its Web page, send an e-mail, or even page maintenance personnel. Because there is no show without video and since video is a linear, time-based media, the consultant decides that the video server, after being commanded to go by the Show Controller, should generate a time code for all the other show systems, which will be distributed over the network using MTC over RTP MIDI.

The consultant creates a system block diagram and user screen:

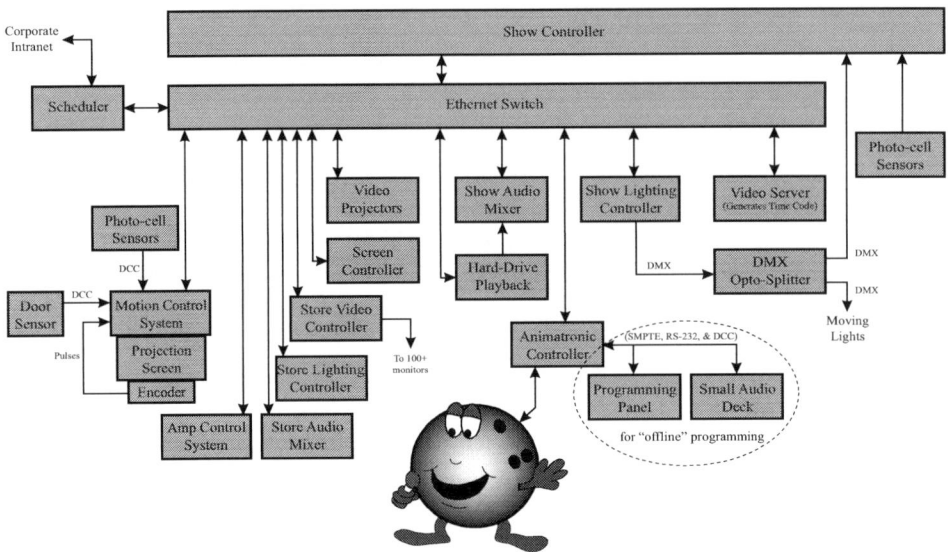

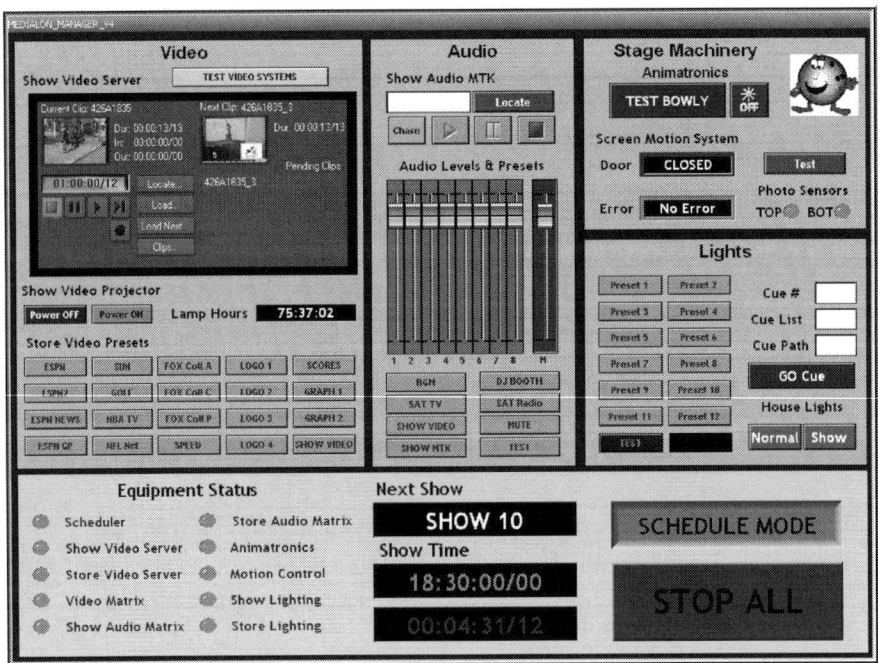

Example Medialon Manager User Screen Created by Alan Anderson, Medialon

The show control consultant generates a show control script, one segment of which is shown below.

Trigger	Event	Action
Five minutes before show time	Five minute warning	Scheduler sends message to Show Controller (SC)
SC receives message		Start Show Pretest
Show pretest	SC tests screen	SC sends message to screen system
Screen replies	Screen OK or not OK	SC sends "Screen OK" (or "show abort") message to scheduler
Screen test complete	SC tests Show Video	SC sends message to video server SC sends message to projectors
Video server/ projectors reply	Video/projectors OK or not OK	SC sends "Show Video OK" (or "show abort") message to scheduler
Show video test complete	SC tests Store Video	SC sends command to store video
Store Video replies	Store Video OK or not OK	SC sends "Store Video OK" (or "not OK") message to scheduler
Store Video test complete	SC tests Show Audio	SC sends message to show sound matrix, SC sends message to amp controller
Matrix and amp controller reply	Matrix/Amp Controller OK or not OK	SC sends "Show Audio OK" (or "show abort") message to scheduler
Show Audio test complete	SC tests Store Audio	SC sends message to store audio
Store Audio replies	Store Audio OK or not OK	SC sends "Store Audio OK" (or "not OK") message to scheduler
Store Audio test complete	SC tests Animatronics	SC sends message to animatronic controller
Animatronic Controller replies	Animatronics OK or not OK	SC sends "Animatronics OK" (or "not OK") message to scheduler
Animatronic Test complete	SC tests Lighting	SC sends MSC message to show lighting
Show Lighting replies	Show Lighting contact closure on or off	SC sends "Show Lighting OK" (or "not OK") message to scheduler
Show Lighting Test complete	SC tests Store Lighting	SC sends message to store lighting system
Store Lighting Test complete	SC evaluates if critical systems (Screen, Show Video, Show Audio) are OK	SC sends "Standing By for Show" message to Scheduler or sends "Critical System Failure, Show Aborting," and cancels show
Show Time	Show 12 Go	SC starts Video Segment 14
Start of show	Video starts rolling	Time Code is generated
`12:00:00:00`	Time Code rolls	

Trigger	Event	Action
12:01:00:00	Bowly starts talking	TC triggers animatronics preshow
12:04:00:00	Store audio starts to fade down	SC sends message to store audio
12:04:30:00	Store lighting starts to fade down	SC sends message to store lighting
12:04:55:00	Screen lowers	SC sends message to screen system
12:04:58:00	Bowly says "Here we go!"	
12:05:00:00	Show starts	
Various	Show lighting/sound chase time code	
12:10:01:00	Show ends	
12:10:02:00	Screen retracts	SC sends message to screen system
12:10:06:00	Store lighting starts to fade up	SC sends message to store lighting
12:10:07:00	Store audio starts to fade up	SC sends message to store audio
12:10:10:00	Bowly finishes speaking	SC sends "Rest" command to Bowly
12:11:06:00	End of time code	SC sends "Show Complete" message to Scheduler, and systems reset as necessary for next show

The Results

The first StrikeTown opens in the world's largest mall in Minnesota and is a huge success. The only problem is that the maintenance issues are too much for one person to handle, so a second full-time person, who deals only with show systems, store audio, and store video ,is added. The self-tests are still used by the new show maintenance person, but he now has time to schedule and coordinate maintenance since he doesn't also have to clear pin-jams in the bowling alley. Big Pete announces future StrikeTown locations in Las Vegas, Orlando, and Dubai.

Chapter 32: Comfortably Rich

It's a crazy night in Massive Stadium. During "Comfortably Rich," the band's biggest hit, images of a flying poodle, edited perfectly to the song's beat, are showing on a giant circular screen. Video walls built into the set switch back and forth from prerecorded poodles to live views of the band. Throughout the song, sound effects and barks from the poodle footage emanate from the massive quad sound system. Then, suddenly, the drummer's riser lifts 20 feet into the air, with massive spots tracking the riser from every direction; this is the moment the crowd has been waiting for—the drum solo. As the riser's ascent slows, the poodle images disappear and the solo starts. As the drummer hits each of his drums, the huge array of moving lights dances precisely to the beat. At the end of his solo, the drummer plays the famous opening beats from "Young Rust," and as the drum riser returns slowly to earth, images of rust and money appear on the screens.

THE MISSION

Elderly rockers Purple Floyd want this spectacle recreated nightly for their final tour (the band had its first "final" tour ten years ago).

DESIGN CONSIDERATIONS

Let's evaluate this application using the design process I laid out in "My Show Control Design Process," on page 362. Of course, we will also incorporate the principles I explained in the "System Design Principles" section, on page 111.

Question 1: What are the safety considerations?

The movement of the mechanized drum riser is potentially dangerous, and this needs to be taken into account.

Question 2: What type of show is it?

It's mostly a linear show, since the band will play (more or less) the same sequence of songs each night.

Question 3: Is the show event-based, time-based, or a hybrid?

Some sort of time code will likely be needed in conjunction with the media, but there will probably be a few event-based triggers as well.

Question 4: What is the control information source?
In this case, it is the band itself. Ideally, the systems should sync to the band, but there must be some general constraints to achieve the synchrony the band desires.

Question 5: What is the type of user interface required?
All the system operators on the tour are skilled technicians, so they want very sophisticated and complete user interfaces.

Question 6: What devices must be controlled/connected?
Let's look at each in detail.

THE SYSTEMS

The band's longtime production manager is in charge of contracting the various people and companies that will supply and operate gear for the tour. With all the recent changes in show technology, he decides to bring in a show control consultant to ensure that the critical sequences work correctly.

Lighting
The tour is designed entirely with moving and LED lights, and since there are no conventional fixtures, there are no dimmer racks; the only connection between the front-of-house lighting-control console and the moving lights is a data cable, which communicates with the fixtures themselves using streaming ACN (sACN). The moving-light controller has a MIDI port—which can accept MIDI Show Control commands; it can also accept ASCII console commands over Ethernet. The following looks have been programmed into the console for the drum solo:

Drum/Note Number	Control Message
Kick/C3	Go Cue 1, List 10
Snare/D3	Go Cue 2, List 10
High Tom/E3	Go Cue 3, List 10
Mid Tom/F3	Go Cue 4, List 10
Low Tom/G3	Go Cue 5, List 10

Sound
Purple Floyd's massive quad sound system is controlled using a large digital console, which has a control surface out in the house and a processing rack backstage. Interconnection is via a proprietary console network and an Ethernet-based audio distribution system. A fiber-optic snake, with a redundant back up system, has been run, and this snake can—in addition to audio—carry SMPTE, MIDI, or

Ethernet signals. Audio for the video segments is provided by the video department.

Video

Live video from the cameras and playback sources is controlled by a video director located backstage with all the video gear. Hard-disk based video servers provide high-definition video to high-power video projectors for the big screen and also drive the content on the video walls, switching back and forth with the live feeds. The video servers can accept DMX or ACN control, and also chase time code.

The poodle video segment is six minutes, 23 seconds, and four frames long; the rust segment is three minutes, 25 seconds, and 20 frames long. Each prerecorded video segment in the show will be assigned a different time code hour: the poodle segment is the third in the show, so it uses hour 3; the rust segment is next, so it uses hour 4.

To make sure everything can sync up, 10 seconds of video black is added on to the head of each video segment and onto the tail (end) of each segment, to fill out to the next largest time-code minute. Here's an example from the poodle segment:

Time Code	Event
03:00:00:00	10 seconds preroll video black
03:00:10:00	Start of poodle segment
03:06:33:04	End of poodle segment
03:07:00:00	End of video black
04:00:00:00	10 seconds pre-roll video black
04:00:10:00	Start of rust segment
04:03:35:20	End of rust segment
04:04:00:00	End of video black

Stage Machinery

The drum riser will lift to a preset position on receipt of an ASCII string "Up" over an Ethernet connection; a "Down" string tells the lift to retract. A special deadman's "authorize" button, which is downstream of any network control, is located near the lift control. The backline drum technician will press and hold the authorize button when the drummer and all crew are in a safe position, enabling the effect. Finally, the Stage Machinery system can broadcast its current position in a UDP packet every 100 ms.

Musical Synchronization

While the show control consultant advocates for a more sophisticated solution of chasing "musical time," the band and the crew feel most comfortable with time code. However, for everything to work seamlessly, the musicians somehow must lock themselves to the time code related to the media, so the sound engineer records a **click track**[1] on the audio track of the video server, which is then played into the drummer's in-ear monitors.

Once synced to a video segment, the drummer and the rest of the band can "float" a little bit within the timing of the segment, as long as they mostly stay in sync. To make the first transition from a time-based to a spontaneous segment, the drummer simply begins his solo as the "Comfortably Rich" video segment ends. When the drummer wants to end his solo, he signals the operators to start the next segment by playing a special beat pattern, which he never plays except at the end of the solo. The backstage operators roll the next segment, and the drummer is able to resync himself (and therefore the band) by listening to the click track. A similar approach is taken to get into and out of guitar solos and the like.

To synchronize the lighting with the electronic drum kit, MIDI Note messages are taken from the drum set and must be processed to trigger corresponding MIDI Show Control Messages.

SHOW CONTROL SCRIPT

While the video cue lists tell part of the synchronization story, the whole sequence is getting quite complicated now, so the production manager generates a script.

Trigger	*Event*	*Action*
Manual	Comfortably Rich preroll	Start video segment
8 beats before first frame	Start of click track	Click track to drummer

---

1. A click track is an audio track used to tell musicians when to start a song. The track is traditionally a series of metronome clicks starting a measure or two before the band should start; hence, the name click track. Click tracks can also be vocal count-offs—"1 2 3 4 1 2 3 4 Go"—or drum patterns.

Trigger	Event	Action
03:00:10:00	Comfortably Rich segment start	
03:05:00:00 (Approx)	Check riser safety	Enable riser movement (Drum Tech)
03:06:15:00	Drum riser rises	
03:06:33:04	End of song/start of solo	
Manual	Enable MIDI drum lights	Trigger light looks
03:06:35:00	Cue video for Young Rust	Cue DVD for next segment
Manual	Disable MIDI drum lights	
Drummer's special beat pattern	Young Rust Pre-Roll	Start video segment
Manual	Lower riser	Disable riser movement (Drum Tech)
8 beats before first frame	Start of click track	Click track to drummer
04:00:10:00	Young Rust segment start	
04:03:35:20	End of song/segment	Cue DVD for next segment

APPROACH 1

The band has used the same video director for years, so, they decide that he will be the one who pushes the "play" button, starting the video segments at the correct times. When he does so, SMPTE Time Code is sent out through a time code distributor/analyzer to the lighting console, so that it can be cued tightly with the video; to the sound console, which does some automated panning; and to a show control system, which sends out network commands to the drum riser.

To get the drum solo lighting to trigger, the consultant configures the show controller to generate appropriate messages for the lighting console when it receives corresponding MIDI note messages from the drum kit. To have the lighting system accurately track the movement of the drum riser, the UDP position packets broadcast by the scenic automation system are mapped by the show control system into X/Y coordinates using some software written by the consultant. These coordinates are then sent to the lighting console, which has some custom software to accept these coordinates and track any of the moving lights. The show control consultant creates this block diagram and user screen:

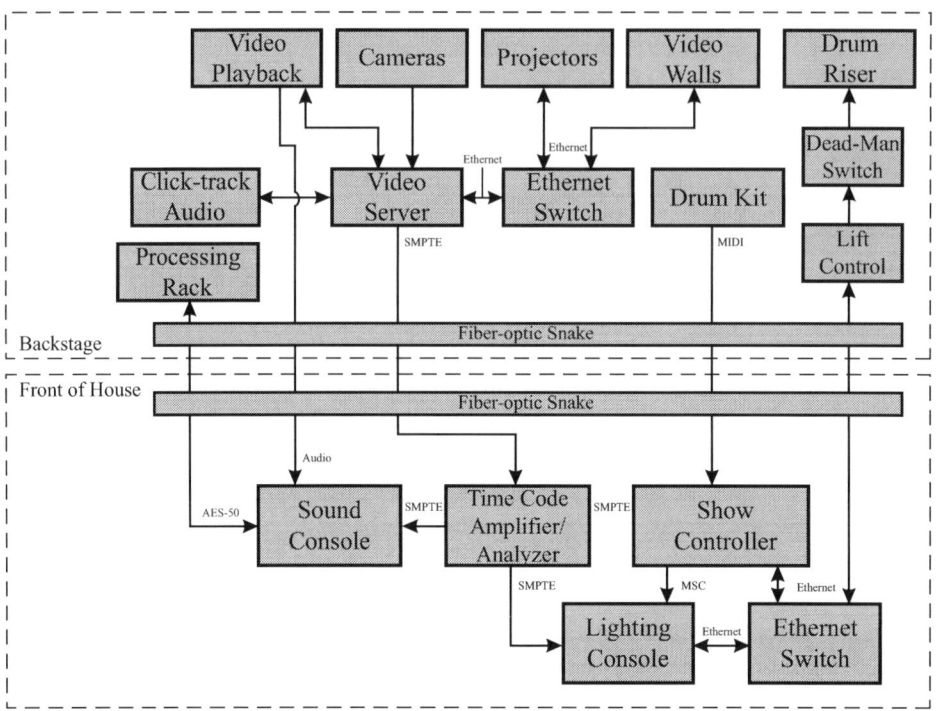

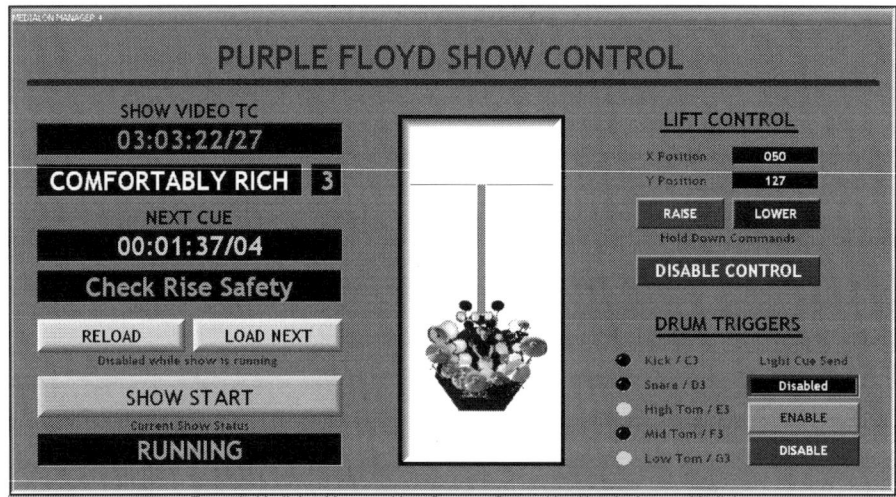

Example Medialon Manager User Screens Created by Alan Anderson, Medialon

The Results

The production manager wisely arranges a large-scale test about two months before the tour is to begin rehearsals. The band has set aside two entire days of rehearsal to work on integrating themselves with the prerecorded segments. Everything's working fine, but on day two, the keyboard player storms offstage yelling, "Why do we have to lock ourselves to those)_(*_)#@^# machines? Why can't they chase us?" The production manager runs to the show control consultant, who says, "Remember, that musical time approach I suggested at the outset? Let's try it." The band agrees, and everyone decides to come back in a month for another test.

APPROACH 2

With the new approach, the consultant will use "musical time" to drive the system. The group decides that the backup keyboard player, who is very comfortable with sequencers and other musical software, will also take on responsibility for the musical time system. All he has to do is "conduct" the "musical time" system which, in this case, won't actually play any musical parts, but will provide some sound effects backing tracks. The musical time system takes the songs and breaks them down into an ordered musical structure of measures and beats.

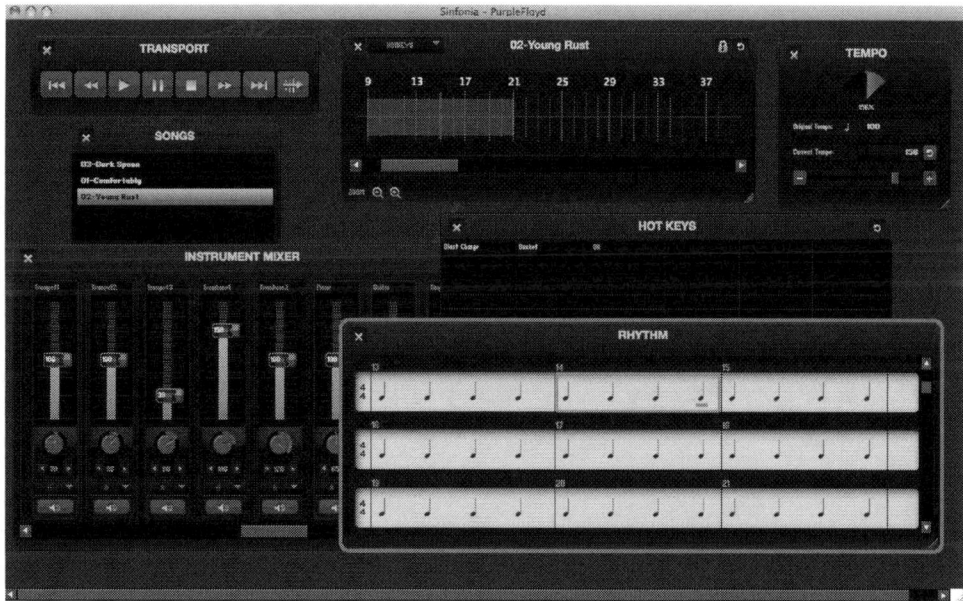

Example Sinfonia® Screen Created by Dr. David B Smith, Realtime Music Solutions

However, unlike a standard MIDI sequencer, this system can be conducted, allowing the keyboard player to vary the tempo, or even conduct each beat individually.

This musical time (song name, measure, beat) information is sent over the network to the show controller, which takes the current data—along with a next beat prediction from the musical time system—and compares it all with a "normal" performance, which is simply a performance with a typical tempo structure on which the band has agreed. The show controller then, in real time, can figure out how much the current performance is deviating from the "normal" performance, and it can vari-speed the video server to keep it locked to the performance. The backup keyboard player simply follows the current performance, and the rest of the systems chase his actions.[2]

Because so much of the system was already connected to the show controller, the only changes in the system layout are to bring the video server on to the network, and get it talking to the show controller via IP in order to accept the start and vari-speed commands, and to run a MIDI line to the sound console. The existing connections to the lighting console, drum kit, and drum riser all stay as they were, and those event-based cues are also triggered by the show control system at the appropriate measure and beat. The time code system is left in place for preshow applications where video is rolling and the band is not yet playing.

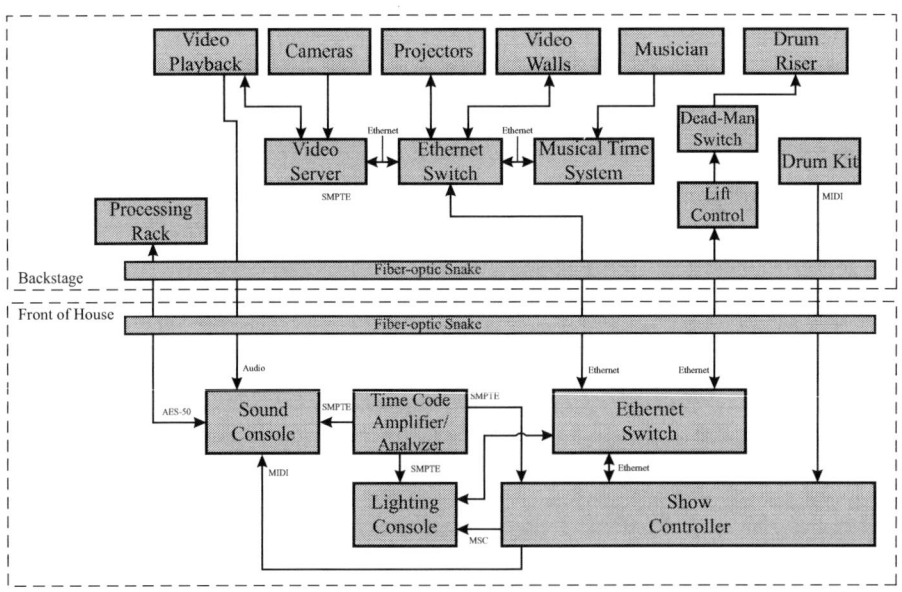

---

2. Note that this musical time approach is not in widespread use, but could be.

The show control programmer updates the main user screen:

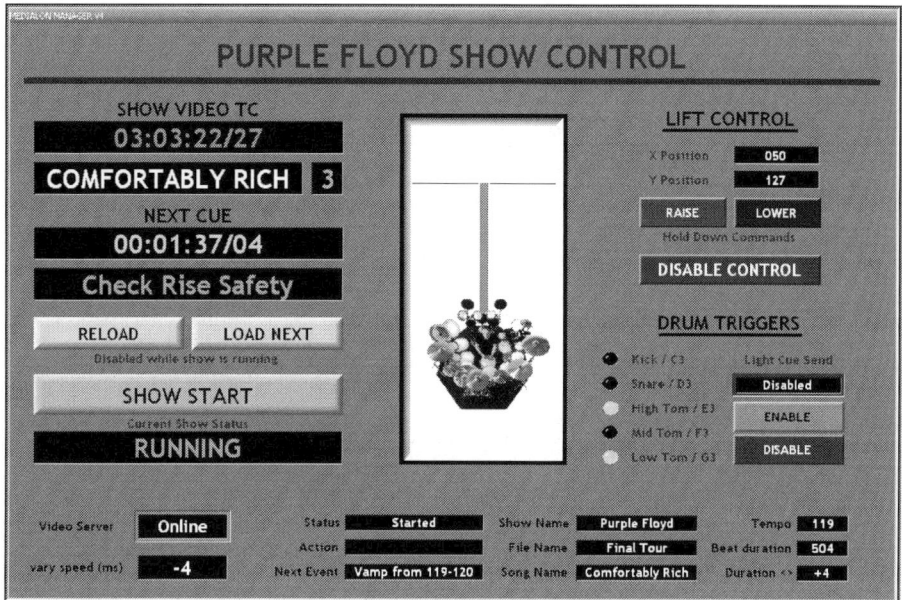

The Results

The band loves it! For the first time in their careers, they are able to simply play, and with a little extra work by the backing keyboard player, the entire system stays perfectly in sync.

Chapter 33: **It's an Itchy World after All**

Itchy and Scratchy wander on to the huge outdoor stage from opposite ends. Itchy, proclaiming his friendship, gives a bouquet of flowers to Scratchy, who is purringly happy. But, seconds later, Scratchy gives his trademark scream as Itchy shoots him off the stage with a large-caliber machine gun. Scratchy staggers back onstage to get a drink of water from a well, but Itchy sneaks in from the side and kicks Scratchy down the opening. The manic mouse then throws a hand grenade down after the crazy cat, and a huge explosion bellows out, spraying the audience with water. Moments later, a charred Scratchy appears at the mouth of the well, staggers across the stage, and lays down under a ledge to get a badly needed cat nap; little does he know that he is sleeping on a bowling alley. Itchy, however, is happy to point this out, and quietly arranges nine warhead bowling pins around the napping cat. Next, he rolls a bomb/bowling ball down the alley and there is a huge explosion. "Cut" yells the "director," as the smoke clears, and the Itchy and Scratchy performers come back onstage. "Now, audience," he asks, "what should we do next?"

THE MISSION

These are the first three segments of the Itchy and Scratchy Epic Stunt Spectacular, which will purportedly show how action movies are made, using indestructible cartoon characters Itchy and Scratchy to demonstrate various stunts. This is to be the main attraction at Duff Gardens, the internationally known theme park located near Orlando, Florida. Unlike most theme park shows, this one has a twist: the audience will select which stunts they want to see, and in which order.

Mr. Burns, Duff's CEO, has taken a personal interest in the show, and so everyone is working hard to ensure that everything goes smoothly. Itchy and Scratchy are Duff's corporate icons, so no expense is spared. However, while the attraction's construction is lavish, its operating costs will be kept as low as possible, because the park management intends the show to run for many years. So, the entire show will be run by one Technical Director (TD), one Sound Operator, and several pyro and special effect technicians.

DESIGN CONSIDERATIONS

Let's evaluate this application using the design process I laid out in "My Show Control Design Process," on page 362. Of course, we will also incorporate the principles I explained in the "System Design Principles" section, on page 111.

Question 1: What are the safety considerations?
There are many dangerous effects in this show, so all proper precautions must be taken.

Question 2: What type of show is it?
It's a nonlinear show, since the audience selects which stunts they want to see and in which order.

Question 3: Is the show event-based, time-based, or a hybrid?
The show will be a loose, event-driven collection of flexible time-based sequences.

Question 4: What is the control information source?
Sometimes a time base will supply control information; other times the human operators will. The audience has a voting system to select and sequence segments, and, in critical safety situations, the actors onstage will actually provide additional control information themselves.

Question 5: What is the type of user interface required?
Skilled technicians are running the show and many elements of the show are unpredictable, so they will need very comprehensive user interfaces.

Question 6: What devices must be controlled/connected?
Let's go through each system.

THE SYSTEMS

Equipment for the production is being supplied by a number of subcontractors, all working for WDI (Wacky Duff Illusions), Duff's in-house design and engineering department, which is designing and providing general contracting services for the show.

Show Control
Park management has standardized all operations in the park on one commercially available show control system to facilitate maintenance, training, and updates. Park management systems will interface with the show controller through an Ethernet link, and this will allow the park's central control system to tell the show

controller whether to run day or night versions of the show, and also do things such as turning on worklights at specific times for cleaning, and so on.

Lighting

The lighting system will be built around a rack-mounted version of a large control console. A full-sized console will be used for the programming period and then the show will be downloaded into the rack-mounted unit for show operation. Either version of the console can be driven via commands over ACN, which will also be used to connect all the conventional dimmers, moving lights, and LED fixtures. The ACN network also accepts commands from a special processor which provides interfacing for emergency lighting systems, allowing certain looks to be brought up independently of the show control system in the event of a park-wide or other emergency.

Video

High-power video projectors are used to generate massive images for the nighttime shows, projecting preshow images on a nearby water tower shaped like Itchy's mouse ears. Two other video outputs drive daylight-viewable LED screens for sequence titles, voting results, and other show imagery. Each projector is driven by a PC, and all the PCs are networked and controllable via Ethernet and connected to the show control system.

Sound

The Sound Operator actually runs three systems: an audio matrix mixer, a sound FX playback triggering system, and a music generation system—all of which can accept various control messages over Ethernet. The matrix system handles all routing and gain manipulation for the system, and receives audio inputs from the sound FX playback system, several wireless mics, and the music system. In addition to a mouse and monitor, the matrix is controlled by a hardware fader console. The system can simultaneously run in a cue-based mode like a lighting console and accept manual input from the physical faders, which the operator uses to make show-to-show adjustments and mix mics. In addition, the system can take commands over the network from the show control system, as well as from park-wide management. The park-wide connection enables special emergency paging lines to be made active at any time and incorporates park background music and a general paging system.

The performers playing the characters do not actually speak; instead, each individual phrase (or scream) is triggered manually from the sound FX playback system by the Sound Operator using a push button panel. This approach ensures proper synchronization regardless of what happens onstage, and in rehearsal, the actors learn certain moves to indicate to the operator when to trigger the sound. This system also plays back other prerecorded sounds, such as the voice-over used to open the show, which are triggered from the show control system over the network.

The Itchy and Scratchy theme song and underscoring music for each segment is generated in real time using the virtual orchestra system. The system generates a fixed-tempo musical sound track, but the virtual orchestra gives flexibility to "vamp" on any measure of the music, allowing the Sound Operator to stretch each segment indefinitely if something goes wrong. When running, the music system provides time code to the network to control other show elements; in vamp mode, the last valid time-code frame is repeated indefinitely.

Mr. Burns went to a competitor's theme park and was disturbed when a character's dialogue appeared to come out of a speaker stage left when the performer was stage right. He asked his sound designer to prevent this problem at Duff Gardens, so the designer adopted a feature of the sound matrix system to "localize" sound images anywhere on the massive stage by varying the arrival time and gain characteristics in zones covering the audience; this allows a laugh from Itchy to sound like it comes from Itchy, no matter where in the audience you are sitting. For this production, the blocking is very predictable, so each localization cue for the actor's moves is simply programmed in and triggered at specific times.

Effects

There are extensive pyrotechnic and other effects in this production, which will be controlled using a computer-based pyro controller that works in conjunction with a special safety-rated Programmable Logic Controller (PLC), the human TD. and the performer.

The PLC determines whether the effect is safe to go based on the state of a variety of detectors and, most importantly, TD and actor-operated safety switches. For dangerous effects, in addition to the TD "authorizing" the effect, the performer must press and hold a button or buttons, indicating that he or she is in a safe position, or the effect will not be allowed to operate. Until the effect is safe to go, the firing contacts to the actual effect are shorted by the PLC, ensuring that no matter what happens upstream in the pyro controller, the effect cannot fire until all is safe. An Ethernet connection from the PLC is used to report status back to the master show controller.

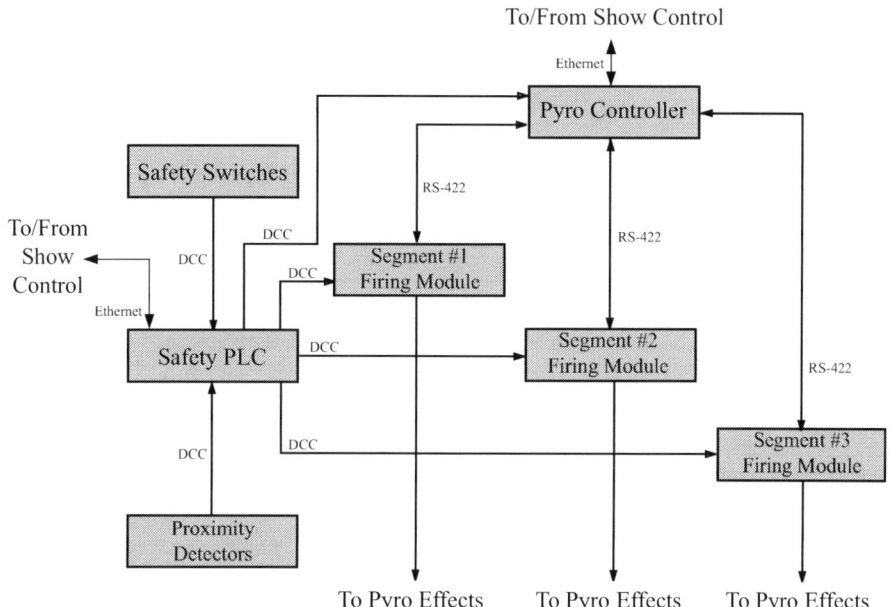

For safety reasons (and to save money), pyro substitutes are used whenever possible; the substitutes are also fired by the pyro control system because some of them still present a danger to the performers. The well explosion, for example, is really a pneumatic "air cannon" effect, which blasts a tremendous amount of water out of the well in conjunction with a sound effect. The machine-gun bullet hits are also pneumatic.

The machine-gun effect uses a special system to give the performer maximum flexibility and to make the effect as believable as possible, it enables the performer to actually control the machine-gun effects directly. When the trigger is squeezed on Itchy's machine gun, a radio signal is sent to a receiver, which generates a message over Ethernet to the show control system. When enabled by the TD through show control, the pneumatic bullet-hit effects are then triggered, along with localized machine-gun sound effects. After extensive testing, the machine-gun burst firing interval has been agreed upon and is programmed into each system.

Cue Light System

Since the actors perform the show to the music—which can change somewhat if something goes wrong—a cue light system is used to indicate when certain actions, such as entrances, should occur. Large monitors backstage indicate when each event is predicted to happen; red lights next to the monitor come on for a warning and turn off to indicate that the event should take place. A similar system is implemented over the TD booth in the front of the house: A pair of green lights (one for backup in case of a burn out) mounted on top of the TD booth indicates to the performer playing Itchy that Scratchy is safely in place for certain effects, such as the grenade and the machine gun. The cue light control system is a standard PC accepting custom ASCII commands over Ethernet.

Audience Voting System

WDI contracts with a company that does audience voting systems to provide a small touch screen for every seat in the venue, all connected via a network. The main controller for the audience voting system connects to the show control system over the show network to indicate the results of the voting using some custom ASCII messages; the show controller then handles the scheduling of the segments.

THE APPROACH

Since so much money is being spent on the attraction, WDI does extensive testing in advance to ensure that as few changes as possible are made after construction begins.

The show is broken up into segments, each of which is time-based, but staying flexible on any life-threatening or time-critical elements. Fifteen segments are available, but only nine (to represent Scratchy's nine lives) are to be selected by the audience for each show in order to keep the show interesting and generate repeat business. As the audience files into the venue, they select the first three segments they want to view, and in which order. After these segments, the "director" comes out and guides the audience through the selection of the next three.

Each segment is started by the technical director (TD) when everything is in place, safe, and operational. The show controller issues a go command to the music system, which then generates time code back on to the network, which in turn triggers the actors' cuing system, light cues, sound cues, and localization cues for that segment.

The system is a large network, with the systems connected as shown here:.

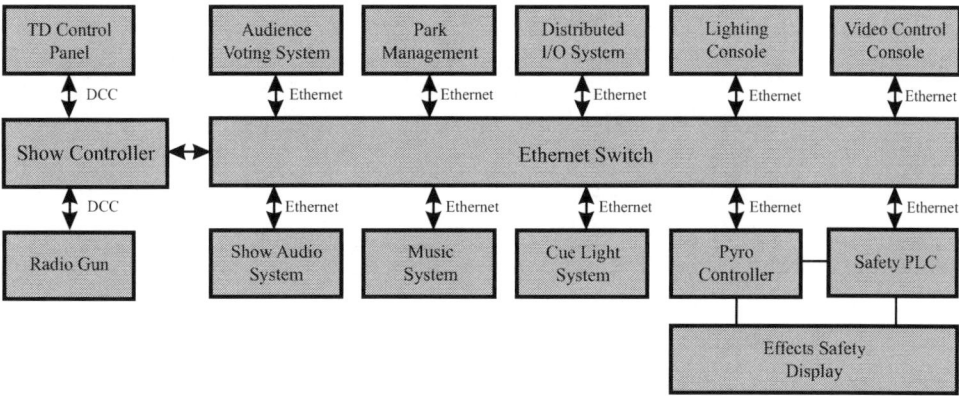

The show control programmers design this user interface:

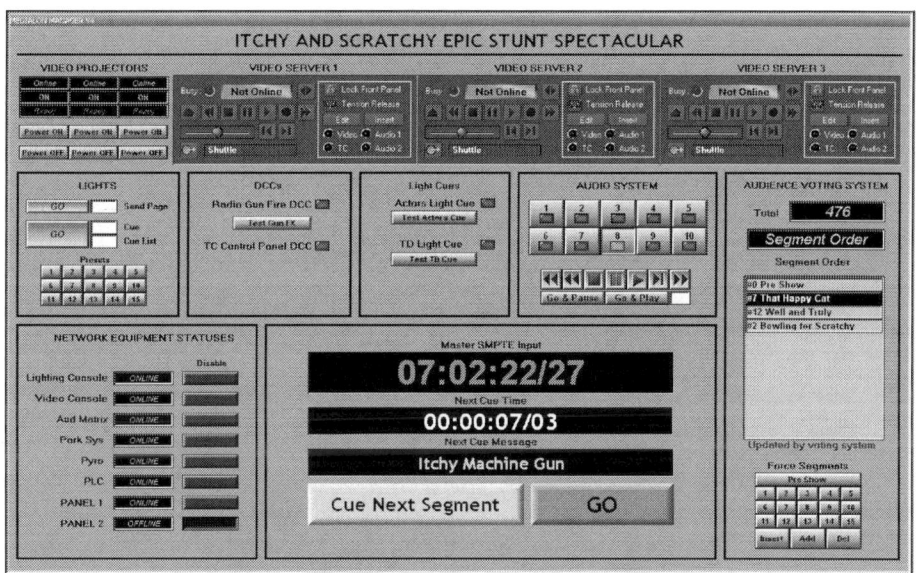

Example Medialon Manager User Screen Created by Alan Anderson, Medialon

Show Control Script

WDI comes up with a detailed cue script for show planning. In addition to trigger sources and show control messages, the list indicates who executes a particular effect. "TD" denotes that the technical director executes the effect; "TC" indicates that the effect is executed automatically based on the time code; Snd means that the Sound Operator takes the cue; and Hs Mgr is the house manager in charge of the audience. This script applies to a nighttime show, including all lighting and image commands; for simplicity, however, cue-light cues, system acknowledgments, and other messages are not included here. For this example, the audience has selected segments 7, 12, and 2.

Time Code	Event	Actions	Ex By	Control Messages
Segment #0	**Preshow**			
Event	TD starts pre-show	Park backgrnd music up Sound localization preset Preshow light look Preshow images up	TD	Sound Go Q000 Localiz.Q000 Lights Q000 Image Q000
Event	10 minutes before show	"Please take your seats, voting is about to begin"	Hs Mgr	Sound Q001
Event	2 minutes before show	"Please select the first three segments"	Hs Mgr	Sound Q002
Event	TD starts show	Itchy and Scratchy theme song	Snd	Sound Q005
00:00:10:13	2nd measure of music	Houselights to half	TC	Lights Q001
00:00:20:02	4th measure of music	Houselights out	TC	Lights Q002
00:00:24:28	Itchy in music	Itchy graphic up/ Preshow out	TC	Image Q001
00:00:32:07	Scratchy in music	Scratchy graphic up	TC	Image Q002
00:01:02:23	Start first segment		TC	
Segment #7	**That Happy Cat**			
07:00:00:00	Segment starts	Flowers music Go "That Happy Cat" image Sound localization cue	TC TC TC	Music Q701 Image Q701 Localiz. Q701
07:00:10:00	Scratchy enters	Lights for Scratchy up	TC	Lights Q701
07:00:18:00	Scratchy moves center	Sound localization cue Scratchy whistling	TC Snd	Localiz. Q702 Sound Q702
07:00:32:00	Itchy enters with flowers	Lights for Itchy up Itchy squeaking	TC Snd	Lights Q702 Sound Q703
07:01:18:00	Itchy and Scratchy move center	Sound localization cue	TC	Localiz. Q703
07:02:03:00	Scratchy moves SR	Sound localization cue	TC	Localiz. Q704

414 • *Part 5: Show Control*

Time Code	Event	Actions	Ex By	Control Messages
07:02:05:00	Itchy exits	Sound localization cue	TC	Localiz. Q705
07:02:30:00	Itchy enters with machine gun	Sound localization cue TD authorizes machine gun	TC TD	Localiz. Q706
Event	Scratchy safely in place	Scratchy authorizes gun	Scr	
07:02:32:00	Scratchy safely in place	Lights for Scratchy in place	TC	Lights Q703
Event	Itchy squeezes trigger	Machine gun bullet hits Machine gun sounds Machine gun flashes Scratchy screaming	Itch Itch Itch Itch	Pyro Q701 Sound Q704 Lights Q704 Sound Q705
Event	Scratchy breaks photocell beam	Flowers music fade out Lights for Scratchy off	TD TD	Sound Q706 Lights Q706
Event	Itchy releases trigger	Machine gun hits stop Machine gun sounds stop Machine gun flashes stop	Itch Itch Itch	Pyro Q702 Sound Q707 Lights Q705
Event	Itchy character laughs	Itchy laughing sound FX	Snd	Sound Q708
07:03:00:00	Itchy exits	Lights fade down	TC	Lights Q707
Segment #12	**Well and Truly**			
Event	Scratchy's head around corner	Well music go	TD	Sound Q1201
12:00:00:00	Scratchy staggers onstage	Lights for Scratchy up "Well and truly" image up	TC TC	Lights Q1201 Image Q1201
12:00:10:00	Scratchy moves towards well	Sound localization cue	TC	Localiz.Q1202
Event	Scratchy moaning	Moaning sound FX	Snd	Sound Q1202
Event	Takes drink from well	Slurping sound FX	Snd	Sound Q1203
12:00:30:00	Itchy enters	Lights for Itchy up	TC	Lights Q1202
Event	Itchy kicks Scratchy down well	TD authorizes explosion Falling sound FX	TD TD	Sound Q1204
12:00:45:00	Scratchy down well	Lights focus on well	TC	Lights Q1203
Event	Itchy laughs	Itchy Laughing sound FX	Snd	Sound Q1205
Event	Scratchy in safety position	Scratchy authorizes explosion	Scr	

Time Code	Event	Actions	Ex By	Control Messages
Event	Itchy throws grenade	TD fires explosion Lights for explosion Sound for explosion Well music out	TD TD TD TD	Pyro Q1201 Lights Q1204 Sound Q1206 Sound Q1206
12:02:00:00	Itchy walks off-stage	Sound localization cue	TC	Localiz.Q1203
Event	Itchy laughs	Itchy laughing sound FX	Snd	Sound Q1207
12:02:05:00	Itchy walks off-stage	Lights fade down	TC	Lights Q1205
Segment #2	**Bowling for Scratchy**			
Event	Scratchy's head peers out well	Bowling music go	TD	Sound Q201
02:00:00:00	Scratchy climbs out of well	Lights for Scratchy "Bowling for Scratchy" image Sound localization cue Scratchy moaning	TC TC TC Snd	Lights Q201 Image Q201 Localiz.Q201 Sound Q202
02:00:10:00	Scratchy moves towards alley	Sound localization cue	TC	Localiz.Q202
02:00:20:00	Scratchy lies down	Lights focus on Scratchy Sound localization cue Purring sound FX Go	TC TC TC	Lights Q202 Localiz.Q203 Sound Q203
02:00:30:00	Itchy enters	Lights up on Itchy	TC	Lights Q203
02:00:33:00	Itchy moves towards alley	Sound localization cue	TC	Localiz.Q204
Event	Itchy arranges bowling pins	Standby for explosion	TD	
02:01:45:00	Itchy arranges bowling pins	Lights focus on Itchy	TC	Lights Q204
Event	Scratchy moves to safety	Scratchy authorizes explosion	Scr	
Event	Itchy rolls bomb down alley	Bowling sound effects	Snd	Sound Q204
Event	Bomb ball hits pins	TD fires explosion Lights for explosion Sound for explosion Purring sound FX stop	TD TD TD TD	Pyro Q201 Lights Q205 Sound Q205 Sound Q206
02:03:00:00	Itchy walks off-stage	Sound localization cue Itchy laughing sound FX	TC TC	Localiz.Q205 Sound Q207

Before each show, the pyrotechnicians load all the pyro equipment and walk through the attraction with the stunt performers to do a safety check. When the show is ready for the audience, the TD executes a cue on the controller that starts preshow mode. In this mode, special park background music is piped through the system, along with any park-wide paging or announcements. If the night show is being run, a preshow light look is brought up, and the Itchy and Scratchy image is projected on the water tower. The house manager keeps in touch with the TD as the audience files in, and the Sound Operator triggers one announcement 10 minutes before show time: "Ladies and gentlemen, please take your seats. The Itchy and Scratchy Stunt Show will start in 10 minutes. You can start voting now for your first three stunts." Two minutes before show time, the house manager checks again and the Sound Op triggers another recorded announcement. At this point, the TD gets verbal confirmation over the headset, closes the voting (which loads up the first three segments), confirms that the actors are ready, and calls places. When the house manager indicates that the audience is seated, the Sound Op starts the introduction sequence—the ever-popular Itchy and Scratchy theme.

As the houselights dim, strains of the theme song can be heard over the sound system: "They fight, they fight, they fight, they fight, they fight! Fight-fight-fight, fight-fight-fight! The Itchy and Scratchy Epic Stunt Spectacular!" The music system generates time code, which is sent back to the master controller, which in turn triggers the image presentation system, and the recorded announcer introduces the popular animated characters. At time code `00:01:02:23`, on the final downbeat, the first audience selected segment (#7 in this case) is automatically triggered; the show controller also switches to monitor the music system for all subsequent time code.

The first sequenced music cue begins, the image system brings up the title of the first segment—"That Happy Cat"—and a localization preset cue is issued. At time `07:00:05:00`, Scratchy's cue light comes on; at `07:00:10:00` the light goes off and Scratchy enters. The system takes an automatic localization cue as he moves center, and the Sound Op triggers a whistling cue as the performer turns his head toward the audience. Itchy is cued to enter and comes onstage with his bouquet of flowers. The Sound Op executes a dialogue cue, and Itchy walks offstage. Several localization cues are then taken automatically.

When Itchy enters with the machine gun and the TD sees that Itchy is in the right position and that everything is safe, he authorizes the machine gun sequence. Scratchy moves to the correct position to go flying through a breakaway wall when hit by the bullets; to signal that he is ready, he stands on two switches mounted in the deck, one for each foot. These switches are also wired into the pyro PLC, and when both switches are closed, the PLC closes a contact to the pyro controller, which is indicated on the TD's control panel and received by the show controller. When the master show controller receives this message, the green lights on top of the booth come on, indicating to Itchy that everything is ready. As soon as Itchy pulls the trigger, a radio signal is sent, which (through the show control system) triggers the machine-gun sound effects, special lighting, a scream sound for Scratchy, and the bullet hits. Scratchy crashes through the wall, and the Sound Op executes a cue that fades the flowers music out (although time code is still generated). The time code turns Scratchy's light look off, and executes a localization cue. When Itchy releases the trigger, all the machine-gun effects stop. Itchy moves his head in a way to signal the Sound Op to play a sadistic laugh cue, and then walks offstage. A light cue is taken, leaving only a spotlight on the hole in the wall where Scratchy crashed through.

When the audience sees Scratchy's bullet-ridden head peer around the corner, they give a big cheer, and the Sound Op starts the second segment (#12 in this case), starting the sequenced "well" music; time code from the music system then starts the other elements. Scratchy moves toward the well, several light and localization cues are triggered automatically, and the Sound Op triggers a moan sound. As Scratchy leans into the well, the Sound Op triggers several manual slurping sounds then Itchy's cue light goes out, he enters and kicks Scratchy down the well. The TD executes a cue that puts the explosion effect into standby and the Sound Op triggers a falling and splash sound effect.

The performer playing Scratchy, of course, only falls about six feet and climbs into a special bunker to protect him from the explosion. Itchy dances around the top of the well to cover the time Scratchy needs to get into the safety position. When in place, Scratchy presses and holds an authorize switch in his bunker underneath the well, which signals the pyro controller, the TD, and the show controller, which then turns on the green lights on top of the booth. Itchy sees the green ready light and tosses the hand grenade down the well; the TD waits a moment and then executes a cue triggering the water-cannon effect, a light cue, an explosion sound effect, and a cue that fades out the "well" music. Sound localization and light cues are then taken automatically, the Sound Operator triggers a laughing cue, and Itchy walks offstage. The lights fade down, and finally Scratchy's charred head is seen over the top of the well; the TD starts the next sequence.

The Sound Op triggers appropriate moaning cues, and other cues are taken automatically from the time code as Scratchy staggers over to the bowling alley. The Sound Op executes a purring cue as Scratchy lays down; Itchy's cue light goes off, and he enters. Itchy's act of arranging the large bowling pins is actually a diversion as Scratchy is moving to an explosion-proof bunker behind the alley. The TD executes a cue telling the pyro system to stand by, Scratchy presses and holds a safety switch, and the green cue light comes on. Itchy sees the light and rolls the ball down the alley. As the ball hits the pins, the TD triggers a command firing the huge explosion cue, the sound effect, light cue, and a sound cue that kills the purring sound. As Itchy walks offstage, the Sound Operator triggers another laughing cue.

The Results

We've covered only the first three of nine segments of the show, but the attraction is a rousing success, and people line up all day to get in! Mr. Burns says the show is "Ex-cellent!"

Conclusion

Storytelling is our business, and technology won't change this. Live entertainment has survived a couple thousand years, and all kinds of threats that were supposed to do us in: movies, television, the Internet, etc. I believe that live performance will be around as long as humans are around, because it clearly meets a human need: the desire to be together and share stories. The best example I know to support this hypothesis is a "LAN Party" like DreamHack, where thousands of people bring their own computers to a large hall and network them together to play games for several days; for a typical event, they have more than *ten thousand computers* on the network. Clearly, there is no technical reason for this—with the Internet, millions of people play video games together from the comfort of their homes. But that's not what Dream-Hack is about—in addition to playing games, they have live concerts, shows, and parties

Photo Petri 'Mumin' Hänninen Blomberg, Dream Hack Press Image

around the clock. They are taking this essentially solitary activity and turning it into a live show.

Many aspects of the early 1990s world I wrote about in the preface (page xv) are gone. Back then, most shows used simple control protocols to link a few devices; more complex, powerful systems had to be custom engineered, and this meant that these technologies were available in limited ways only to the "big boys": theme parks and other shows and attractions with larger budgets and resources. I'm not a nostalgic person, and so I say good riddance to those old days. The victory of Ethernet over its contenders, coupled with amazing computer power and cheap memory, has now levelled the playing field, distributing these amazing control technologies to ever smaller kinds of production. The little low budget shows I see around New York these days have creation power and interaction capabilities in a laptop that would have only been available to theme park designers 20 years ago, and I would say that we are no longer significantly restricted by technology, but instead are only limited by the laws of nature and our creative imaginations. This is something I couldn't have written in the first edition of this book in 1994, so the future seems very bright to me.

But that doesn't mean that we don't need to keep moving forward; keep in mind that many of our legacy backbone technologies—things like DMX, MIDI, and SMPTE Time Code—have existed longer than many of my students have been alive. If these old technologies do the job, then fine, let's keep using them. But I personally think that many are holding us back creatively. So get out there, experiment (in advance of the show when you have time to learn from your mistakes), and push this whole thing forward. Maybe it will take a generation to really reach our potential, but there's no reason we need to wait that long!

CONTACT INFO AND BLOG

If you have any (polite) comments, questions, or corrections, I'd like to hear them. Please check my website at: `http://www.controlgeek.net`. You can e-mail me from there, and I have a blog, book errata, and many other online resources.

Thanks!

John Huntington
Brooklyn, New York City

July, 2012

Acknowledgments

My thanks to the following (in order of their help):

- John C. Huntington Jr. for help with funding this self-published edition.
- Mike Saltzman, Esq. for help with getting out of my Focal contract.
- My self publishing team (see page xix).
- Philip Nye of Engineering Arts again for great suggestions and examples.
- Dan and Glenn Birket of Birket Engineering for reviewing the Pyro, Fog/Smoke/Fire and Animatronics chapters.
- Jeff Berryman of Bosch for help with the OCA section.
- Dan Antonuk of ETC for reviewing the ACN information.
- Steve Terry of ETC for reviewing the lighting chapter.
- Ellen Juhlin of Meyer Sound for a lot of help with the OSC chapter.
- Scott Blair, Revolution Display; Paul Kleissler of City Theatrical, and Simon Newton of the Open Lighting Project for help with the RDM chapter.
- Alan Hendrickson of the Yale School of Drama, for reviewing the Stage Machinery chapter.
- Chris Ashworth, of Figure 53 for providing a screen capture.
- Wayne Howell, of Artistic Licence, for reviewing the Art-Net section.
- Josh Weisberg for reviewing the Image Presentation chapter.
- Benoit Bouchez of Kiss-Box for reviewing the RTP-MIDI section.
- Mike Lay of Philips Color Kinetics and Myles Ambrose for reviewing the networking chapters.
- Jason Potterf of Cisco for providing a HUGE amount of very helpful feedback on the networking chapters.
- Norman Ballard of Laser Production Services, and Tony Clynick for reviewing the laser chapter.
- Glenn Birket, David Boevers, Richard Cadena, Jonathan Deans, Gary Fails, Scott Fisher and Steve Terry for generously providing endorsement quotes.
- Special additional thanks to Jonathan Deans for writing the foreword.
- Everyone who reads my blog at www.controlgeek.net.

Third Edition (Published 2007)

Dedication: *In memory of show control guru and pioneer George Kindler, my friend and mentor, who died at age 57 while I was at work on this edition.*

- Cara Anderson, my acquisitions editor at Focal Press, for all her help and guidance throughout the process.
- Alan Rose, who handled the production and pre-press work for Focal Press, and his associate Lauralee Reinke who copy edited the text and cleaned up the layout.
- Aaron Bollinger, who created all of the fantastic new graphics for this edition.
- Andrew Gitchel, who acquired or shot all the new photos for this edition.
- Philip Nye, for help categorizing shows and for extensive help on the ACN chapter.
- Dan Antonuk for a ton of help getting me through the ACN material, and for providing screen captures of ACN discovery.
- Benoit Bouchez for extensive help with the MIDI-RTP chapter.
- The members of the show control email list, who vetted some of the book's concepts.
- Carlton Guc at Stage Research, for creating custom SFX program screens for the practical examples section, and Alan Anderson at Medialon, for creating custom Manager program screens for the practical examples section.
- Loren Wilton for capturing the Soundman-Server transactions.
- Mark Reilly at Crestron, for locating the Laser scan head section.
- Alan Hendrickson for providing information on NFPA 79.
- Peter Willis for providing the RDM discovery transaction.
- Special thanks to everyone who did a technical review of a section of the book (listed alphabetically): Dan Antonuk, Electronic Theatre Controls; Rick Baxter, Production Electrician; François Bergeron, Thinkwell Design and Production; Glenn Birket, Birket Engineering, Inc.; Scott Blair, Director of Digital Lighting Development, High End Systems; Benoit Bouchez, Kiss-Box; Steve Carlson, Executive Secretary, IEEE 802.3 Working Group; Travis Gillum, LifeFormations, Inc./MasterMind Control; Mitch Hefter, Sr. Project Engineer at Entertainment Technology, a Genlyte company; Alan Hendrickson, Professor of Technical Design and Production, Yale School of Drama; Wayne Howell, Artistic Licence; Jim Janninck, TimberSpring; Sierk Janszen, Rapenburg Plaza; John Lazzaro, CS Division, UC Berkeley; Charlie Richmond, Richmond Sound Design Ltd.; Karl G. Ruling, ESTA Technical Standards Manager; Elizabeth Simon, Magenta Art Projects; Dr. David B. Smith, Chair, Entertainment Technology Department, City Tech; Ian W. Smith, Production Resource Group, L.L.C.; Steve Terry, Vice President Research & Development, Electronic Theatre Controls; George Tucker, Manager, Applications Engineering, Crestron; Stephane Villet, Medialon; Ryan Waters, CTO, Microlight Systems

Second Edition (Published 2000)

Dedication: *To my mother, who didn't live to see this edition.*

- Jerry Durand, Philip Nye, and Dave Barnett for finding typos in the first edition.
- Caroline Bailey for putting up with me during the whole second edition process, for the "Ten Pin Alley" section title.
- Tom Lenz and Barbara Wohlsen for assisting me on this edition. Tom did and/or redid many of the line drawings and Barbara found and gathered most of the photos or screen captures.
- Marie Lee, Lauren Lavery, Terri Jadick, Charles McEnerney, and everyone at Focal Press.
- JoAnne Dow of Wizardess Designs, Tom DeWille of PyroPak/LunaTech, Kevin Gross of Peak Audio, Robert Harvey of White Rabbit/RA Gray, Chuck Harrison of Far Field Associates, Dennis Hebert of Gilderfluke, Alan Hendrickson of the Yale School of Drama, Jim Janninck of TimberSpring, Michael Karagosian of MKPE, George Kindler and Kevin Ruud of Thoughtful Designs, Mike Lay and Steve Terry of Production Arts, Jeff Long of Granite Precision, Lars Pedersen of Scharff/Weisberg, Mike Rives of TDA, Charlie Richmond of Richmond Sound Design, David Scheirman of JBL Professional, and David Smith of the New York City Technical College, for reviewing sections of the draft manuscript. Extra thanks to Chuck Harrison for his knowledge and inspiration for the first edition all the way back in 1987 at Associates & Ferren, and Steve Terry for the nice foreword and encouragement along the way.
- Everyone at all the companies who provided me with information on their systems.
- Gardiner Cleaves at Production Arts Lighting, for converting the first edition FastCAD drawings to DXF.
- Jim Kellner and Mike Fahl of Dataton, for providing information on many video and projection standards.
- Sierk Janszen of Avenger, for providing information on IR protocols.

First Edition (Published 1994)

Dedication: To my parents, for making it all possible. (And for making me possible!)

- Cy Becker at SMPTE, for information on the latest time code standard.
- David Bertenshaw, Tony Brown, and Andy Collier at Strand UK, for information about Strand's European standards.
- Dr. John Bracewell of Ithaca College, for guidance and inspiration.
- Margaret Cooley at Audio Visual Labs, for information on AVL products and protocols.
- Sound designer Jonathan Deans, for help on the thesis.
- Tom DeWille of Luna Tech and Ken Nixon of PyroDigital Consultants, for their guidance on the pyro chapter.
- Anders Ekval at Avab, for information about Avab's protocol.
- Bran Ferren, for inspiration and information.
- Tony Gottelier for giving me the European perspective.
- Richard Gray of R. A. Gray, for information about SDX and for reading my thesis.
- Alan Hendrickson, Ben Sammler, and the rest of the faculty at the Yale School of Drama, for their help and guidance on the thesis and throughout graduate school.
- Mike Issacs at Lone Wolf, for up-to-the-minute information about Media-Link.
- George Kindler of Thoughtful Designs, for checking the book for completeness.
- Bob Moses of Rane, for information on the AES-24 effort and for reviewing the manuscript, and for helping me with the term "meta-system."
- Pat MacKay, publisher, and Karl Ruling, technical editor, at *TCI* and *Lighting Dimensions*, for their help.
- Charlie Richmond, for the MIDI Show Control standard, answers to countless questions, continued support, and for reviewing the manuscript with a fine-tooth comb.
- Paul Shiner at Crest, for information on NexSys.
- Lou Shapiro of Erskine-Shapiro, for help early on in the process.
- Karen Speerstra, Sharon Falter, John Fuller, and everyone at Focal Press, for all their help.
- Steve Terry of Production Arts Lighting, for his support, encouragement, employment, information about DMX512, and for reviewing the manuscript. Also thanks to John McGraw of Production Arts.
- Laurel Vieaux at QSC, for information on QSControl.
- Ralph O. Weber of Digital Equipment Corporation, for information on Two-Phase Commit MIDI Show Control.
- Kevin Kolczynski at Universal Studios, Florida, for supplying the cover photograph.

Appendix: Decimal/Hex/Binary/ASCII Table

The following table shows numbers in several formats and the corresponding ASCII characters (up to 7 bits only):

Decimal	Hex	Binary	ASCII Character
000	00	00000000	NUL – Null Character
001	01	00000001	SOH – Start of Header
002	02	00000010	STX – Start of Text
003	03	00000011	ETX – End of Text
004	04	00000100	EOT – End of Transmission
005	05	00000101	ENQ – Enquiry
006	06	00000110	ACK – Acknowledge
007	07	00000111	BEL – Bell
008	08	00001000	BS – Backspace
009	09	00001001	HT – Horizontal Tab
010	0A	00001010	LF – Line Feed
011	0B	00001011	VT – Vertical Tab
012	0C	00001100	FF – Form Feed
013	0D	00001101	CR – Carriage Return
014	0E	00001110	SO – Shift Out
015	0F	00001111	SI – Shift In
016	10	00010000	DLE – Data Link Escape
017	11	00010001	DC1 – Device Control 1
018	12	00010010	DC2 – Device Control 2
019	13	00010011	DC3 – Device Control 3
020	14	00010100	DC4 – Device Control 4
021	15	00010101	NAK – Negative Acknowledgement
022	16	00010110	SYN – Synchronous Idle
023	17	00010111	ETB – End of Transmission Block
024	18	00011000	CAN – Cancel
025	19	00011001	EM – End of Medium

Decimal	Hex	Binary	ASCII Character
026	1A	00011010	SUB – Substitute
027	1B	00011011	ESC – Escape
028	1C	00011100	FS – File Separator
029	1D	00011101	GS – Group Separator
030	1E	00011110	RS – Record Separator
031	1F	00011111	US – Unit Separator
032	20	00100000	Space
033	21	00100001	!
034	22	00100010	"
035	23	00100011	#
036	24	00100100	$
037	25	00100101	%
038	26	00100110	&
039	27	00100111	'
040	28	00101000	(
041	29	00101001	)
042	2A	00101010	*
043	2B	00101011	+
044	2C	00101100	,
045	2D	00101101	-
046	2E	00101110	.
047	2F	00101111	/
048	30	00110000	0
049	31	00110001	1
050	32	00110010	2
051	33	00110011	3
052	34	00110100	4
053	35	00110101	5
054	36	00110110	6
055	37	00110111	7
056	38	00111000	8
057	39	00111001	9
058	3A	00111010	:

Decimal	Hex	Binary	ASCII Character
059	3B	00111011	;
060	3C	00111100	<
061	3D	00111101	=
062	3E	00111110	>
063	3F	00111111	?
064	40	01000000	@
065	41	01000001	A
066	42	01000010	B
067	43	01000011	C
068	44	01000100	D
069	45	01000101	E
070	46	01000110	F
071	47	01000111	G
072	48	01001000	H
073	49	01001001	I
074	4A	01001010	J
075	4B	01001011	K
076	4C	01001100	L
077	4D	01001101	M
078	4E	01001110	N
079	4F	01001111	O
080	50	01010000	P
081	51	01010001	Q
082	52	01010010	R
083	53	01010011	S
084	54	01010100	T
085	55	01010101	U
086	56	01010110	V
087	57	01010111	W
088	58	01011000	X
089	59	01011001	Y
090	5A	01011010	Z
091	5B	01011011	[

Decimal	Hex	Binary	ASCII Character
092	5C	01011100	\
093	5D	01011101	]
094	5E	01011110	^
095	5F	01011111	_
096	60	01100000	`
097	61	01100001	a
098	62	01100010	b
099	63	01100011	c
100	64	01100100	d
101	65	01100101	e
102	66	01100110	f
103	67	01100111	g
104	68	01101000	h
105	69	01101001	i
106	6A	01101010	j
107	6B	01101011	k
108	6C	01101100	l
109	6D	01101101	m
110	6E	01101110	n
111	6F	01101111	o
112	70	01110000	p
113	71	01110001	q
114	72	01110010	r
115	73	01110011	s
116	74	01110100	t
117	75	01110101	u
118	76	01110110	v
119	77	01110111	w
120	78	01111000	x
121	79	01111001	y
122	7A	01111010	z
123	7B	01111011	{
124	7C	01111100		

Decimal	Hex	Binary	ASCII Character
125	7D	01111101	}
126	7E	01111110	~
127	7F	01111111	DEL
128	80	10000000	
129	81	10000001	
130	82	10000010	
131	83	10000011	
132	84	10000100	
133	85	10000101	
134	86	10000110	
135	87	10000111	
136	88	10001000	
137	89	10001001	
138	8A	10001010	
139	8B	10001011	
140	8C	10001100	
141	8D	10001101	
142	8E	10001110	
143	8F	10001111	
144	90	10010000	
145	91	10010001	
146	92	10010010	
147	93	10010011	
148	94	10010100	
149	95	10010101	
150	96	10010110	
151	97	10010111	
152	98	10011000	
153	99	10011001	
154	9A	10011010	
155	9B	10011011	
156	9C	10011100	
157	9D	10011101	

Decimal	Hex	Binary	ASCII Character
158	9E	10011110	
159	9F	10011111	
160	A0	10100000	
161	A1	10100001	
162	A2	10100010	
163	A3	10100011	
164	A4	10100100	
165	A5	10100101	
166	A6	10100110	
167	A7	10100111	
168	A8	10101000	
169	A9	10101001	
170	AA	10101010	
171	AB	10101011	
172	AC	10101100	
173	AD	10101101	
174	AE	10101110	
175	AF	10101111	
176	B0	10110000	
177	B1	10110001	
178	B2	10110010	
179	B3	10110011	
180	B4	10110100	
181	B5	10110101	
182	B6	10110110	
183	B7	10110111	
184	B8	10111000	
185	B9	10111001	
186	BA	10111010	
187	BB	10111011	
188	BC	10111100	
189	BD	10111101	
190	BE	10111110	

Decimal	Hex	Binary	ASCII Character
191	BF	10111111	
192	C0	11000000	
193	C1	11000001	
194	C2	11000010	
195	C3	11000011	
196	C4	11000100	
197	C5	11000101	
198	C6	11000110	
199	C7	11000111	
200	C8	11001000	
201	C9	11001001	
202	CA	11001010	
203	CB	11001011	
204	CC	11001100	
205	CD	11001101	
206	CE	11001110	
207	CF	11001111	
208	D0	11010000	
209	D1	11010001	
210	D2	11010010	
211	D3	11010011	
212	D4	11010100	
213	D5	11010101	
214	D6	11010110	
215	D7	11010111	
216	D8	11011000	
217	D9	11011001	
218	DA	11011010	
219	DB	11011011	
220	DC	11011100	
221	DD	11011101	
222	DE	11011110	
223	DF	11011111	

Decimal	Hex	Binary	ASCII Character
224	E0	11100000	
225	E1	11100001	
226	E2	11100010	
227	E3	11100011	
228	E4	11100100	
229	E5	11100101	
230	E6	11100110	
231	E7	11100111	
232	E8	11101000	
233	E9	11101001	
234	EA	11101010	
235	EB	11101011	
236	EC	11101100	
237	ED	11101101	
238	EE	11101110	
239	EF	11101111	
240	F0	11110000	
241	F1	11110001	
242	F2	11110010	
243	F3	11110011	
244	F4	11110100	
245	F5	11110101	
246	F6	11110110	
247	F7	11110111	
248	F8	11111000	
249	F9	11111001	
250	FA	11111010	
251	FB	11111011	
252	FC	11111100	
253	FD	11111101	
254	FE	11111110	
255	FF	11111111	

Glossary

1000-BASE-T: 1000 mbit/s (1 gbit/s) Ethernet over copper. Also called "Gigabit" Ethernet.

100-BASE-T: 100 mbit/s Ethernet over copper. Also called, "Fast" Ethernet.

10-BASE-T: 10 mbit/s Ethernet over copper.

802.11: The widely used, wireless Ethernet standard.

802.3: The IEEE committee that standardizes Ethernet.

8P8C Connector: Also called (somewhat incorrectly) an "RJ45" connector, has eight positions and accommodates eight conductors. The connector is very widely used in Ethernet applications with Category 5 cable.

Absolute: Independent data or measurement, which can be reconstructed without referencing preceding or following data points.

Address Resolution Protocol (ARP): A protocol which resolves a physical address, such as an Ethernet MAC address, to an IP address.

American Standard Code for Information Interchange (ASCII): A numeric coding system for alpha-numeric (and other) characters.

Analog: A data representation where the value is measured or expressed as part of continuous scale.

Animatronics: Otherwise known as "Character Animation"; describes a show with a robotic character synchronized with an audio track.

Art-Net: A proprietary scheme for sending lighting control data over Ethernet that has been put into the public domain.

Audio Video Bridging (AVB): An IEEE standard that uses Ethernet and special switches for the transmission of audio and video.

Authority Having Jurisdiction (AHJ): An authority, such as a local or state government, which has jurisdiction over something. For example, here in NYC, the local city government is the AHJ for electrical code issues and, while we use the National Electric Code, the AHJ can impose other laws, rules, and regulations.

Balanced Transmission: See "Differential Transmission".

Bandwidth: A measurement of how much information can be transmitted over a communications link, typically in bits per second.

Base 10: a numbering system with digits 0-9.

base 16: A numbering system with 16 digits, 0-F.

Base 2: See binary.

Baud Rate: The number of symbols or signalling units sent over a data link in one second.

Binary: A numbering system with only two options for digits: 0 and 1.

Binary-Coded Decimal: A numbering scheme where each decimal (Base 10) digit is encoded with a binary word.

Bits Per Second (BPS or bit/s) The number of bits transmitted down a communications link in one second.

Bitwise: Operates on binary information directly at the level of the the individual bits. Can operate at a very low level in a computer system to work very fast.

Bluetooth: A wireless transmission system used mostly for peripheral devices.

Bonjour: Apple's cross-platform service discovery system, which allows services to be discovered on a network without knowing the IP address in advance.

Bridge: A technique to connect two network segments without a router.

Broadcast Domain: A piece of a network (or a whole network), with all nodes in the segment being reachable by broadcast messages.

Broadcast IP Address: Typically 255.255.255.255; packets sent to this IP address will be forwarded to every station in the broadcast domain.

Broadcast Storm: When a redundant pathway exists in a switched network, broadcast traffic can end up looping around the system and repeating, causing a broadcast storm. The Spanning Tree Protocol (STP) was designed to prevent this problem.

Broadcast: A packet forwarded to every other node on the network.

Byte: A group of eight bits. More accurately called an "Octet."

Cable: An assembly made up of two or more wires.

Cat 5e Cable: A widely used networking cable with four twisted pairs and no sheild.

Checksum: A simple form of error detection that adds a block of transmitted data values together, and then transmits that check, along with the data, for checking by the receiver.

Click track: Typically an audio track used to tell musicians when to start a song and what tempo to keep. The track is traditionally a series of metronome clicks starting a measure or two before the band should start; hence, the name click track. Click tracks can also be vocal count-offs—"1 2 3 4 1 2 3 4 Go"—or drum patterns.

Client: A system or program that accesses a services offered by a "server".

Coaxial Cable: A cable type with a center conductor and an overall shield, separated by an insulating "dielectric."

Colon Hexadecimal: A way of representing IPv6 addresses using hex and colons.

Connectionless: A protocol or communication technique that simply addresses and sends out its data to the remote system without ensuring first that a connection exists.

Connection-Oriented: A protocol or communication technique that establishes an end-to-end connection to the remote system before any data is sent.

Console: A controller for lighting, sound, etc. Typically meant for human interfacing with the machine.

Contact Closure: The actual or simulated closing of contacts (typically, two pieces of metal physically touching, or a transistor) to complete or break a circuit.

Contactor: Typically, a large, high power relay.

Cross-Fade: A transition between two presets or "looks," typically done in a time-based way.

Crossover Cable: A cable, like a null modem cable, that crosses over the transmit and receive lines in an Ethernet cable to properly connect transmit and receive lines.

Cue: The basic building block of entertainment control; represents a state or a change in state.

Current-loop Interface: A signalling connection where the receiver determines the "1" or "0" based on the level of the current flowing through the receiver.

Cyclic redundancy check (CRC): A sophisticated and highly reliable error detection scheme, which takes a block of data and divides it by a special divisor. The result of that division is transmitted along with the data for checking by the receiving system.

Data Circuit-terminating Equipment (DCE): In a serial interface DCE is, historically, the modem or other communications equipment. See also DTE.

Data Rate: The rate of data transmission down a communications link, typically measured in Bits Per Second (BPS).

Data Terminal Equipment (DTE): In a serial interface, DTE is the computer or terminal. See also DCE.

Datagram: The basic data unit of a packet-switched network; it contains a header and then the data payload.

De facto Standard: An informal standard based on generally accepted practice.

Default Gateway: The IP address to which a host, network, or VLAN will forward packets to which it can't otherwise find a route.

Deterministic Network: A network that can deliver its data in a predictable amount of time.

Differential Transmission: In electrical data communications links, a connection where a the signal is sent in normal and opposite polarity. The receiver can use the difference signals to cancel out noise in these systems.

Digital Signal Processing (DSP): Computer-based processing of digital signals, typically through dedicated DSP hardware.

Digital: A data representation where the value is measured or expressed as a series of discrete numbers or samples.

Dimmer-Per-Circuit: A lighting control system where each circuit in the system is directly connected to its own dimmer. Historically, circuits had to be hard patched to larger dimmers.

Dimmers: Electrical devices that vary aspects of the the electrical feed to a luminaire in order to dim it.

Direct Current (DC): Electricity that flows only in one direction (from negative to positive).

DMX Address: In DMX, the starting slot to which the addressed device should listen.

DMX512-A: The widely used lighting control protocol based on RS-485.

Domain Name System (DNS): A system that translates URLs like controlgeek.net to IP addresses.

Dotted Decimal: The numbering format, of three decimal numbers separated by periods, used in IPv4 addressing.

Double-Precision Variable: A variable stored (typically) with 64 bits.

Dynamic Host Configuration Protocol (DHCP): A system that automatically assigns an IP address, Default Gateway, and other settings to a host.

Electrical Isolation: Isolating one power source or voltage from another. This is typically done in control systems to protect low current, low voltage devices, like computers.

Electro-Magnetic Interference (EMI): Interference induced onto an electrical line by a magnetic field. Typically, this happens in the form of unwanted noise in a communications link.

Emergency Stop or E-Stop: A special control circuit that operates by disconnecting power to the lowest-level components of the system, allowing them to be shut off no matter what.

Enabling System: A control system that integrates human and computer monitoring to safely fire effects. The computer can enable the system for the human, or vice versa.

Encapsulation: A process where a data packet has additional information appended, so that it can be passed onto another layer or process.

Encoder: A device that takes a physical position, and communicates that position or data in a way that can be read by machines.

Entertainment Control: Any of the control systems that are used in a single discipline (lighting, sound, etc).

Error detection: A scheme to detect errors when they occur during transmission down a communications link.

Ethernet: A widely used networking technology, standardized under IEEE 802.

Fail-Safe Design: A system design that will, upon failure of one or more components, fail into a "safe" state.

Failure Mode and Effects Analysis (FMEA): A method of identifying and analyzing potential failure modes.

Fiber-Optic Cable: A cable, typically made of either glass or plastic, which is used to carry light to communicate data.

File Transfer Protocol (FTP): A protocol used to transfer files over networks. It ensures that the files can be reassembled correctly.

FireWire: A high performance, short haul serial transmission system for isochronous transport of bits.

Flooding: When frames or packets are sent to every interface on a device except the incoming interface.

Flow Control: A technique used to regulate the flow of data over a communications link.

Forwarding: When a packet or frame is sent onto another interface or system.

Frame Check Sequence (FCS): The CRC error detection result transmitted in the Ethernet frame.

Frame: A data packet (from Layer 3) to which frame synchronization information has been added for transmission on Layer 2.

Full-Duplex switch: A switch that can operate in full-duplex mode, where each connection can send and receive frames simultaneously.

Full-Duplex: A communications link where any party can talk at any time.

Half-Duplex: A communications mode where either party in a conversation can talk, but not at the same time.

Hexadecimal: See Base 16.

Highest Takes Precedence (HTP): A control approach (typically in lighting) where the highest of all the control inputs is accepted and acted upon.

Host ID: The parts (bits) of an IP address which identify a particular host.

Host: A computer connected to the network and assigned an IP address. A device like a hub, which doesn't have any innate intelligence, is a node on the network but is not a host, even though it's connected to the network.

Hub: A mostly obsolete device for connecting multiple Ethernet devices together. The hub is simply a repeating device, passing on every frame to every connected device.

Human Interface Guidelines (HIG): Guidelines for developing user screens and other interfaces with humans.

Human Machine Interface (HMI): Anything that allows a human to interface with a machine. In industrial control, this typically means a touch screen.

Hyper Text Transfer Protocol (HTTP): A widely used protocol that can transfer "hypermedia," including text and graphics. It is the protocol used on most Web pages.

IGMP Snooping: A feature of some switches that allows the switch to snoop into Layer 3 IGMP traffic to determine whether or not a particular interface on the switch should receive multicast traffic.

Insulated Gate Bipolar Transistor (IGBT): A solid-state device that can allow a small current to switch a large load.

Internet Assigned Numbers Authority (IANA): The organization which assigns, globally, IP addresses, port numbers, etc.

Internet Group Management Protocol (IGMP): A protocol used to establish and manage multicast group memberships.

Internet Protocol (IP): A protocol that offers a standardizing addressing scheme to allow computers to be connected on a network.

Internet: A network of networks.

Interrupt Driven: A system that can be alerted when a sub system needs attention.

Intranet: An internal or private network within an organization.

Latest Takes Precedence (LTP): A control approach (typically in lighting) where the last (not necessarily the highest) changed of all the control inputs is accepted and acted upon.

Light Amplification by Stimulated Emission of Radiation (LASER): A device which creates a coherent beam of light at a single (or very small range of) wavelengths.

Linear Media: Media, such as video or audio, which only makes sense when it's playing, frame after frame, or sample after sample.

Linear Show: A show with a single, fixed, storyline, which is normally performed in the same sequence each performance.

Line-Rate: A switch that has enough internal speed to accomodate every interface running at full capacity simultaneously.

Link-Local Address: An IP address that is only used inside a subnet for local purposes. It can be automatically configured by hosts without a DHCP server.

Local Area Network (LAN): A network connecting hosts all in a relatively small geographic area and typically owned and operated by a single organization.

Localhost Address: See "loopback address."

Logical Link Control (LLC) Layer: The uppermost layer within Ethernet.

Loopback Address: A special address (127.0.0.1 in IPv4, and ::1 in IPv6) that loops back to the local host. This can be used for testing and other purposes.

MAC Address: The physical address used in Ethernet. Typically, this is a globally unique ID number burned in at the factory and is not changeable by the user.

MAC Layer: The layer in Ethernet which handles access control.

Managed Switch: A switch with a management interface, allowing configuration of things like VLANs.

Medium Dependent Interface X Crossover (MDIX): A technology that auto-configures Ethernet ports to correctly connect transmit and receive lines. Most modern Ethernet connections use Auto MDIX so that crossover cables are not necessary.

MIDI Machine Control (MMC): A protocol, based on MIDI, designed to control media players in a play/stop/locate way.

MIDI Show Control (MSC): A protocol based on MIDI, which controls show equipment using commands such as "Go."

MIDI Time Code (MTC): A way to send SMPTE Time Code time "addresses" over a MIDI link.

Modem: A MODulator-DEModulator. Converts a digital communications signal into something else, like an audio signal, and back.

Motion Control: The industrial term for the technologies that underlie scenic and other automation systems.

Motion Profile: A plot of a device's velocity against time. Typically, a motion profile will have three phases: acceleration, peak velocity, and deceleration.

Multicast: A packet able to be received by a group of recipients simultaneously.

Multidrop Connection: A data connection from one machine to one or more others. DMX uses a multidrop connection.

Multilayer Switch: A switch that offers some functionality at Layer 3 or higher.

Multiplex: Sending multiple channels of data down a single communications link.

Musical Instrument Digital Interface (MIDI): A current loop serial interface used to send musical events like "note on" between musical devices.

Neighbor Discovery Protocol (NDP): A protocol in IPv6 that replaces ARP and allows a device to find out information about the network.

Network Address Translation (NAT): A technique used by routers to translate internal, private, non-routeable IP addresses out to the public internet and back. NAT is widely used now because of IPv4 address space depletion.

Network ID: The parts (bits) of an IP address which identify a particular subnet.

Network Interface Card (NIC): The connection to an Ethernet network. This used to be a card physically installed in a machine, hence the name.

Network: A collection of "nodes" which can communicate with each other.

Nibble: a piece of a byte.

Node: A connection point on a network.

Nonlinear Show: Multiple, seperate components or segments of a show can run independently, in a dynamic order, or even simultaneously.

Nonrouteable IP Addresses: A range of IP addresses allocated for "private networks". Packets with these addresses are not passed onto the Internet by routers, so they are often called nonrouteable or private IP addresses.

Normally Closed: A contact that is closed, or completing the circuit, in its "normal" state. Typically used in something like an Emergency Stop circuit, where the machine is "normally" operating until an emergency, when the circuit needs to be broken to stop the machine.

Normally Open: A contact that is open, or breaking the circuit, in its "normal" state.

Null Modem Cable: A cable that flips the transmit and receive lines to properly connect transmitter and receiver. Also see Crossover Cable.

Octet: A group of eight bits.

Open Control Alliance (OCA): An interoperable, network based audio and video control system under development in 2012.

Open Sound Control (OSC): An open, message-based protocol for communications between syntheziers, computers, interface and other kinds of devices. Typically it runs on Ethernet or USB over IP.

Open Standards: Standards that are free to use and open to anyone.

Optical Isolator: A solid-state device that isolates voltages or circuits from each other, typically by having an LED turn on, which in turn activates a light-activated switch.

Packet: A "unit" of data used in a packet-switched (Layer 3) network. At layer 2, a packet is packaged into a Frame.

Parallel Communications Interface: An interface where multiple data lines form the link, allowing multiple bits to be sent simultaneously, in parallel.

Parity Checking: A simple form of error detection that appends a bit to a digital word during transmission.

Peer-to-Peer: A system where a number of equals can communicate and, possibly, control each other.

Physical layer (PHY): The physical hardware layer of OSI, including cable, connector, standardized voltages, etc.

Ping Command: Sends out a series of messages to test a connection to a host.

Point to Point Connection: A data connection from one machine to one other.

Port Mirroring: A feature of managed switches to copy all traffic from one interface to another, where it can be monitored.

Port: In IP networking, a number assigned to a process inside a computer; allows more than one process to share a common IP address.

Power over Ethernet (PoE): A technique to supply a small amount of power (typically around 15 watts maximum) to a connected Ethernet device using the Cat 5 cable.

Private IP Addresses: See nonrouteable IP addresses.

Professional Lighting and Sound Association (PLASA): The primary trade association for North America and Europe.

Proprietary Standard: A standard owned or controlled by one organization.

Quality of Service (QoS): A way of prioritizing traffic on a network. Typically used with things like Voice Over IP.

Real Time Protocol (RTP) Payload Format for MIDI; a standardized way of carrying MIDI over a network.

Relative: Dependent data or measurement, which can only be reconstructed by referencing preceding or following data points.

Relay: An electro-mechanical device that allows a small current or voltage to control a large (or isolated) current or voltage.

Reliable: When referring to networks, reliability describes the criteria of the network for ensuring that a packet will be delivered.

RJ45 Connector: More correctly referred to as an 8P8C connector, with eight positions, since the actual RJ45 specification refers to a telephone company standard. The connector is widely used in network wiring.

Router: A device that can connect networks together.

Routing Matrix: Also called a "crosspoint matrix", a device that takes a number of inputs and selectively connects them to one or more outputs.

Routing Table: A table kept by a router indicating where to find a path to a specific network.

RS-232: An unbalanced serial interface standard used for short-haul communications.

RS-422: A balanced serial interface standard used mostly in industrial data communications.

RS-485: A balanced serial interface standard used mostly in industrial data communications, and in DMX.

Sensor: A device that measures or models a physical property or some state in the world, and sends that information on to another system. For example, a temperature sensor measures the temperature and passes that information on in some form.

Serial Communications Interface: An interface where data bits are sent serially, one after another.

Server: A computer or program which serves the requests of other programs, called "clients".

Service Set IDentifier (SSID): The name broadcast by a Wireless Access Point (WAP).

Shielded, Twisted Pair (STP): A cable type with one or more pairs of wire, and an overall shield.

Show Control: Connecting more than one entertainment control system together.

Silicon-Controlled Rectifier (SCR): A solid-state device which can allow a small current to switch a large load.

Simple Network Management Protocol (SNMP): An IP-based protocol for managing devices like switches and routers over a network.

Simplex: A communications mode where the talker can only talk and the listener can only listen.

Single Failure Proof Design: A design that can safely accommodate the failure of a single component.

Single-Ended Transmission: In electrical data communications links, a connection where a simple signal and a ground reference are sent. The receiving system has no way to differentiate between the signal and noise in these systems, so these interfaces are not good for long-distance transmission.

Single-Precision Variable: A variable stored (typically) with 32 bits.

Sinking: A transistor circuit arrangement where the transistor is between the load and the ground.

Snubber: Typically a diode/capacitor combination, or a "transorb," which can absorb and/or dissipate voltage transients created when the electro-magnetic field from a DC coil collapses (as in a DC relay being turned off).

Society of Motion Picture and Television Engineers (SMPTE) Time Code (TC): A way to transmit or record digital time information on an audio format.

Socket: A combination of an IP address and a port.

Sourcing: A transistor connection arrangement where the load is between the transistor and the ground.

Spanning Tree Protocol (STP): A protocol used to manage redundant pathways and ensure that loops do not occur in switched networks. Loops could cause Broadcast Storms.

Standard: A useful working practice, protocol, measurement, or connection method agreed upon by a group.

Streaming ACN: ANSI E1.31 – 2009, a lightweight, simplified version of ACN that allows simple DMX512 universes and slots to be transported over a network in a way compliant with and understood by "full" ACN.

Subnet Mask: A way of indicating which part of an IPv4 address is the network, and which is the host.

Subnet: A piece of a larger network.

Switch: A device for connecting multiple Ethernet devices together, and capable of forwarding only the appropriate frames to each connected device.

System: A group of elements connected together into a complex whole.

T568A: A standard for which wires to put to which pins on an 8P8C connector. T568B is more widely used.

T568B: A widely used standard for which wires to put to which pins on an 8P8C connector.

Terminal Emulator: A piece of software that emulates old hardware "dumb terminals" on modern computers.

Touch Screen: A computer screen that can be operated by touching the screen. Often called a "Human-Machine Interface" in the industrial control world.

Transmission Control Protocol (TCP): A protocol that typically is used in conjunction with the Internet Protocol (IP) to offer "reliable" delivery.

Unbalanced Transmission: See "Single-Ended Transmission".

Unicast: A process where a packet is forwarded to only one other node on the network.

Unicode: A character encoding scheme similar to ASCII, but accommodating of international characters.

Unique Local Addresses (ULA): A block of IPv6 addresses that are not routeable and can be used on local networks.

Universal Serial Bus (USB): A high speed, short haul interface used primarily for connecting peripherals to computers.

Universe: In DMX, a way to indicate when more than 512 slots are needed, and physically separate runs, or "universes" of DMX are needed.

Unmanaged Switches: A switch configured permanently at the factory and without the ability to configure things like VLANs.

Unreliable: An unreliable network will send a packet, but does not guarantee that it will be delivered. Unreliable networks are often used where speed of delivery is more important than data integrity.

Unshielded, Twisted Pair (UTP): A cable type with one or more pairs of wire, and no overall shield. Cat 5e cable is an example of UTP.

User Datagram Protocol (UDP): a protocol that typically is used in conjunction with the Internet Protocol (IP) to offer simple, "unreliable" delivery of packets.

UTF-8: A variable width Universal Character Set encoding standard that can encode the entire Unicode set, and is also backwards compatible with ASCII.

Variable: A memory storage location, which can be accessed and/or changed.

Video Server: A computer system that plays back video files on command.

Video Switchers: Mixes multiple video sources into one (or a few) outputs.

Virtual Local Area Network (VLAN): A group of hosts that have a common broadcast domain, but can share a physical infrastructure with other VLANs.

Virtual Network Control (VNC): Is an open source, free desktop sharing system that allows a computer to be controlled from a remote location just as if the user were sitting in front of the machine.

Virtual Private Network (VPN): A security technique that, basically, extends a private network out through the public Internet.

Voltage-Loop Interface: A signalling connection where the receiver determines the "1" or "0" based on the level of the voltage received.

Wide Area Network (WAN): A network connecting large distances (like a city or country), typically through the use of a common carrier, like a phone company.

Wi-Fi: see 802.11.

Wire: A single electrical conductor.

Wireless Access Point (WAP): A wireless bridge to a hardwired Ethernet system that allows portable computers to connect, using radio waves, to the network.

Wireshark: A free, open source network analysis and troubleshooting tool that is widely used.

Index

Numerics

0A, see Line Feed
0D, see Carriage Return
1000BASE-LX, 168
1000BASE-SX, 168
1000BASE-T, 168
100BASE-FX, 168
100BASE-T, 168
100BASE-TX, 168
10BASE2, 167
10BASE5, 167
10BASE-FL, 167
10BASE-T, 167
232, see RS-232
4-20mA analog current loops, 133
422, see RS-422
485, see RS-485
48VDC phantom power, 174
4B5B Encoding, 168
5-4-3 rule, 216
568, see ANSI/TIA/EIA-568-B
802, see IEEE 802 or Ethernet
8P8C, see RJ45

A

A/V Stumpfl, 53, 54
absolute encoder, 91
absolute, defined, 74, 75
AC servo drives, 46
AC zero-crossing, 17
Academy of Production Technology, i
acceleration, 44, 52
Access Point (AP), see Wireless Access Point
access switches, 216
ACN, see Architecture for Control Networks
active sensing, 287
Address Resolution Protocol, 217
Address Resolution Protocol (ARP), 162, 200, 202, 237
address, DMX, 242
Adobe® Framemaker®, xvi
AES, see Audio Engineering Society
AES-24, 7, 335, 341, 343, 426
AES-24-1-1999, 342
AES-50, 33
Allen-Bradley, 345
Alternate Start Code (ASC), 251, 259
Amazon.com, xv
Amazon's Createspace™, xvi
Ambrose, Myles, 423
American National Standards Institute (ANSI), 8, 322
American Standard Code for Information Interchange (ASCII), 31, 38, 126, 132, 145, 152, 177, 207, 305, 346, 347, 349, 382, 389, 399
amplifiers, 30
AMX, 242
analog, 3
analog encoding, 42
and (logical operator), 87
Anderson, Alan, 383, 394, 402, 413, 424
Anderson, Cara, 424
animatronics, 51
ANSI E1.11-2008 Entertainment Technology USITT DMX512-A, Asynchronous Serial Digital Data Transmission Standard for Controlling Lighting Equipment and Accessories, 20, 241

ANSI E1.17 - 2010, Entertainment Technology — Architecture for Control Networks (ACN), 21, 269
ANSI E1.20-2010, Entertainment Technology - RDM, Remote Device Management Over DMX512 Networks, 259
ANSI E1.27-1-2006 Entertainment Technology Standard for Portable Control Cables for Use with ANSI E1.11 (DMX512-A) and USITT DMX512/1900 Products, 245
ANSI E1.27-2 – 2009, Entertainment Technology -- Recommended Practice for Permanently Installed Control Cables for Use with ANSI E1.11 (DMX512-A) and USITT DMX512/1990, 245
ANSI E1.3 - 2001 (R2011), Entertainment Technology 0 to 10V Analog Control, 20
ANSI E1.31 – 2009, Entertainment Technology – Lightweight streaming protocol for transport of DMX512 using ACN, 255, 280
ANSI E1.37-1 - 201x, Additional Message Sets for ANSI E1.20 (RDM) – Part 1, Dimmer Message Sets, 266
ANSI, see American National Standards Institute
ANSI/TIA/EIA-568-B, 136
Antheil, George, 143
Antonuk, Dan, 270, 423, 424
APIPA, see Automatic Private IP Addressing
Apple Computer, 156
Application layer, 125
Arabic, 99
Architecture for Control Networks (ACN), 21, 258, 269, 399, 409
ARP cache, 201, 202
ARP request, 201
ARP, see Address Resolution Protocol
Artistic Licence, 252, 424
Art-Net II (Art-Net), 252

ASC, see Alternate Start Code
ASCII, see American Standard Code for Information Interchange
Ashworth, Chris, 423
AS-Interface (AS-i), 49, 345
Associates & Ferren, 37, 425
asynchronous, 149, 361
asynchronous communication, 147
Audio Engineering Society (AES), 6, 7, 328, 341, 342
Audio Video Bridging (AVB), 32
Audio Visual Labs, 426
authority having jurisdiction, 111
authorizing effects, 113, 399
automata, 357
automated rigging systems, 41
Automatic Private IP Addressing (APIPA), 189
Avab, 242, 426
AVB, see Audio Video Bridging
Avenger, 425
axis, 41

B

backbones, 139
backoff delay, 165
BACKUP YOUR DATA, 117
Baier, Michael, 53
Bailey, Caroline, 425
balanced signal transmission, see differential signal transmission
Balanced Transmission, 134
Balanced, see "Differential"
Ballard, Norman, 423
Ballet Mécanique, 143
bandwidth, 127
Barnett, Dave, 425
base 10, 99
base 16, 101
base two, 100
Basic Service Set (BSS), 176

Baud rate, 127
Baxter, Rick, 424
BCD, see Binary Coded Decimal
Becker, Cy, 426
Bennett, Keith, 360
Benoit Bouchez, 423
Bergeron, François, 424
Berryman, Jeff, 423
Bertenshaw, David, 426
best effort transmission, 182
binary search tree, 263
binary, defined, 100
binary-coded decimal (BCD), 103
bi-phase modulation, 327
Birket Engineering, Inc., i, 62, 424
Birket, Dan, 423
Birket, Glenn, 423, 424
bit/s, see bits per second
bits per second, 127, 153
bitwise operations, 104, 194
Blair, Scott, 259, 424
Bluetooth, 156
Blu-Ray, 36
BNC connector, 33, 167
Boevers, David, 423
boilers, 58
Bollinger, Aaron, xix, 424
Bonjour, 297
Bouchez, Benoit, 424
bps, see bits per second
Bracewell, Dr. John, 426
Brainstorm Electronics, 329
Bridging, 173
broadcast address, 185
broadcast domain, 215
broadcast IP address, 186
broadcast storm, 218
broadcast transmission, 83, 150, 232
Brooks, Evan, 322
Brown, Tony, 426
BSR E1.33, Extensions to E1.31 for Transport of ANSI E1.20 (RDM), 257

BSR E1.37-2 Additional Message Sets for ANSI E1.20 (RDM) – Part 2, IPv4 & DNS Configuration Messages, 268
BSR E1.45 - 201x, Transport of IEEE 802 data frames over ANSI E1.11 (DMX512-A), 257
BSS, see Basic Service Set
burner-management systems, 58
Burns, Montgomery, 407
bus topology, 84
bytes per second, 127
bytes, defined, 100
Bytes/second, see Bytes Per Second

C

cable, defined, 134
Cadena, Richard, 423
CAN, see Controller Area Network
canned systems, 359
Carlito's Way, 293
Carlson, Steve, 270, 424
Carnegie Mellon University School of Drama, i
Carriage Return (CR), 153, 382
carrier, 140
Carrier Sense, Multiple Access, Collision Detection (CSMA/CD), 163, 165, 179
Cat 5 cable, 136
Cat 5, see Category 5e
categories, 136
Category 1, 136
Category 2, 136
Category 3, 137
Category 4, 137
Category 5, 137
Category 5e, 137
Category 6, 137
cathode-ray tube (CRT), 328
CD 80, 242
CD, see Collision Detection

centralized system, 80
character animation, 51
checksum, 131
choke, 17, 18
Chunky Move, 359
CID, see Component IDentifiers
CIDR, see Classless Inter-Domain Routing
CIP, see Common Industrial Protocol
circuit, 135
Cirque du Soleil, ii, 358
Cisco, 230
Cisco Small Business, 227
City Tech, xvi
City Theatrical, ii, 264
Citytech, 226
clapper, 322
Classless Inter-Domain Routing (CIDR), 184
Clear To Send (CTS), 149
Cleaves, Gardiner, 425
CLI, see Command Line Interface
click track, 400
client, 82
client-server, 82, 206
closed-loop, 78, 79
coax, 136
coaxial cable, 136
Cobra Firing Systems, 62
Cobranet, 33
coil, 347
collector pin, 96
Collier, Andy, 426
Collision Detection (CD), 163
collision domain, 164, 165
colon-hexadecimal, 234
color scrollers, 19
Colortran, 242
Colortran CMX, 242
com ports, 153
command line interface (CLI), 172
command-response protocol, 81
common carrier, 161
Common Industrial Protocol (CIP), 346
Component IDentifiers (CID), 272

components, 270
concussion mortars, 61
connectionless, 183
connection-oriented, 182
console, 13
contact arrangements, 95
contact closure, 31, 38, 54, 58, 63, 93, 94, 95, 96, 293, 361, 376, 391, 393
contactor, 95
contention, 133
Control Geek, see www.controlgeek.net
controlled stop mode, 113
Controller Area Network (CAN), 345
convergence, 19
converting number bases, 105
Cooley, Margaret, 426
core switches, 216
corn chips, xv
CPWG, see PLASA Control Protocol Working Group
CR, see Carriage Return
CRC, see Cyclic Redundancy Check
Createspace, iv
Crest, 426
Crestron, 424
crimping, 138
critical thinking, 86
cross-fade, 13
crossover, 151
crossover cable, 169
crosspoint matrix, 36
CRT, see Cathode Ray Tube
CSMA, see Carrier Sense, Multiple Access
CTS, see Clear To Send
cue light, 68, 129
cue, defined, 68
current-loop interfaces, 133
Cyclic Redundancy Check (CRC), 132, 166, 328, 345

456 • Index

D

DA, see Directory Agent or Distribution Amplifier
daisy chain, 84
Dante, 33
Dark Operate, 92
Data Circuit-terminating Equipment (DCE), 148, 149
data collisions, 163, 260
data rate, 127
Data Terminal Equipment (DTE), 148, 149
data word, 153
datagrams, 183
Dataton AB, 36
David Boevers, i
Dawkins, Richard, 147
day mode, 72
DC motor, 46
DC servo amplifier, 46
DC voltage, 20, 43, 97
DCC, see Dry Contact Closure
DCE Ready, 149
DCE, see Data Circuit-terminating Equipment
DCID, see Device Class IDentifier
DDL, see Device Description Language
deadman's switch, 114, 382
Deans, Jonathan, v, 360, 423, 426
deceleration, 44, 52
Default Gateway, 186
default gateway, 192, 219
delimiters, 305
demodulates, 140
de-multiplexer, 128
Desire Under the Oaks, 369
Detecting Network Attachment in IPv4 (DHC-DNA), 256
deterministic, 4
Device Class IDentifier (DCID), 275
Device Description Language (DDL), 272, 277
device ID, 302
Device Management Protocol (DMP), 272, 273
DeviceNet, 49, 345
DeWille, Tom, 425, 426
DHC-DNA, see Detecting Network Attachment in IPv4
DHCP server, 187
DHCP, see Dynamic Host Configuration Protocol
dielectric, 136
differential transmission, 134, 150, 245
Digidesign, 322
Digigram, 33
digital, 3
digital encoding, 42
Digital Equipment Corporation (DEC), 162, 314, 426
Digital MultipleXing 512 Slots (DMX512A), xv, 6, 20, 25, 38, 54, 59, 84, 150, 241, 259, 361, 381, 399
Digital Signal Processing (DSP), 27
Digital Video Recorder (DVR), 229
dimmer class data, 251
dimmer-per-circuit, 16
dimmers, 13
DIN 180° connectors, 282
Diode-Pumped Solid-State (DPSS), 23
DIP, see Dual Inline Package
direct connection cable, 169
Directory Agents, 276
discovery, 259, 260, 263, 276
Disney, 357
Disney's Hollywood Studios®, 357
Disneyland, 357
Disney-MGM Studios, 357
distributed dimming systems, 18
distributed system, 80
Distribution Amplifier (DA), 330
distribution switches, 216
Distripalyzer, 329
DMP, see Device Management Protocol
DMX slot number, 243
DMX universes, 243

Index • 457

DMX, see Digital MultipeXing 512 Slots
DNS, see Domain Name System
Domain Name Server, 237
Domain Name System (DNS), 183, 190, 203, 232
Don't Panic!, 119
doorbell, 135
Doremi, 360
dotted decimal, 183
double precision, 109
Dow, JoAnne, 425
DPSS, see Diode-Pumped Solid-State
Drama Book Shop, xv
DreamHack, 421
drive racks, 48
Drop-Frame-NTSC Time Compensated Time Code, 325
Dry Contact Closure (DCC), 31, 93, 375
DSP, see Digital Signal Processing
DTE to DTE Connections, 151
DTE, see Data Terminal Equipment
Dual Inline Package (DIP), 243
Duff Gardens, 407, 410
dumb computer terminals, 148, 153
Dumb devices, 73
Durand, Jerry, 425
duty cycle, 18
DVR, see Digital Video Recorder
Dynamic Host Configuration Protocol (DHCP), 173, 185, 200, 203, 237, 256, 275

E

EAP, see Extensible Authentication Protocol
Echelon, 346
EECO, 322
effect, 41
EFX (show), 357
EIA Recommended Standards Comparison, 150
EIA serial connection, 38
EIA, see Electronic Industries Alliance
EIA-232, see RS-232
EIA-422, see RS-422
EIA-485, see RS-485
Einstein, 118
Ekval, Anders, 426
electrical isolation, 97
electrical noise, 18
Electro Controls, 242
Electro Magnetic Interference (EMI), 17, 18, 134, 150
electro-magnet, 46
Electronic Industries Alliance (EIA), 136, 147, 148
Electronic Theatre Controls, 302, 424
elegance, 111
Ellen Juhlin, 423
emergency stop, 94, 112
EMI, 282
EMI, see Electro Magnetic Interference
enabling system, 113, 382
encapsulated, 159
encoder, 45, 90
encryption, 178
End of System Exclusive (EOX), 287
End Of Travel (EOT), 43
enhanced function (DMX), 251
entertainment control. defined, 1
Entertainment Services and Technology Association (ESTA), 241, 424
Entertainment Technology, a Genlyte company, 424
EOT, see End Of Travel
EOX, see End Of system eXclusive
EPCOT®, 357
EPI-29 (Revised Rules for Allocation of Internet Protocol Version 4 Addresses to ACN Hosts), 256
error correction, see error detection
error detection, 129, 131, 153
Erskine-Shapiro, 426
ES-Bus, 315

ESTA, see Professional Lighting and Sound Association
E-Stop, see emergency stop
ETC, 14
EtherCAT, 49, 346
Ethercon, 138
Ethernet, 4, 6, 32, 49, 98, 132, 138, 141, 162, 179, 181, 182, 216, 274, 361, 389
EtherNet/IP™, 346
Ethersound, 33
evolution, 1, 99
exclusive or (logical operator), 86
Extensible Authentication Protocol (EAP), 178
eXtensible Markup Language (XML), 272

F

Fahl, Mike, 425
Fails, Gary, 423
fail-safe design, 63, 112, 372
Failure Mode and Effects Analysis (FMEA), 115
Falter, Sharon, 426
Far Field Associates, 425
FCTN, see Fixed Component Type Name
Ferren, Bran, 426
FHSS, see Frequency Hopping Spread Spectrum
fiber-optic cable, 33, 142
field bus, 48
filament "sing", 17
File Transfer Protocol (FTP), 181, 203
fire and smoke, 97
FireWire, 156
fireworks, 61
Fisher Technical Services, ii
Fisher, Scott, 423
five-pin XLR, 246
Fixed Component Type Name (FCTN), 275
fixed IP address, 189

fixture address, 243
fixture libraries, 245, 280
flashpots, 61, 62
floating point, 85
flood, 216, 232
flow control, 132, 153
fly, 41
FMEA, see Failure Mode and Effects Analysis
Focal Press, xvi, 423, 425, 426
fog, defined, 57
FOH, see Front of House
foil-tape tabs, 37
form C contacts, 95
form schedule, 95
forward, 216
Forward Phase Control (FPC), 17, 18
Foster, Fred, 242
FPC, see Forward Phase Control
frame, 159, 166
frame check sequence, 166
free topology, 85
Freed, 335
Freed, Adrian, 335
freewheeling, 326
Frequency Hopping Spread Spectrum (FHSS), 176
Frequency Shift Keying (FSK), 141
frequency-division multiplexing, 128
front of house (FOH), 29
front-end, 44
Frozen Yellow Garden Hose, 167
FSK, see Frequency Shift Keying
FTP, see File Transfer Protocol
full duplex, 129
full-duplex switches, 165, 171
Fuller, John, 426

G

galvanic, 98
galvanic isolation, 250

Gary Fails, ii
gas sniffers, 58
gateway, 190
general MIDI, 290, 292
General Purpose Interface (GPI), 31, 38, 93
generator, 43
gen-locked, 39
George Lucas Super Live Adventure, 357
Gigabit Ethernet, 168
Gilderfluke, 425
Gillum, Travis, 424
Gitchel, Andrew, 424
Glenn Birket, i
global unicast, 236
Glow, 359
glycol, 57
GNOME Human Interface Guidelines, 364
Gobo rotators, 19
Golder Group, 360
Google Book Search, xv
Gottelier, Tony, 426
GPI, see General Purpose Interface
graceful abort, 114
Granite Precision, 425
Gratuitous ARP, 186, 202
Gravesend Inn, *226*
Gray, Richard, 426
Great Moments with Mr. Lincoln, 357
Gross, Kevin, 425
ground loop, 97
Guc, Carlton, 371, 424
guru, 361

H

half-duplex, 129, 259
hard patched, 16
hardware address, 200
hardware flow control, 132
harmonic load, 19
Harrison, Chuck, 425
Harvey, Robert, 425

Hebert, Dennis, 425
Hefter, Mitch, 242, 424
Henderson, J.O., 360
Hendrickson, Alan, 115, 423, 424, 425, 426
Hergen, James, 53
hex, see hexadecimal
hexadecimal, 101
HIG, see Human Interface Guidelines
High End Systems, 259, 424
Highest Takes Precedence (HTP), 16
HMI, see Human-Machine Interfaces
home run, 83
horizontal cables, 139
host, 3
host identifier, 183
hot backup, 115
hot-patchable, 154
house sync, 39
Howell, Wayne, 252, 424
HTP, see Highest Takes Precedence
HTTP, see Hyper Text Transfer Protocol
hub, 170
hub and spoke, 170
hum, 97
Human Interface Guidelines (HIG), 364
Human-Machine Interfaces (HMI), 45, 90
Huntington, John C, Jr., 423
hydraulic, 46, 47
Hyper Text Transfer Protocol (HTTP), 181, 203
HyperMac, 33

I

I/O (input/output), 48
IANA, see Internet Assigned numbers Authority
IBM, 149
ICANN, see Internet Corporation for Assigned Names and Numbers
ICMP, see Internet Control Message Protocol
idle, 148

IEC, see International Electro-technical Committee
IEEE 1394, 156
IEEE 802.11 wireless Ethernet, 175
IEEE 802.11i, 178
IEEE 802.15.7 "Visible Light Communication", 257
IEEE 802.1Q, 226
IEEE 802.3, see Ethernet
IEEE 802.3af, 174
IEEE, see Institute of Electrical and Electronic Engineers
IETF, see Internet Engineering Task Force
if-then (logical operator), 86
IGBT, see Insulated Gate Bipolar Transistor
IGMP Snooping, 232
IGMP, see Internet Group Management Protocol
ILDA, see International Laser Display Association
IMAP, see Internet Message Access Protocol
incremental, 74
Indiana Jones™ Epic Stunt Spectacular!, 357
Information Technology (IT), 4, 389
infra-red, 92
infrastructure BSS, 177
initialized, 91
input, defined, 3
input/output (I/O), 48
Institute of Electrical and Electronic Engineers (IEEE), 32, 162, 168, 178
Insulated Gate Bipolar Transistor, 96
Intel, 162
interlaced, 324
International Electro-technical Committee (IEC), 8
International Laser Display Association (ILDA), 25
International Standards Organization, (ISO), 157
Internet, 118, 161
Internet Assigned Numbers Authority (IANA), 203
Internet Control Message Protocol (ICMP), 190
Internet Corporation for Assigned Names and Numbers (ICANN), 235
Internet Engineering Task Force (IETF), 8, 256, 276, 296
Internet Group Management Protocol (IGMP), 185, 232, 257
Internet Message Access Protocol (IMAP), 182, 203
Internet Protocol (IP), 161, 162, 181, 183, 233, 389
Internet Protocol Version 6 (IPv6), 234
Internet Service Provider (ISP), 161, 235
Internet Society (ISOC), 8
Internet Standards (STD), 8
Interoperability, 270
interrupt-driven, 82, 361
inter-VLAN routing, 230
Intranet, 161
inverting, 87
IP address, 183, 192
IP classes, 183
IP, see Internet Protocol
ipconfig command, 186
ipMIDI, 298
IPv6, 185, 234
ISBN, iv
ISO 7498, 157
ISO, see International Standards Organization
ISOC, see Internet Society
isochronous, 155, 156
isolation, galvanic, 97
isolation, optical, 97
ISP, see Internet Service Provider
Issacs, Mike, 426
IT, see Information Technology
Itchy and Scratchy, 407
Ithaca College, 426

J

Jadick, Terri, 425
jam-sync, 328, 330
Janninck, Jim, 424, 425
Janszen, Sierk, 424, 425
Japanese MIDI Association, 302
JBL Professional, 425
Jillette, Penn, 145
Jonathan Deans, ii
Josh Weisberg, 423

K

KÀ (Cirque du Soleil), 358
Karagosian, Michael, 425
kbit/s, see kilobits per second
Kellner, Jim, 425
kilobits per second, 127
Kindler, George, 360, 424, 425, 426
Kiss-Box, 424
Kleissler, Paul, 423
Kliegl, 242
Kolczynski, Kevin, 426
KX International, 51

L

La Damnation de Faust, 358
ladder logic, 45
Lamarr, Hedy, 143
LAN, 161
LAN Parties, 421
LAN, see Local Area Network
Las Vegas, 358
laser, 23
Latest Takes Precedence (LTP), 16
Lavery, Lauren, 425
Lay, Mike, 423, 425
layered communications system, 125
layers, 157
Lazzaro, John, 296, 424

LCCN, iv
LCS, 360
LDI, see Lighting Dimensions International
lease, 185
LED displays, 37
Lee, Marie, 425
Lenz, Tom, xviii, 425
Lepage, Robert, 358
LF, see Line Feed
LifeFormations, Inc/MasterMind Control, 424
Light Operate, 92
Lighting Dimensions International (LDI), 302
Lighting Dimensions magazine, xv, 426
Lighting Methods Incorporated (LMI), 242
Lightware USA, 36
limit switch, 42, 90
Line Feed (LF), 153, 382
linear encoder, 43
linear media, 70
linear show, 67
Linear Time Code (LTC), 327
line-rate, 171
Link-Local, 189
link-local, 236, 297
liquid nitrogen, 57
Live Design magazine, xv
LLC, see Logical Link Control
Local Area Network (LAN), defined, 161
local echo, 153
localhost, 185
Logical Link Control (LLC), 162, 163
logical operators, 86, 104
Lone Wolf, 426
Long, Jeff, 425
LonWorks, 346
Look Solutions USA, Ltd., 57
loopback, 185
lower explosive limit, 58
LTC, see Linear Time Code
LTP, see Last Takes Precedence
Luna Tech, 426

M

MAB, see Mark After Break
MAC address, 166, 170, 200, 202, 260
MAC, see Media Access Control
MacKay, Pat, 426
MADI, 33
Magenta Art Projects, 424
magic tricks, xv
maintained contact, 95
maintenance mode, 72
malware, 117
Mama Mia, 360
managed switch, 172
Management Information Base (MIB), 344
manchester encoding, 141, 327
mark, 147, 148, 247
Mark After Break (MAB), 247
mask, subnet, 194
Master-Slave, 81
Mbit/s, see Megabits per second
McEnerney, Charles, 425
McGraw, John, 426
MDIX, 170
Mechanical Design for the Stage, 115
mechanized props, 41
Media Access Control (MAC), 162, 163
media control systems, 361
media converter, 173
media servers, 19
Media-Link, 7, 426
Medialon Manager, 72, 383, 394, 402, 413, 424
Megabits per second (Mbit/s), 127
Meid, Dorian, 36
Melanie, Netta, xix, 423
Meldrum, Andy, 302
Metal-Oxide Silicon Field-Effect Transistors (MOSFET), 17
Metropolitan Opera, 358
Meyer Sound Laboratories, 30, 31, 346
Meyer, Chris, 322
MIB, see Management Information Base

Micro 2, 242
Microlight Systems, 424
Microsoft® Excel, 335
Middle C, 292
MIDI Machine Control (MMC), 296, 315
MIDI Manufacturer's Association (MMA), 281, 288, 301, 302, 315
MIDI note, 371
MIDI relay, 375
MIDI Show Control (MSC), 7, 55, 63, 290, 296, 299, 301, 317, 371, 375, 384, 398
MIDI splitter, 294, 376
MIDI Time Code (MTC), 286, 290, 296, 299, 305, 310, 315, 321, 322, 330, 344, 389, 390, 391
MIDI Visual Control (MVC), 343
MIDI, see Musical Instrument Digital Interface
midspan, 174
mission critical, 4
MMA, see MIDI Manufacturers Association
MMC, see MIDI Machine Control
Modbus, 346
Modbus/TCP, 346, 391
Modbus-IDA, 346
modems, 148
modes, 142
Modicon, 346
modulate, 140
momentary contact, 95
Mortal Engine, 359
mortars, 62
Moses, Bob, 426
motion control, 41, 43, 44
motion detectors, 92
motion profile, 44
motors, 46
MSC, see MIDI Show Control
MTC, see MIDI Time Code
Multicast, 160
multicast, 273, 296
multicasting, 236

multidrop, 133, 150
multilayer switch, 173
multi-mode optical fiber, 142, 167, 168
multiplex, 127
multiplexing, frequency-division, 127
multiplexing, time division, 127
multipoint, 133
Murphy's Law, 115
Musical Instrument Digital Interface (MIDI), xv, 21, 25, 32, 38, 54, 134, 271, 281, 322, 361, 398
musical time synchronization, 72, 400, 403
MVC, see MIDI Visual Control
Mystere (Cirque du Soleil), 360

N

NAT, see Network Address Translation
National Aeronautics and Space Administration (NASA), 322
National Fire Protection Association (NFPA), 58, 61
National Television Standards Committee (NTSC), 324
Native VLAN, 226
NC, see Normally Closed
NDP, see Neighbor Discovery Protocol
needle valves, 52
Neighbor Discovery Protocol (NDP), 237
NETSTAT command, 204
network, 3, 123, 145
network address, 185
Network Address Translation (NAT), 232
network bridge, 175
network diameter, 165
network ID, 185
Network IDentifier, 183
Network Interface Card (NIC), 166, 169
network name, 177
network segment, 165
Network Time Protocol (NTP), 203, 343
networking, 123

Neutrik, 138
New York City, 422
New York City College of Technology (City Tech), xvi, 226
NexSys, 426
NFPA 1123, Code for Fireworks Display, 61
NFPA 1126, Standard for the Use of Pyrotechnics before a Proximate Audience, 61
NFPA 160, Standard for Flame Effects Before a Proximate Audience, 58
NFPA 79, the Electrical Standard for Industrial Machinery, 112, 424
NFPA, see National Fire Protection Association
nibble, 100
NIC, See Network Interface Card
NIC, see Network Interface Card
night mode, 72
Nixon, Ken, 426
NO, see Normally Open
nodes, 3
Non Return to Zero (NRZ), 141
non routable IP addresses, 184
nonlinear show, 67
Normally Closed (NC), 94
Normally Open (NO), 94
not (logical operator), 87
Nowicki, Cindy, xix
NRZ, see Non-Return to Zero
NTP, see Network Time Protocol
NTSC, see National Television Standards Committee
null modem, 151, 169
null start code, 247, 251, 259
numerals, 99
Nye, Philip, 423, 424, 425

O

OCA, see Open Control Alliance
octave problem, 295

octets, 100
ODVA, see Open Device Net Vendors Association
off by one problem, 294
offset, 326
On/Off effects, 51
open collector, 96
Open Control Alliance (OCA), 341
Open Device Net Vendors Association (ODVA), 345
Open Sound Control (OSC), 335, 391
Open Systems Interconnect (OSI), 157, 181
open-loop, 78, 79
operational modes, 72
operator fatigue, 114
optical film sound track, 360
optical isolation, 250, 282
opto splitters, 259
or (logical operator), 86
Organizationally Unique Identifier (OUI), 166
OSC, see Open Sound Control
OSI, see Open Systems Interconnect
OUI, see Organizationally Unique Identifier
output, defined, 3
overflow, 132

P

Packet Forwarding, 160
packets, 159
pairing, 156
Pangolin, 24
parallel transmission, 145
parity, 153
parity checking, 130
passwords, 119
PCAOM, see Polychromatic Acousto-Optic Modulator
PDU, see Protocol Data Unit
Peak Audio, 33, 425
Pedersen, Lars, 425

peer review, 115
peer to peer, 82
peer-to-peer, 82
phantom power, 174
phase, 32
Phillips Vari*Lite, 244
photo-electric sensor, 92, 376
PHY
    See Physical Layer
PHY, see Physical Layer
physical address, 200
physical computing, 155
Physical Layer (PHY), 162
pilot tone, 333
ping command, 190
Pink Floyd, 37
Pioneer®, 347
Plain old Telephone Service (POTS), 136
PLASA, 241, 251, 259, 269
PLASA Control Protocol Working Group (CPWG), 241, 268
PLC, 230
PLC, see Programmable Logic Controller
plenum cable, 139
plug and play, 154, 276
plugfests, 268
pneumatic, 46, 47, 52, 89
PoE, see Power over Ethernet
point to point, 123, 133, 145
polling, 81, 260
Polychromatic Acousto-Optic Modulator (PCAOM), 23
polynomial, 132
porn, 6
port, 170, 203, 261, 297
port mirroring, 205
positional feedback, 42
post-production, 53
potentiometers, 43
POTS, see Plain Old Telephone Service
Potterf, Jason, 423
Power over Ethernet (PoE), 174, 229
PowerPoint, 35

precision, 109
pre-fire time, 63
prefix, 235
pre-roll time, 323
preset, 13
primary-secondary, 81
printer, 145
private networking, 184
probe request, 177
process, 3
process control, 57
ProcessField Bus (Profibus), 347
processors, 2, 30
producer-consumer model, 83
Production Arts, 293
Production Arts Lighting, 7, 242, 425, 426
Production Resource Group, 424
productivity, labor, 360
Professional Lighting and Sound Association (PLASA), 8, 20, 57, 241, 259, 261, 263, 269
Profile for Interoperability (EPI), 273
Programmable Logic Controller (PLC), 45, 48, 58, 230, 346, 391, 410, 418
propagation velocity, 164
proportional effects, 52
proportional valve, 47
Protocol Data Units (PDU), 273
protocol stack, 181
protocol suite, 181
proximity switches, 92
proxy, 260
Pulse-Width Modulation, 18
punched in, 53
Purple Floyd, 397
PuTTY, 206, 207
PWM, see Pulse-Width Modulation
PyroDigital Consultants, 426
PyroPak/LunaTech, 425
pyrotechnics, 61

Q

QoS, see Quality of Service
quadrature, 91
Quality of Service (QoS), 233

R

R. A. Gray, 426
Radio Frequency (RF), 175
Radio Frequency IDentification (RFID), 93
radio modems, 143
Radio-Frequency Interference (RFI), 18
RAID, see Redundant Array of Inexpensive Drives/Disks
random-access, 28
Rane, 426
Rapenburg Plaza, 424
Rapid Spanning Tree Protocol (RSTP), 218
RD, see Received Data
RDM, see Remote Device Management
RDMnet, 268
Real Time Protocol (RTP), 296
Realtime Music Solutions, 403
Received Data (RD or RX), 149
recommended minimum sets, 306
recommended practice, 6
recovery journal, 296
redundancy, 115
redundancy check, 131
Redundant Array of Inexpensive/independent Drives/disks (RAID), 117
refresh rate, 248
regenerate, 330
Regional Internet Registry (RIR), 253
registered jack, 138
Reilly, Mike, 424
Reinke, Lauralee, 424
relative encoder, 91
relative, defined, 74
relay, 46, 95, 347
relay logic, 45
reliability, 116

reliable, 182
reliable transmission, 272
remote controls, 90
Remote Device Management (RDM), 251, 259
Renkus-Heinz, 346
repeating hub, 170
Request to Send (RTS), 149
Requests For Comments (RFC), 8
Reset Sequence, 247
resolver, 43
responder, 260
resync, 147
retro-reflective, 92
Reverse Phase Control (RPC), 17, 18
RF, see Radio Frequency
RFC 1157, 344
RFC 2608, 276
RFC 3550, 296
RFC 4122, 275
RFC 4695, 296
RFC 761, 181
RFC 768, 181
RFC 791, 181
RFC, see Request For Comments
RFI, see Radio Frequency Interference
RFI, see Radio Frequency Interference, 18
RFID, see Radio Frequency IDentification
Richard Cadena, i
Richmond Sound Design, 207, 302, 424, 425
Richmond, Charlie, 302, 308, 314, 360, 424, 425, 426
Ring Cycle, 358
ring topology, 84
RIR, see Regional Internet Registry
rise time, 17
riser cables, 139
Rives, Mike, 425
RJ11, 138
RJ45 (8P8C), 138, 139, 168, 174
RJ45 (8P8C) pin out, 139, 246
Rockwell Automation, 345
Roland, 281, 343

Root Layer Protocol (RLP), 273
Rose, Alan, 424
rotary shaft encoder, 43
router, 173, 211
routes, 83
routing matrix, 36
routing table, 220, 231
RPC, see Reverse Phase Control
RS-232, 31, 148, 226, 346, 347, 391
RS-422, 31, 150, 351
RS-485, 150, 252, 346
RSTP, 218
RSTP, see Rapid Spanning Tree Protocol
RTP Payload Format for MIDI, 296
RTP, see RTP Payload Format for MIDI
RTS, see Request To Send
Ruling, Karl G., 424, 426
running status (MIDI), 291
Ruud, Kevin, 425
RX, see Received Data

S

SA, see Service Agents
sACN, see streaming ACN
safety, 48, 111, 362
Saltzman, Mike, 423
Sammler, Ben, 426
sample lock, 32
scanning, wireless connection, 177
scenic automation systems, 41
Scharff Weisberg, Inc., 425
Scheirman, David, 425
Scott Blair, 423
Scott Fisher, ii
SCR
SCR, see Silicon Controlled Rectifier
Scratchy, 407
SDP, see Session Description Protocol
SDT, see Session Data Transport
sensors, 37, 89
Sequential Circuits, 281, 322

serial communications, 146
server, 82
Service Location Protocol (SLP), 276
Service Set IDentifier (SSID), 177
session, 158
Session Data Transport (SDT), 272, 273
Session Data Transport protocol, 256
Session Description Protocol (SDP), 297
Session Initiation Protocol (SIP), 297
SFX (software), 371, 424
Shapiro, Lou, 426
Shielded Twisted Pair (STP), 136
Shiner, Paul, 426
show action equipment, 41
show business, xv
show control email list, 424
show control, defined, 2
show controller, 361
show enabled, 72
show must go on, 4
signal connection methods, 2
signal ground, 149
signalling units, 127
signature resistance, 175
Silicon Controlled Rectifier (SCR), 17, 20, 96
Simon Newton, 423
Simon, Elizabeth, 424
Simple Mail Transfer Protocol (SMTP), 181, 203
Simple Network Management Protocol (SNMP), 33, 204, 344, 391
simplex, 129
sine wave, 17
sine wave dimming, 18
Sinfonia, 72
single failure-proof design, 112
Single Pole, Double Throw (SPDT), 95
single precision, 109
single-ended transmission, 134, 150
sinking, 96
sinusoidal, 18
SIP, see Session Initiation Protocol

slate, 322
slides, 37
slope, 44
slots, 241
SLP, see Service Location Protocol
Smith, Dr. David B, 403
Smith, Dr. David B., 72, 424, 425
Smith, Ian W., 424
smoke, defined, 57
SMPTE 12M, SMPTE Standard for Television, Audio, and Film—Time and Control Code, 321
SMPTE, see Society of Motion Picture and Television Engineers
SMTP, see Simple Mail Transfer Protocol
snake, 373
SNMP, see Simple Network Management Protocol
snubber, 96
Society of Motion Picture and Television Engineers (SMPTE) Time Code (TC), xv, 6, 25, 38, 49, 55, 63, 315, 321, 344, 353, 360, 390, 391, 392, 398, 401
socket, 203
soft-patching, 16
software flow control, 132
solenoid, 46, 47, 372
solenoid valves, 51
solid-state relay, 93, 96
Song Position Pointers (SPP), 286
Sony 9-Pin, 350
SoundMan-Server, 207, 424
sourcing, 96
space, 148
Spanning Tree Protocol (STP), 218
SPDT, see Single Pole, Double Throw
Speerstra, Karen, 426
SPP, see Song Position Pointer
spread spectrum radio transmission, 143, 176
spyware, 119
SSID, see Service Set IDentifier
stage elevators, 41

stage equipment, 41
stage machinery, 41
stage manager, 68, 129, 360
Stage Research, 28, 371, 424
stand by, 68
standard, de facto, 5
standard, defined, 4
standard, open, 5
standard, proprietary, 5
star topology, 83, 167
start bit, 147
start code, 247
stateless address autoconfiguration, 189
static IP address, 189
stepper motor, 46
Steve Carlson, 174
Steve Terry, 423
stop bit, 147, 153
STP, see Shielded Twisted Pair
STP, see Spanning Tree Protocol
Strand Lighting, 14, 16, 242, 302
stratums, 343
streaming ACN (sACN), 252, 255, 256, 398
strobe, 145
stuck note, 287
submasters, 14
subnet mask, 194
subnets, 194
Sunn, 302
switches, 89, 159, 170, 216
switching fabric, 171
switching hubs, 83, 170
sync pulse, 333
synchronous, 146, 149
synthesizer, 281
SysEx, see System Exclusive messages
system common messages, 285
System Exclusive (SysEx) messages, 285, 302
system, defined, 2

T

T568A, 139
T568B, 139
tach pulse, 333
tachometer, 43, 46
tagging, 226
target position, 42, 90
TCI magazine, 426
TCP, see Transmission Control Protocol
TD, see Technical Director or Transmitted Data
Technical Director (TD), 114, 407
Telecommunications Industry Association (TIA), 136, 147, 148
TELetype NETwork (Telnet), 206
Teller, 146
TELNET command, 206
Tera Term, 153
terminal, 206
terminal emulator, 152, 206
terminated, 249
Terry, Steve, 242, 423, 424, 425, 426
the show must go on, 111, 115
Theatre Crafts magazine, xv
things that go boom, 61
Thinkwell Design and Production, 424
Thoughtful Designs, 425, 426
thru-beam, 92
TIA, see Telecommunications Industry Association
TIA/EIA-568B, 138
TimberSpring, 424, 425
time code drop outs, 326
time code slate, 322
time-division multiplexing, 127
Times Square, xv
token-passing protocols, 84
Tomlinson, J.T., 360
Tony Clynick, 423
topology, 83
Toresdahl, Todd, 360
toroidal inductor, 17

torque, 46
touch screen, 90
touch-screen, 362
TRACEROUTE command, 208
tracking, 13
transformer, 43, 98, 166
Transmission Control Protocol (TCP), 37, 162, 181, 182, 203, 204, 206, 273, 296
Transmitted Data (TD or TX), 149
transparent, 260
Transport layer, 125
trapezoidal profiles, 44
Troika Ranch, 359
trojans, 117
troubleshooting, 119
truth table, 86
Tucker, George, 424
tune request, 286
twisted-pair cable, 135
Two-Phase Commit (2PC), 314
TX, see Transmitted Data

U

UA, see User Agents
UACN, see User Assigned Component Name
UC Berkeley, 296, 424
UCS, see Universal Character Set
UDP, see User Datagram Protocol
UID, see Unique ID
ULA, see Unique Local Address
unbalanced transmission, 150
Unbalanced, see "Single-Ended"
unicast, 296
Unicode, 126
Uniform Resource Identifier (URI), 277
Uninterruptable Power Supply (UPS), 116
Unique ID (UID), 260, 261, 263
Unique Local Addresses (ULA), 236
United States Institute for Theatre Technology (USITT), 241, 242, 302

Universal Character Sets (UCS), 126
Universal Real-Time System Exclusive ID, 301
Universal Real-Time System Exclusive MIDI messages, 316
Universal Resource Locator (URL), 232
Universal Serial Bus (USB), 154
Universal Studios, Florida, 426
Universally Unique IDentifier (UUID), 256, 274
universes, 243
unmanaged switch, 172, 216
unmutes, 263
unreliable, 183, 274
unresolved address, 200
unshielded, 135
Unshielded Twisted Pair (UTP), 135, 167, 168
untwisted, 135
upper layers, 157
UPS, see Uninterruptable Power Supply
URI, see Uniform Resource Identifier
URL, see Universal Resource Locator
USB Type A connectors, 155
USB, see Universal Serial Bus
User Agents (UA), 276
User Assignable Component Name (UACN), 275
User Datagram Protocol (UDP), 181, 183, 186, 203, 204, 256, 273, 274, 296, 343, 344, 399
user interface, 361
USITT Callboard, 302
USITT, see United States Institute for Theatre Technology
UTF-8, 126
UTP, see Unshielded Twisted Pair
UUID, see Universally Unique IDentifier

V

value engineering, 366

valves, 46
vamp, 410
Vari*Lite, 19, 244, 302
variable types, 85
variable, defined, 109
Variables, 85
vari-speed, 323
velocity feedback, 43, 46
velocity loop, 43, 46
vertical blanking interval, 328
vertical cables, 139
Vertical Interval Time Code (VITC), 328
video, 35
video projectors, 37
video server, 19, 35, 38
video switchers, 36
video sync, 39
Vieaux, Laurel, 426
Villet, Stephane, 424
virtual circuit, 182
Virtual Lan (VLAN), 231
Virtual Local Area Network (VLAN), 172, 224, 226
Virtual Network Control (VNC), 344
virtual orchestra, 410
Virtual Private Network (VPN), 233
virus checkers, 117
viruses, 119
VITC, see Vertical Interval Time Code
VL3500Q, 244
VLAN, see Virtual Local Area Network
VLAN. see Virtual Local Area Network
VNC, see Virtual Network Control
Voice Over Internet Protocol (VOIP), 174
Voice Over IP (VOIP), 233
VOIP, see Voice Over Internet Protocol
voltage-loop interfaces, 133
VPN, see Virtual Private Network

W

walkie-talkies, 129

WAN, see Wide Area Network
WAP, see Wireless Access Point
Waters, Roger, 37
Waters, Ryan, 424
Wawrzynek, John, 296
Wayne Howell, 423
WDI (Wacky Duff Illusions), 408
Weber, Ralph O., 314, 426
Wendlandt, Bill, 360
WEP, see Wired Equivalent Privacy
White Rabbit/RA Gray, 425
Whois, 203
whole numbers, 85
Wide Area Network (WAN), 161
WiFi Protected Access (WPA), 178
WiFi, see IEEE 802.11 wireless Ethernet
Willis, Peter, 424
Wilton, Loren, 207, 424
Wings Platinum, 53
wire, 134
Wired Equivalent Privacy (WEP), 178
Wireless Access Point (WAP), 173, 175, 176, 177
wireless Ethernet, see IEEE 802.11 wireless Ethernet
Wireshark, 204, 205
Wizardess Designs, 425
Wohlsen, Barbara, 425
Wootten, Damon, 360
word clock, 32
World's Fair, 357
WPA, see Wi-Fi Protected Access
Wright, Matt, 335
www.controlgeek.net, xvii, xix, 82, 183, 232, 422, 423

X

Xerox, 162
XLR, 328, 373
XML, see eXtensible Markup Language

Y

Yale School of Drama, xv, 424, 425, 426
Yamaha, 281

Z

zero-crossing, 17
Zeta Instrument Processor Interface (ZIPI), 335
ZIPI, see Zeta Instrument Processor Interface

Made in the USA
San Bernardino, CA
09 October 2016